Peter Michler, Simone Luca Portalupi

Semiconductor Quantum Light Sources

Also of Interest

Polarization of Light.
In Classical, Quantum, and Nonlinear Optics
Chekhova, Banzer, 2021
ISBN 978-3-11-066801-8, e-ISBN (PDF) 978-3-11-066802-5,
e-ISBN (EPUB) 978-3-11-060509-9

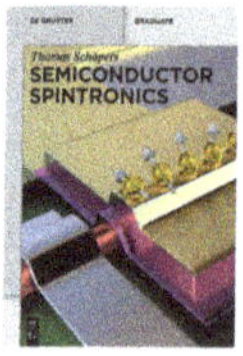

Semiconductor Spintronics
Schäpers, 2021
ISBN 978-3-11-063887-5, e-ISBN (PDF) 978-3-11-063900-1,
e-ISBN (EPUB) 978-3-11-063932-2

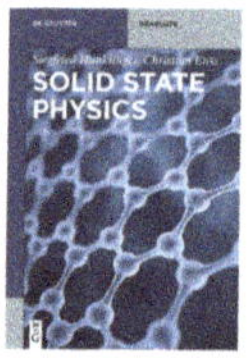

Solid State Physics
Hunklinger, Enss, 2022
ISBN 978-3-11-066645-8, e-ISBN (PDF) 978-3-11-066650-2,
e-ISBN (EPUB) 978-3-11-066708-0

Quantum Technologies.
For Engineers
Müller, Greinert, 2023
ISBN 978-3-11-071744-0, e-ISBN (PDF) 978-3-11-071745-7,
e-ISBN (EPUB) 978-3-11-071750-1

Quantum Crystallography
Matta, Huang, Massa, 2023
ISBN 978-3-11-056566-9, e-ISBN (PDF) 978-3-11-056667-3,
e-ISBN (EPUB) 978-3-11-056571-3

Peter Michler, Simone Luca Portalupi

Semiconductor Quantum Light Sources

Fundamentals, Technologies and Devices

DE GRUYTER

Authors
Prof. Dr. Peter Michler
Universität Stuttgart
Institut für Halbleiteroptik und
Funktionelle Grenzflächen
Allmandring 3
70569 Stuttgart Vaihingen
Germany
p.michler@ihfg.uni-stuttgart.de

Dr. Simone Luca Portalupi
Universität Stuttgart
Institut für Halbleiteroptik und
Funktionelle Grenzflächen
Allmandring 3
70569 Stuttgart Vaihingen
Germany
s.portalupi@ihfg.uni-stuttgart.de

ISBN 978-3-11-070340-5
e-ISBN (PDF) 978-3-11-070341-2
e-ISBN (EPUB) 978-3-11-070349-8

Library of Congress Control Number: 2023945074

Bibliographic information published by the Deutsche Nationalbibliothek
The Deutsche Nationalbibliothek lists this publication in the Deutsche Nationalbibliografie; detailed bibliographic data are available on the Internet at http://dnb.dnb.de.

Cover image: Entangled photon pairs generated by a circular Bragg grating cavity embedding a semiconductor quantum dot / Sascha Kolatschek
Typesetting: VTeX UAB, Lithuania
Printing and binding: CPI books GmbH, Leck

www.degruyter.com

To my wife Silke and my sons Jan, Dennis, and Tim
P. M.
To my mother Lucia and my father Angelo
S. L. P.

Preface

Semiconductor quantum light sources based on quantum dots, such as triggered single- and entangled photon sources, are fascinating devices since they provide the ultimate control over the light emission process. Moreover, they deliver light states with superior optical and quantum optical properties. This includes high single-photon emission probabilities (brightness), low multiphoton probabilities (single-photon purity), and identical photon wave packets (indistinguishability), i. e., ultimately Fourier transform-limited photons. Furthermore, in recent years, triggered polarization- and time-bin entangled photon pair states have been demonstrated with very high fidelities and photonic cluster state generation. In the meanwhile the covered spectral range extends from the ultraviolet up to the telecom bands. From a more practical point of view, alternatively to optical excitation, the sources can be also electrically driven and therefore be very compact and robust. Many quantum photonic implementations will strongly benefit from their excellent properties, e. g., quantum communication, photonic quantum sensing, photonic quantum simulation, and optical quantum computing.

The aim of this textbook is to give an extensive experimental view onto the fundamental properties of quantum light states, their characterization methods, and their generation by quantum-dot-based light sources. The first two topics are discussed in a broader sense and are therefore relevant for all kinds of quantum light emitters, whereas for the photon generation process, we restrict ourselves on semiconductor-based quantum dots. This book targets physics and quantum engineering students at undergraduate and graduate levels, as well as research scientists, physicists, and engineers in academia and industry. It offers practical bases for understanding quantum light sources and for evaluating their performances. This book is written by experimentalists for experimentalists, while keeping rigor in the explanation of the theory necessary to understand the experiments. It is partly based on lecture notes of the *Semiconductor Quantum Optics* course and on the *Physics and Technology of Nanostructures for Quantum Optics* course for bachelor/master physics and photonic engineering students at the University of Stuttgart.

The book is organized in three parts and 14 chapters. Part I (Chapters 1–6) contains fundamentals of quantum light states and their classification, and it is therefore useful for understanding quantum light from all different kinds of quantum emitters. Besides a classical description of linewidth broadening effects, first- and second-order correlation functions (Chapter 1) and quantum specific aspects are also introduced. This includes a short introduction into the quantum theory of radiation and field quantization in open space and cavities, as well as a quantum description of the correlation functions necessary to adequately describe nonclassical light states such as photon number states (Chapters 3 and 4). In Chapter 5, all relevant single emitter properties such as linewidth, brightness, single-photon purity, photon coherence, photon indistinguishability, and degree of entanglement are covered together with experimental settings to benchmark these properties. In Chapter 6, cavity quantum electrodynamics (c-QED) is treated with

https://doi.org/10.1515/9783110703412-201

a special emphasis on the interaction of a two-level system with a light field discussing also the so-called weak and strong coupling regimes of cavity QED.

Part II (Chapters 7–11) is devoted to semiconductor quantum dot specific topics. It starts with an introduction of the basic physical properties of quantum dots, including their electronic and optical properties (Chapter 7). Different excitonic configurations and their specific role in quantum light generation are discussed. Chapter 8 is dedicated to the interaction of the electronic excitations with phonons and possible electric and magnetic field fluctuations. These interactions lead to dephasing, spectral diffusion, and therefore to linewidth broadening, which can seriously impact the quality of quantum light. In Chapter 9, several reversible and permanent techniques for tuning the electronic and optical properties of QDs are presented, including temperature, electric and magnetic fields, and strain. Chapter 10 gives an overview of different optical and electrical excitations schemes and discusses their pro and cons. Finally, Chapter 11 is dedicated to an extensive discussion on the photon indistinguishability. We will discuss two-photon interference experiments of two different types, i. e., for photons originating from one and the same source and from two remote sources.

Part III (Chapters 12–14) is devoted to quantum-dot-related technologies and their devices. First, in Chapter 12, the most commonly utilized nanofabrication techniques are discussed, from semiconductor growth to lithography and etching. Particular care will be given to the so-called deterministic lithography approaches, which play a central role in the realization of high-performance quantum light sources. Chapter 13 provides a detailed overview of the photonic structures, which can be utilized for enhancing the light extraction and modifying the emission properties, such as the lifetime. The most common experimental approaches for the characterization of the key properties are also presented. The last chapter goes into the details of single- and entangled photon sources, their operation, design, and achievable performances, in particular, when coming to photonic quantum devices. Finally, photonic integrated circuits are discussed with focus on the main components required for on-chip quantum operations as linear quantum computing.

We thank Sascha Kolatschek, Julian Maisch, and Nam Tran for providing numerous figures for the book. We especially thank our PhD students (in alphabetic order) Florian Hornung, Raphael Joos, Julian Maisch, Ulrich Pfister, and Tim Strobel and our colleagues Matteo Galli, Dario Gerace, Michael Jetter, and Marc Sartison for critically reading parts of the manuscript and for many useful comments to improve and clarify explanations and deviations. The cover picture of the book is based on graphics by Sascha Kolatschek. We also thank the publisher, De Gruyter, especially Kristin Berber-Nerlinger, Melanie Götz, and Ria Sengbusch for their friendly support.

Stuttgart, July 2023

Peter Michler
Simone Luca Portalupi

Contents

Abbreviations

ARP	adiabatic rapid passage
BS	beamsplitter
CBG	circular Bragg grating
CCD	charge-coupled device
CdSe	cadmium selenide
CdTe	cadmium telluride
cQED	cavity quantum electrodynamics
CW	continuous wave
DBR	distributed Bragg reflector
DNSP	dynamic nuclear spin polarization
EID	excitation-induced dephasing
F_P	Purcell factor
FPI	Fabry–Perot interferometer
FSS	fine-structure splitting
FWHM	full width at half maximum
GaAs	gallium arsenide
HBT	Hanbury–Brown and Twiss
HOM	Hong–Ou–Mandel
ICP-RIE	inductively coupled plasma reactive ion etching
InAs	indium arsenide
InGaAs	indium gallium arsenide
InP	indium phosphide
MBE	molecular beam epitaxy
MCA	multi channel analyzer
MOVPE	metalorganic vapor-phase epitaxy
MZI	Mach–Zehnder interferometer
NA	numerical aperture
NRC	nonresonant QD-cavity coupling
OH	Overhauser field
PCFS	photon-correlation Fourier spectroscopy
PBS	polarizing beamsplitter
PSB	phonon side bands
PD	pure dephasing
PhC	photonic crystal
PhCWG	photonic crystal waveguide
PIC	photonic integrated circuit
PL	photoluminescence
μ-PL	microphotoluminescence
QD	quantum dot
QCSE	quantum confined Stark effect
RIE	reactive ion etching
SD	spectral diffusion
SIL	solid-immersion lens
SNSPD	superconducting nanowire single-photon detector
SPLED	single-photon light emitting diode
SPS	single-photon source
TAC	time-to-amplitude converters

https://doi.org/10.1515/9783110703412-202

TCSPC	time-correlated single-photon counting
TIR	total internal reflection
TLS	two-level system
TPE	two-photon excitation
TPI	two-photon interference
X	exciton
X^+, X^-	positive, negative trion (or positively, negatively charged exciton)
XX	biexciton
ZnSe	zinc selenide
ZnTe	zinc telluride
ZPL	zero-phonon line

Part I: **Fundamentals of optics and quantum optics for single- and entangled photon sources**

1 Chaotic light and correlation functions

1.1 Introductory remarks

The goal of this chapter is to classically describe the optical properties of so-called *chaotic light sources*. Chaotic points up the fact that the emitters in such a source radiate independently, with no relationship among the phases for independent emitters (e. g., atoms, quantum dots). The light from a single spectral line of a discharge lamp and the filament lamp are examples of chaotic light sources. In the following, we will investigate their emission dynamics, coherence properties, and photon statistics. We will see that some of these classical concepts carry over to the quantum theory of light, whereas specific properties of so-called nonclassical light beams need a full quantum mechanical description (see Chapter 2). The treatments of the theory in this chapter include some shortcuts and omissions of more advanced derivations, which can be found, e. g., in the book of Loudon [118]. Nevertheless, we will focus on the basic concepts for understanding single-emitter properties.

1.2 Model: chaotic light

1.2.1 Homogeneous broadening

The discussion can start by considering a *single emitter* (e. g., an atom, ion, molecule, defect center, quantum dot, etc.) emitting with frequency ω_0 (classical picture). The emitter radiates electromagnetic radiation steadily until it suffers a collision (e. g., atom-atom or electron–phonon in a solid-state matrix, i. e., a collision process is assumed to be short with respect to the radiative lifetimes of the emission processes involved). After the collision, the radiation continues with a different phase, which is unrelated to the phase before the collision. In the following, we will discuss the influence of collision-induced *random phase changes* on the emission properties. Let us now introduce the following assumptions:

- There is no *phase relation* before and after the collision. Each phase change $\Delta\varphi$ is completely *randomly distributed* within $[0, 2\pi]$.
- The phase $\varphi_i(t) = \varphi_i$ remains constant during the *mean collision-free time* τ_0 between two collisions.
- The electric field amplitude of the wave from the single emitter (atom, QD, etc.) can be expressed as

$$E_i(t) = E_0 \exp(-i\omega_0 t + i\varphi_i(t))$$

  with $\omega_0 = 2\pi\nu_0$ as the angular frequency and $\varphi_i(t)$ as the time-dependent phase of the electric field.

https://doi.org/10.1515/9783110703412-001

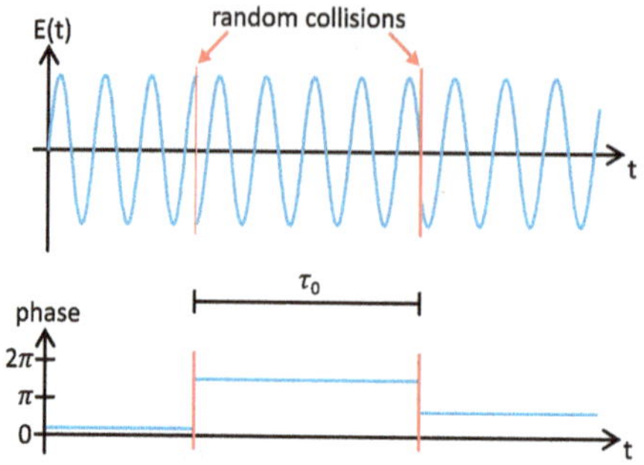

Figure 1.1: Time evolution of the electric field of a *single emitter* with angular frequency $\omega_0 = 2\pi\nu_0$, where the vertical lines represent random collisions, and the phase $\varphi_i(t)$ varies statistically $\Delta\varphi_i \in [0, 2\pi]$. The phase $\varphi_i(t) = \varphi_i$ stays constant during the *mean collision-free time* τ_0.

- The time evolution of the electric field $E_i(t)$ and the corresponding phase $\varphi_i(t)$ is schematically shown in Figure 1.1.

In the next step, we consider an *ensemble of emitters* under the following assumptions:
- Each emitter emits with a *random time-dependent* phase $\varphi_i(t)$ $(i = 1, 2, \dots, \nu)$.
- The emission from different emitters form a plane parallel light beam, and the light has a *fixed polarization* for all ν emitters. These simplifications allow an easy discussion, and the light fields can be added algebraically.

The total *electric field* of all ν emitters (all with equal angular frequency ω_0) is then given by a scalar sum of the individual fields:

$$\begin{aligned} E(t) &= \sum_{i=1}^{\nu} E_i(t) \\ &= E_1(t) + E_2(t) + \cdots + E_\nu(t) \\ &= E_0 \exp(-i\omega_0 t)\{\exp(i\varphi_1(t)) + \exp(i\varphi_2(t)) + \cdots + \exp(i\varphi_\nu(t))\} \\ &= E_0 \exp(-i\omega_0 t) a(t) \exp(i\varphi(t)). \end{aligned} \tag{1.1}$$

In the last expression the quantities $a(t)$, as the resultant time dependent *amplitude*, and $\varphi(t)$, as the *phase* of the ensemble field, have been introduced (see Fig. 1.2).

The discussed collisions result in a broadening of the linewidth, which takes the overall form of a Lorentzian distribution.

Fluctuations of similar forms occur in the presence of other line-broadening processes, which also lead to Lorentzian frequency distributions (homogeneous broadening), e. g., in the case of radiative broadening.

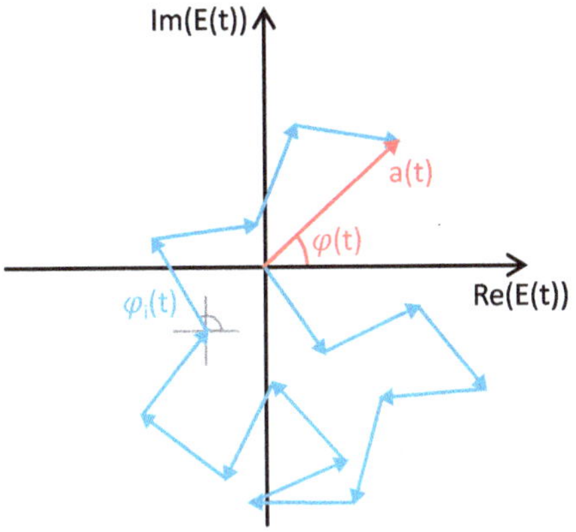

Figure 1.2: The diagram displays the amplitude $a(t)$ and the phase $\varphi(t)$ of the resultant electric field vector. The individual arrows represent unit vectors, each of which has a random phase angle.

1.2.2 Inhomogeneous broadening

Now we will discuss an inhomogeneous broadened chaotic light source with a spread in emission frequencies, which is described by a *Gaussian distribution*. This is, for example, the case for a Doppler broadened ensemble of atoms or an ensemble of solid-state emitters dominated by spectral diffusion.

Differently from the homogeneous broadened case, where each atom had the same angular frequency, ω_0, here let us consider an *ensemble* of ν emitters with frequencies ω_i laying within a Gaussian distribution centered at ω_0, phase φ_i, and the following properties:

- Each emitter radiates with a *random fixed phase* φ_i = const. and a *random fixed frequency* ω_i = const. Different emitters have different frequencies ω_i. For example, for an atomic ensemble with Doppler broadening, the frequency shift from ω_0 is given by the respective atomic velocities, and for solid-state emitters, the spectral shift is given by the strength of nearby fluctuating electric or magnetic fields (spectral diffusion).
- All emitters radiate with the same *fixed polarization*.
- The emission of each emitter is described by a *plane wave*.

The total electric field of the ensemble emission results from the sum of all individual emitters:

$$E(t) = E_0 \sum_{i=1}^{\nu} \exp(-i\omega_i t + i\varphi_i) \tag{1.2}$$

with fixed field amplitude E_0 of each individual emitter.

Average intensity

The *average of the intensity* $\bar{I}(t)$ over a cycle ("cycle average over one period T of the field") of oscillations is given by

$$\bar{I}(t) = \frac{1}{2}\varepsilon_0 c|E(t)|^2 = \frac{1}{2}\varepsilon_0 c E_0^2 a(t)^2 \tag{1.3}$$

with $a(t)$ defined as in Eq. (1.1). Note that besides the *temporal average* over a period T, later it will be used in an expression of the form $\langle \ldots \rangle$, which describes the *statistical average* or *longer time averages* ("long time average"; $t \to \infty$). The random modulation $a(t)$ causes the time dependence of the intensity $\bar{I}(t)$. In the presence of homogeneous broadening, together with large fluctuations on timescales of the coherence time, faster fluctuations are also present and overlapped with the first ones. On the contrary, in the presence of inhomogeneous broadening, only large fluctuations on timescales comparable to the coherence time are visible. A time series of the cycle-averaged intensity $\bar{I}(t)$ for homogeneous and inhomogeneous broadenings can be found, e. g., in [118].

1.3 First-order coherence $g^{(1)}(\tau)$ and temporal field correlations

The unnormalized first-order correlation function of light with stationary statistic is defined as

$$G^{(1)}(\tau) = \langle E^*(t)E(t+\tau)\rangle. \tag{1.4}$$

This expression describes the expectation value of the *temporal correlation* of the electric field and thus the first-order coherence. It only depends on the time delay τ between the two field values. The corresponding normalized version of the first-order temporal coherence is given by

$$g^{(1)}(\tau) = \frac{\langle E^*(t)E(t+\tau)\rangle}{\langle E^*(t)E(t)\rangle}. \tag{1.5}$$

In the following, we consider light sources with stationary statistical properties (i. e., the mechanisms causing the fluctuations do not vary with time). Therefore we can assume the equivalence of time averaging and statistical averaging, which is valid for light beams whose fluctuations are produced by ergodic random processes [118].

Putting Eq. (1.1) into the first-order correlation function, we get the expression

$$\begin{aligned} G^{(1)}(\tau) = E_0^2 \exp(-i\omega_0\tau)\langle\{&\exp(-i\varphi_1(t)) + \exp(-i\varphi_2(t)) + \cdots + \exp(-i\varphi_\nu(t))\} \\ &\times \{\exp(i\varphi_1(t+\tau)) + \exp(i\varphi_2(t+\tau)) + \cdots + \exp(i\varphi_\nu(t+\tau))\}\rangle. \end{aligned} \tag{1.6}$$

For the model of ν independent emitters, the phase angles of *different emitters* have different random values for each moment in time t. Therefore we get *cross terms* of the type

$$\sim \langle \exp(-i\varphi_i(t)) \times \exp(i\varphi_j(t+\tau))\rangle,$$

which give a zero average contribution. The remaining terms with equal indices give

$$G^{(1)}(\tau) = E_0^2 \exp(-i\omega_0\tau) \sum_{i=1}^{\nu} \langle \exp(-i\varphi_i(t)) \exp(i\varphi_i(t+\tau))\rangle \tag{1.7}$$
$$= \nu \cdot \langle E_i^*(t) E_i(t+\tau)\rangle.$$

In the last transformation, we consider that all ν emitters are identical and therefore contribute equally. This leads to an important conclusion: the correlation function for the beam as a whole is determined by the *single-emitter* contributions!

After a collision event (e. g., electron–phonon), the phase angle of each wave train jumps to a random value. This leads to a zero average contribution to the first-order field correlation. Therefore the *single-emitter correlation function* $\langle E_i^*(t)E_i(t+\tau)\rangle$ in Eq. (1.7) is proportional to the probability that the emitter possesses an interaction-free time $\tau_0 > \tau$. Thus we can write the single-correlation function as

$$\langle E_i^*(t) E_i(t+\tau)\rangle = E_0^2 \exp(-i\omega_0\tau)\langle \exp(i\varphi_i(t+\tau) - i\varphi_i(t))\rangle$$
$$= E_0^2 \exp(-i\omega_0\tau) \int_{\tau}^{\infty} d\tau' p(\tau'),$$

where $p(\tau)\,d\tau$ describes the probability that an emitter has an interaction-free time in the time interval $[\tau, \tau + d\tau]$. The random collision processes are discussed in the framework of the kinetic theory of gases.

For a single atom, the probability $p(\tau)\,d\tau$ that an atom has a period of free flight between collisions lasting a length of time in the range τ until $\tau + d\tau$ is

$$p(\tau)\,d\tau = \frac{1}{\tau_0} \exp\left(-\frac{\tau}{\tau_0}\right) d\tau \tag{1.8}$$

with τ_0 = the *interaction-free time*, i. e., here the mean period of free flight (mean collision-free time; see [118]).

Putting the expression of $p(\tau)\,d\tau$ into the equation of the single-correlation function and integrating over $[\tau, +\infty]$, we get

$$\langle E_i^*(t) E_i(t+\tau)\rangle = E_0^2 \exp(-i\omega_0\tau) \exp\left(-\frac{\tau}{\tau_0}\right). \tag{1.9}$$

The normalization of the corresponding ensemble correlation function (Eq. (1.7)) with respect to the atom (emitter) number ν and the time averaged value $\langle E_i^*(t)E_i(t)\rangle = E_0^2$ results in the *normalized single-emitter correlation function*

$$g^{(1)}(\tau) = \exp\left(-i\omega_0\tau - \frac{\tau}{\tau_0}\right)$$
$$= \exp(-i\omega_0\tau - \gamma_{\text{coll}}\tau), \tag{1.10}$$

where $\gamma_{\text{coll}} = (\tau_0)^{-1}$ has been used for the *collision dephasing rate* (or *collision rate*) of the atoms.

In the case that besides collision broadening, also *radiative broadening* plays a role for the dephasing, we can extend the model by introducing an additional multiplicative factor $\exp(-\gamma_{\text{sp}}t)$, which considers the decay of the electric field by radiative recombination. Consequently, this additional factor $\exp(-\gamma_{\text{sp}}\tau)$ appears in the correlation function $G^{(1)}(\tau)$ (Eq. (1.6)).

Then the *total dephasing rate* is

$$\gamma = \gamma_{\text{sp}} + \gamma_{\text{coll}}. \tag{1.11}$$

The total *coherence time* is given by the inverse of the total dephasing rate γ as

$$\tau_c = \frac{1}{\gamma}. \tag{1.12}$$

Therefore we get the normalized single-particle correlation function

$$g^{(1)}(\tau) = \exp\left(-i\omega_0\tau - \frac{|\tau|}{\tau_c}\right) = \exp(-i\omega_0\tau - \gamma|\tau|). \tag{1.13}$$

The modulus was introduced to consider both *negative* and *positive* values of τ.

Figure 1.3 shows the characteristic symmetric exponential decay of the modulus $|g^{(1)}(\tau)|$ for the discussed *homogeneous* broadened emission.

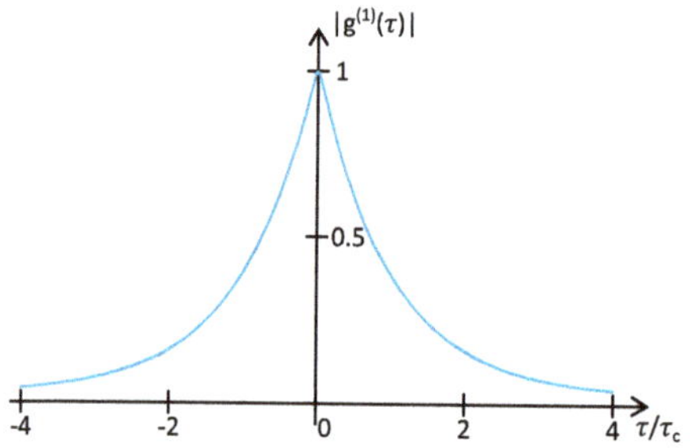

Figure 1.3: $|g^{(1)}(\tau)|$ for a homogeneously broadened light source.

For an *inhomogeneous broadened ensemble*, the first-order correlation function can be calculated as follows. Putting Eq. (1.2) into the first-order correlation function, we get

$$G^{(1)}(\tau) = \langle E^*(t)E(t+\tau)\rangle$$

$$= E_0^2 \sum_{i,j=1}^{\nu} \langle \exp(i\omega_i t - i\varphi_i - i\omega_j \cdot (t+\tau) + i\varphi_j) \rangle.$$

In this case the phases of the emitters are fixed but randomly distributed. Therefore the contributions for $i \neq j$ average to zero, leaving the simplified sum for $i = j$:

$$G^{(1)}(\tau) = E_0^2 \sum_{i=1}^{\nu} \exp(-i\omega_i \tau).$$

To consider the previously discussed inhomogeneous Gaussian distribution in frequency ω_i explicitly, we can replace the sum over the individual emitters into an integral over the normalized Gaussian distribution. Here the Gaussian distribution represents the weight function of the individual frequency components. Because of the normalization, we have to consider an additional factor ν for the number of the emitters.

We therefore get the following expression for the correlation function of the ensemble:

$$G^{(1)}(\tau) = \nu E_0^2 \frac{1}{\sqrt{2\pi}\sigma} \int_0^{\infty} d\omega \exp(-i\omega\tau) \exp\left(-\frac{(\omega-\omega_0)^2}{2\sigma^2}\right)$$
$$= \nu E_0^2 \exp\left(-i\omega_0\tau - \frac{1}{2}\sigma^2\tau^2\right).$$

As a result, we get the *normalized single-emitter first-order correlation function*

$$g^{(1)}(\tau) = \exp\left(-i\omega_0\tau - \frac{1}{2}\sigma^2\tau^2\right)$$
$$= \exp\left(-i\omega_0\tau - \frac{\pi}{2}\left(\frac{\tau}{\tau_c}\right)^2\right). \tag{1.14}$$

For the present case of an *inhomogeneously* broadened ensemble, the *coherence time* τ_c is defined as

$$\tau_c = \frac{\sqrt{\pi}}{\sigma}. \tag{1.15}$$

Figure 1.4 shows the modulus of the correlation function $|g^{(1)}(\tau)|$, where the profile represents also a *Gaussian function*. Note that the width of the function in the time domain possesses the *inverse* value in comparison to the Gaussian distribution in the frequency domain (Fourier pair).

The relationship between the coherence time τ_c and spectral bandwidth $\Delta\nu$ of an emission line (Lorentzian, Gaussian, or Voigt profile) can be generally expressed as

$$\tau_c \sim \frac{1}{\Delta\nu}. \tag{1.16}$$

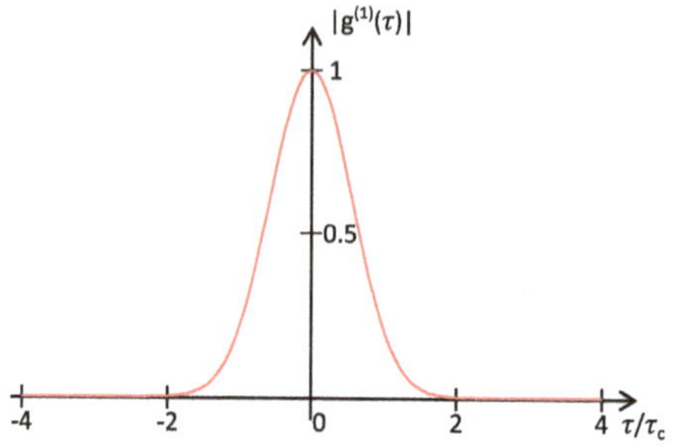

Figure 1.4: $|g^{(1)}(\tau)|$ for a inhomogeneously broadened light source.

The inverse proportionality reflects the property of the Fourier transformation (time vs. frequency domain). Small spectral linewidths correspond to a long coherence time and vice versa. This will be shown explicitly later in this section.

It is obvious from the definition of $g^{(1)}(\tau)$ that $g^{(1)}(0) = 1$ and that the first-order coherence vanishes $g^{(1)}(\tau) \to 0$ for any kind of chaotic light for $\tau \gg \tau_c$.

We can now define the *coherence length* L_c on the basis of the coherence time τ_c accordingly to

$$L_c = \tau_c \cdot c \sim \frac{c}{\Delta\nu} \tag{1.17}$$

with $c = c_0/n$ the velocity of light in a medium with refractive index n (air: $n \approx 1$).

In case that two different line broadening processes affect the emission frequencies of single quantum emitters, the resulting line shape is given by

$$F(\omega) = \int_{-\infty}^{\infty} d\nu F_1(\nu) F_2(\omega + \omega_0 - \nu) \tag{1.18}$$

with the respective two normalized lineshape functions $F_1(\omega)$ and $F_2(\omega)$. Here ω_0 is the central frequency of the two distributions. The integral is invariant under interchange of F_1 and F_2. More than two line broadening mechanisms can be considered by repeatedly applying Eq. (1.18). In many cases, both Lorentzian and Gaussian line broadening processes are active in the solid state. In this case the resulting lineshape is described by a complex error function, and it is named after Voigt [118]. The connection between the correlation function and the emission spectrum will be discussed in the next chapter.

1.4 Michelson interferometry and $g^{(1)}(\tau)$-measurement

In this chapter, we will first discuss the working principle of a Michelson interferometer and how it is used to determine the *coherence time* or rather the *coherence length* of a light source. The Michelson interferometer is discussed from the experimental point

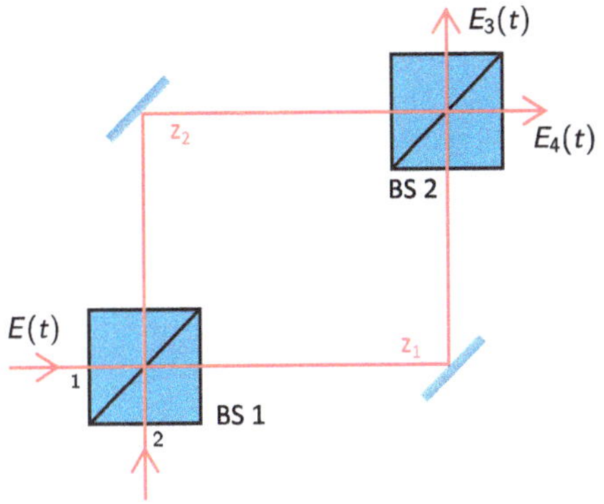

Figure 1.5: Sketch of a Mach–Zehnder interferometer. $E(t)$ and $E_3(t)$, $E_4(t)$ denote the input and output electric fields, respectively. The two internal path lengths of the interferometer between beamsplitter 1 (BS1) and beamsplitter 2 (BS2) are denoted by z_1 and z_2.

of view in Section 5.2.6. Second, we will discuss the fundamental relation between the *coherence properties* of a light source and its corresponding *spectral shape.*

The discussion starts with the treatment of a Mach–Zehnder interferometer with one input field, which is equivalent to a Michelson interferometer but is simpler to illustrate (see Fig. 1.5). In this simplified model experiment, we assume that a parallel light beam (plane wave) falls on the input port 1 of the first beamsplitter (BS1) and no light enters from input port 2. Then $E(t)$ is again the complex electric field of the incident light beam at the input port 1 of the beamsplitter, and both beamsplitters are symmetrical and satisfy the usual beamsplitter relations

$$|R|^2 + |T|^2 = 1, \quad RT^* + TR^* = 0, \tag{1.19}$$

where R and T are the beamsplitter reflection and transmission coefficients, respectively. After the two output ports, the fields are consequently $T\,E(t)$ and $R\,E(t)$. They have in general to travel different lengths z_1 and z_2 before they enter the input ports of the second beamsplitter (BS2) after the times z_1/c and z_2/c. After the output port 4 of the second beamsplitter, the electric field is given by

$$E_4(t) = RT\,E(t_1) + TR\,E(t_2) \tag{1.20}$$

with

$$t_1 = t - (z_1/c) \quad \text{and} \quad t_2 = t - (z_2/c). \tag{1.21}$$

The corresponding intensity of the output light averaged over a cycle of oscillation is given by

$$\bar{I}_4(t) = \frac{1}{2}\,\varepsilon_0 c |E_4(t)|^2$$

$$= \frac{1}{2}\varepsilon_0 c|R|^2|T|^2\{|E(t_1)|^2 + |E(t_2)|^2 + 2\,\mathrm{Re}[E^*(t_1)E(t_2)]\}. \tag{1.22}$$

The typical time resolution of detectors is longer than the coherence times of chaotic light sources. This means that for a proper comparison with the experimental results, it becomes necessary to average $\bar{I}_4(t)$ over an observation period much longer than the coherence time τ_c:

$$\langle \bar{I}_4(t)\rangle = \frac{1}{2}\varepsilon_0 c|R|^2|T|^2\{\langle|E(t_1)|^2\rangle + \langle|E(t_2)|^2\rangle + 2\,\mathrm{Re}\langle E^*(t_1)E(t_2)\rangle\}. \tag{1.23}$$

The output intensity consists of three terms. The first two describe the individual intensities from the two paths (1) and (2). No interference effects arise from these two terms. The interference effects are due to the third contribution, which involves the general first-order correlation function given by

$$\langle E^*(t_1)E(t_2)\rangle = \frac{1}{T}\int_T dt_1 E^*(t_1)E(t_2). \tag{1.24}$$

Note that in this case, there is only a single time variable in the integrand, as t_2 differs from t_1 only by a fixed amount $\tau = (z_2 - z_1)/c$ obtained from Eq. (1.21). Consequently, we can write

$$\langle E^*(t)E(t+\tau)\rangle = \frac{1}{T}\int_T dt E^*(t)E(t+\tau). \tag{1.25}$$

With the definition of the intensity (see Eq. (1.3)), we can now rewrite Eq. (1.23) as

$$\langle \bar{I}_4(t)\rangle = 2|R|^2|T|^2\langle \bar{I}(t)\rangle\{1 + \mathrm{Re}(g^{(1)}(\tau))\}, \tag{1.26}$$

where $\langle \bar{I}(t)\rangle$ is the averaged input intensity. Thus the first-order correlation can be conveniently measured with a Mach–Zehnder or Michelson interferometer, which is equivalent to the above-described Mach–Zehnder interferometer. We can note without proof that $g^{(1)}(-\tau) = g^{(1)}(\tau)^*$.

If we assume a Lorentzian line broadening mechanism, then the output intensity of the Mach–Zehnder can be written in a more explicit form using Eq. (1.13):

$$\langle \bar{I}_4(t)\rangle = 2|R|^2|T|^2\langle \bar{I}(t)\rangle\{1 + \exp(-|\tau|/\tau_c)\cos(\omega_0\tau)\} \tag{1.27}$$

with $\tau = (z_2 - z_1)/c$, where $z_2 - z_1$ is the relative path length difference inside the interferometer. The *visibility* is defined as

$$V = |g^{(1)}(\tau)| = \exp(-\tau/\tau_c). \tag{1.28}$$

Thus the fringes in Eq. (1.27) disappear due to the finite coherence time τ_c of the light source.

Now we will discuss the fundamental relation between the *coherence properties* of a light source and its corresponding *spectral shape*. To determine the average spectrum of random light, we carry out a Fourier decomposition of the electric field $E(t)$:

$$E(\nu) = \int_{-\infty}^{\infty} E(t) \exp(-i2\pi\nu t)\, dt. \tag{1.29}$$

Here $|E(\nu)|^2$ represents the *energy spectral density* in the interval $[\nu, \nu + d\nu]$.

Note that the *complex* electric field amplitude $E(t)$ has been defined so that $E(\nu) = 0$ for negative values of ν. We can now define a truncated Fourier transform $E_T(\nu)$ within a time window $[-\frac{T}{2}, +\frac{T}{2}]$ where the time T is taken to be much longer than the coherence time τ_c,

$$E_T(\nu) = \int_{-\frac{T}{2}}^{\frac{T}{2}} E(t) \exp(-i2\pi\nu t)\, dt. \tag{1.30}$$

With $E_T(\nu)$ and the time window taken to ∞, we can define the *power spectral density*

$$S(\nu) = \lim_{T\to\infty} \frac{1}{T} \langle |E_T(\nu)|^2 \rangle. \tag{1.31}$$

The *intensity* of the light beam can now be expressed with the spectral density $S(\nu)$:

$$I = \int_{0}^{\infty} S(\nu)\, d\nu. \tag{1.32}$$

We can show that the first-order coherence function $G^{(1)}$ (Eq. (1.4)) and the spectral density $S(\nu)$ (Eq. (1.31)) form a *Fourier transform pair*:

$$S(\nu) = \int_{-\infty}^{\infty} G^{(1)}(\tau) \exp(-i2\pi\nu\tau)\, d\tau, \tag{1.33}$$

which is the well-known *Wiener–Khintchine theorem*. It shows that the spectrum is fundamentally connected to the first-order correlation function, i. e., time-dependent fluctuations are directly connected to the shape of light spectrum.

1.5 Second-order coherence $g^{(2)}(\tau)$ and intensity fluctuations

After the discussion of field fluctuations of chaotic light and their role in first-order interference experiments, we will discuss a physical description of intensity fluctuations and their measurement. Here we discuss higher-order interference effects that depend on the correlations of two intensities. The relevant quantity for the description is the *second-order correlation function*, which will be defined below. We will also show that the second-order correlation function is closely connected with some properties of the first-order correlation function. The treatment of the theory in this chapter includes again some shortcuts and omissions of more advanced derivations, which can be found, e. g., in the book of Loudon [118].

For all further discussions on the intensity behavior of chaotic light, we will use the following assumptions:

- We consider a stationary light beam whose fluctuations are produced by ergodic random processes (see also Section 1.3).
- The time-resolution $\Delta t_{\text{Det.}}$ for correlation measurements is much better than the characteristic coherence time τ_c, i. e., instantaneous intensity measurements can be performed.
- It is averaged over a large number of values of the cycle-averaged intensity $\bar{I}(t)$, measured over a period of time much longer than τ_c. In this case the time average can be replaced by a statistical average over the distribution of phase angles.
- The emission frequencies for all emitters are identical, $\omega_i = \omega_0$.
- The emitter phases are statistically distributed in the range $[0, 2\pi]$.

First, we will consider two auxiliary quantities, (a) the long-time average of intensity $\bar{I}$ and (b) the mean square intensity $\langle \bar{I}(t)^2 \rangle$.

(a) The long-time average intensity, by Eqs. (1.1) and (1.3), is given by

$$\begin{aligned}\bar{I} \equiv \langle \bar{I}(t) \rangle &= \frac{1}{2}\varepsilon_0 c E_0^2 \langle |\exp(i\varphi_1(t)) + \exp(i\varphi_2(t)) + \cdots + \exp(i\varphi_\nu(t))|^2 \rangle \\ &= \frac{1}{2}\varepsilon_0 c E_0^2 \langle [\underbrace{1+1+1+\cdots}_{\nu \text{ summands}} + \quad (\text{cross terms}) \quad] \rangle \\ &= \frac{1}{2}\varepsilon_0 c E_0^2 \cdot \nu. \end{aligned} \tag{1.34}$$

For the calculation, we used the ergodic hypothesis that the *long time average* of the total intensity $\bar{I}(t)$ can be replaced by the *ensemble average* $\langle \bar{I}(t) \rangle$ by summing up over all independent emitters. We considered that only the corresponding terms of the type $\exp(i\varphi_j)\exp(-i\varphi_j) = 1$ will contribute to the sum, whereas cross terms with $i \neq j$ will average out to zero.

(b) The mean square intensity $\langle \bar{I}(t)^2 \rangle$ is

$$\langle \bar{I}(t)^2 \rangle = \frac{1}{4}\varepsilon_0^2 c^2 E_0^4 \langle |\exp(i\varphi_1(t)) + \exp(i\varphi_2(t)) + \cdots + \exp(i\varphi_\nu(t))|^4 \rangle. \tag{1.35}$$

The explicit multiplication of this expression is lengthy but straightforward. As already mentioned above, in (a), only the product terms that contain their complex conjugates contribute. There exist ν terms with four equal indices and $2\nu(\nu-1)$ terms that have pairwise the same indices. With the result from (a) we get

$$\begin{aligned}\langle \bar{I}(t)^2\rangle &= \frac{1}{4}\varepsilon_0^2 c^2 E_0^4(\nu + 2\nu(\nu-1))\\ &= \left(2-\frac{1}{\nu}\right)\bar{I}^2.\end{aligned} \tag{1.36}$$

For large emitter numbers $\nu \gg 1$, we get

$$\langle \bar{I}(t)^2\rangle = 2\bar{I}^2. \tag{1.37}$$

Let us now calculate the *variance* $(\Delta I)^2$:

$$(\Delta I)^2 = \langle \bar{I}(t)^2\rangle - \langle \bar{I}(t)\rangle^2 = 2\bar{I}^2 - \bar{I}^2 = \bar{I}^2. \tag{1.38}$$

This means that the size of fluctuation (ΔI) for a chaotic light source is equal to its mean value $\bar{I}$.

Second-order correlation function $g^{(2)}(\tau)$

The second-order correlation function plays an important role in quantum optics. It allows the distinction between classical light, e. g., from LEDs and lasers, and nonclassical light, e. g., from single quantum emitters.

Let us again consider stationary light beams that can be described by plane waves (no transverse effects) and exhibit a single polarization. In this case the second-order correlation can be defined as

$$g^{(2)}(\tau) = \frac{\langle \bar{I}(t)\bar{I}(t+\tau)\rangle}{\langle \bar{I}(t)\rangle^2} = \frac{\langle E^*(t)E^*(t+\tau)E(t+\tau)E(t)\rangle}{\langle E^*(t)E(t)\rangle^2}. \tag{1.39}$$

The order of field factors follows a convention, which will become important in quantum optics. From the symmetry of the definition it follows that

$$g^{(2)}(\tau) = g^{(2)}(-\tau). \tag{1.40}$$

Let us now calculate the second-order correlation function for chaotic light, both for homogeneous and inhomogeneous linewidth broadenings. Therefore we will use the same model for the emitters as already discussed in the previous sections. The electric field is made up of independent emitters, and time averaging can be replaced with ensemble averaging (ergodic hypotheses). Considering ν emitters, we have

$$E(t) = \sum_{i=1}^{\nu} E_i(t),$$

and plugging this term into the numerator of $g^{(2)}(\tau)$, we get

$$\begin{aligned}\langle E^*(t)E^*(t+\tau)E(t+\tau)E(t)\rangle = &\sum_{i=1}^{\nu}\langle E_i^*(t)E_i^*(t+\tau)E_i(t+\tau)E_i(t)\rangle \\ &+ \sum_{i\neq j}\{\langle E_i^*(t)E_j^*(t+\tau)E_j(t+\tau)E_i(t)\rangle \\ &+ \langle E_i^*(t)E_j^*(t+\tau)E_i(t+\tau)E_j(t)\rangle\}.\end{aligned}$$

Here only the terms in which the fields from each emitter are multiplied by their complex conjugates are maintained. The other terms vanish because of random relative phases. If we consider that each emitter contributes equally, then we can write

$$\begin{aligned}\langle E^*(t)E^*(t+\tau)E(t+\tau)E(t)\rangle = &\nu\langle E_i^*(t)E_i^*(t+\tau)E_i(t+\tau)E_i(t)\rangle \\ &+ \nu(\nu-1)\langle E_i^*(t)E_i(t)\rangle^2 + \nu(\nu-1)|\langle E_i^*(t)E_i(t+\tau)\rangle|^2.\end{aligned}$$

In this connection the prefactor ν appears because of the contributions of single emitters, and the factor $\nu(\nu-1)$ results from pair contributions. In the realistic case of a large number of emitters, i. e., $\nu \to \infty$, only the pair contribution is decisive. We therefore get

$$\langle E^*(t)E^*(t+\tau)E(t+\tau)E(t)\rangle \approx \nu^2\langle E_i^*(t)E_i(t)\rangle^2 + \nu^2|\langle E_i^*(t)E_i(t+\tau)\rangle|^2.$$

We can rewrite this last term with the definition of

$$g^{(1)}(\tau) = \frac{\langle E^*(t)E(t+\tau)\rangle}{\langle E^*(t)E(t)\rangle},$$

and by normalization on the number of emitters (division by ν^2) we get

$$g^{(2)}(\tau) = 1 + |g^{(1)}(\tau)|^2 \quad (\text{for } \nu \gg 1). \tag{1.41}$$

Thus the first- and second-order correlation functions of *chaotic light* are fundamentally connected to each other. It is now interesting to discuss the possible range of values of $g^{(2)}(\tau)$ for chaotic light. The allowed values for the first-order coherence are (see the previous section)

$$0 \leq |g^{(1)}(\tau)| \leq 1.$$

Consequently, the allowed values for the second-order coherence are

$$1 \leq |g^{(2)}(\tau)| \leq 2$$

with

$$g^{(2)}(0) = 2 \quad \text{and} \quad g^{(2)}(\tau) \to 1 \quad \text{for } \tau \gg \tau_c.$$

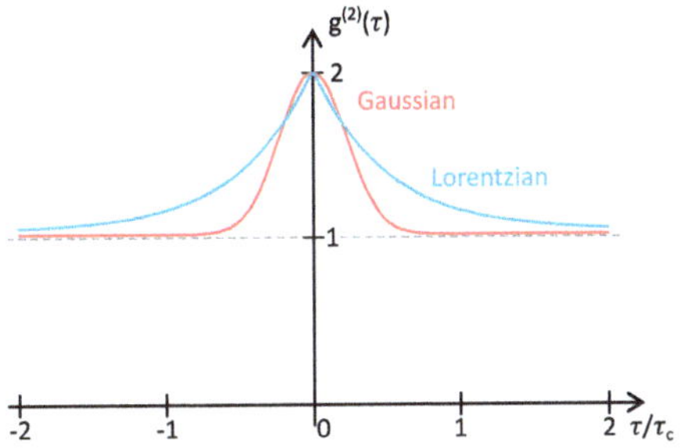

Figure 1.6: Second-order coherence functions of homogeneously (blue) and inhomogeneously (red) broadened chaotic light.

It is important to remark that in a more general treatment based on Cauchy's inequality, there is no upper limit for $g^{(2)}(0)$ for a stationary or nonstationary classical light source, i. e., $1 \leq g^{(2)}(0) \leq \infty$ (see [118]).

The degree of second-order coherence can now be easy calculated for Lorentzian and Gaussian broadened chaotic light (see Fig. 1.6) using Eq. (1.41) and their respective first-order coherence functions (Eqs. (1.13) and (1.14)):

– Homogeneous broadening (Lorentzian broadening):

$$g^{(2)}(\tau) = 1 + \exp(-2\gamma|\tau|) = 1 + \exp\left(-\frac{2|\tau|}{\tau_c}\right).$$

– Inhomogeneous broadening (Gaussian broadening):

$$g^{(2)}(\tau) = 1 + \exp(-\Delta^2\tau^2) = 1 + \exp\left(-\pi\left(\frac{\tau}{\tau_c}\right)\right)^2.$$

1.6 Hanbury Brown–Twiss interferometer and $g^{(2)}(\tau)$-measurement

In this section, we discuss the working principle of the Hanbury Brown–Twiss (HBT) interferometer. The setup is schematically shown in Fig. 1.7, where a symmetric 50:50 beamsplitter and two detectors symmetrically placed with respect to the output ports of the beamsplitter are considered. The experimental use of such an interferometer is discussed in Section 5.2.5.

The HBT interferometer measures the correlation between two intensities expressed in terms of the second-order coherence. It is the working horse in quantum optics to characterize quantum light, e. g., to prove and quantify the purity of single-photon emission (cf. Section 5.2.5), which will be discussed later in the framework of quantum theory.

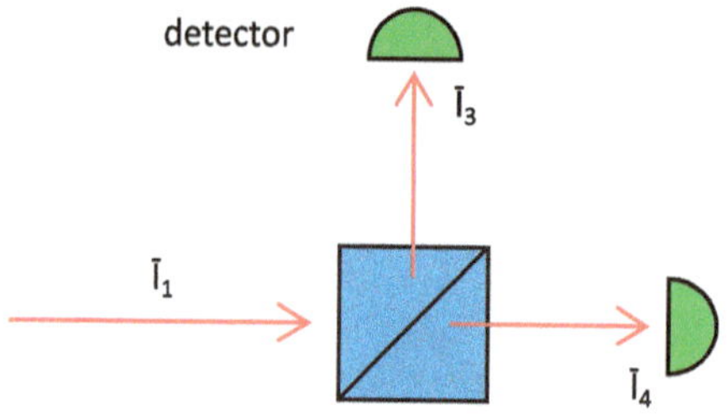

Figure 1.7: Schematic sketch of a Hanbury Brown–Twiss setup with a 50:50 beamsplitter and two detectors symmetrically placed at the output port of the beamsplitter, i. e., at the same linear distance *z* from the light source.

The incoming cycle-averaged intensity $\bar{I}_1(z,t)$ is divided by the 50:50 beamsplitter into two equal beams $\bar{I}_3(z,t)$ and $\bar{I}_4(z,t)$. Because of energy conservation, we can write

$$\bar{I}_3(z,t) = \bar{I}_4(z,t) = \frac{\bar{I}_1(z,t)}{2}, \tag{1.42}$$

and the long-time averages are

$$\bar{I}_3 = \bar{I}_4 = \frac{\bar{I}_1}{2}. \tag{1.43}$$

The intensities $\bar{I}_3$ and $\bar{I}_4$ are measured by detectors, typically avalanche photo-detectors, and are multiplied together in the correlator. The detailed experimental setup will be discussed in Section 5.2.5. The *second-order correlation function* can then be expressed via the temporal correlation of the intensities $\bar{I}_3$ and $\bar{I}_4$:

$$\begin{aligned} g^{(2)}_{3,4}(\tau) &= \frac{\langle \bar{I}_3(t)\bar{I}_4(t+\tau)\rangle}{\bar{I}_3\bar{I}_4} \\ &= \frac{\langle \bar{I}_1(t)\bar{I}_1(t+\tau)\rangle}{\bar{I}_1^2} \\ &= g^{(2)}_{1,1}(\tau). \end{aligned} \tag{1.44}$$

Obviously, the normalized cross-correlation between the two output beams provide a measurement of the second-order coherence function of the input beam.

1.7 Summary

- The model of a chaotic light source is a very useful tool to understand the basic properties of light fluctuations and their consequences for spectral line broadening.

- Homogeneous line broadening leads to Lorentzian frequency distributions, whereas an inhomogeneous broadening mechanism leads typically to Gaussian frequency distributions.
- The Wiener–Khintchine theorem shows that the spectrum of a light source is fundamentally connected to its first-order correlation function.
- The first-order correlation function $g^{(1)}(\tau)$ quantifies the coherence of a light field and can be measured with the Michelson interferometer (MI).
- The second-order correlation function $g^{(2)}(\tau)$ quantifies intensity correlations of a light field and can be measured with the Hanbury Brown–Twiss (HBT) interferometer.

2 Quantum theory of radiation

2.1 Introductory remarks

A consistent theory of quantum light generation with individual quantum emitters must treat the whole system of emitters and radiation in quantum mechanical terms. In this chapter, we introduce the quantization of the light field and discuss its fundamental properties. We summarize the derivation of the field quantization starting from the classical theory of light. A more sophisticated derivation can be found in standard textbooks on quantum optics.

2.2 Classical theory of the electromagnetic field

Maxwell's equations can be transformed into a set of two independent differential equations by introducing the *scalar potential* $\Phi(\vec{r},t)$ and the *vector potential* $\vec{A}(\vec{r},t)$:

$$\nabla^2\Phi(\vec{r},t) = -\frac{1}{\varepsilon_0}\cdot\rho(\vec{r},t), \tag{2.1}$$

$$\nabla^2\vec{A}(\vec{r},t) - \frac{1}{c^2}\frac{\partial^2}{\partial t^2}\vec{A}(\vec{r},t) = -\mu_0\cdot j_{\text{trans}}(\vec{r},t). \tag{2.2}$$

Here $\vec{j}_{\text{trans}}$ is the transverse component of the current density, where $\nabla\cdot\vec{j}_{\text{trans}} = 0$ and $\vec{j} = \vec{j}_{\text{long}} + \vec{j}_{\text{trans}}$ with $\vec{j}_{\text{long}}$ the longitudinal component of the current density with $\nabla\times\vec{j}_{\text{long}} = 0$, $\rho(\vec{r},t)$ is the density of free charges, ε_0 is the electric permittivity, and μ_0 is the magnetic permeability of the vacuum. We use the usual notation for the fields,

$$\vec{E}(\vec{r},t) = -\frac{\partial}{\partial t}\vec{A}(\vec{r},t) - \nabla\Phi(\vec{r},t), \tag{2.3}$$

$$\vec{B}(\vec{r},t) = \nabla\times\vec{A}(\vec{r},t), \tag{2.4}$$

and the *Coulomb gauge*, which satisfies the condition

$$\nabla\cdot\vec{A} = 0. \tag{2.5}$$

In an empty space the current density and the density of charge are zero ($\vec{j}_{\text{trans}} = 0$, $\rho = 0$). Furthermore, in the Coulomb gauge, we can choose $\Phi = 0$. The Maxwell equations therefore merge into a more simple equation for $\vec{A}(\vec{r},t)$, i. e., a homogeneous wave equation, and the fields are given by

$$\nabla^2\vec{A}(\vec{r},t) - \frac{1}{c^2}\frac{\partial^2}{\partial t^2}\vec{A}(\vec{r},t) = 0, \tag{2.6}$$

https://doi.org/10.1515/9783110703412-002

$$\vec{E} = -\frac{\partial \vec{A}}{\partial t}, \tag{2.7}$$

$$\vec{B} = \nabla \times \vec{A}. \tag{2.8}$$

The total energy of the electromagnetic field is given by

$$E = \frac{1}{2} \int_V \mathrm{d}^3 r \left(\varepsilon_0 \vec{E}^2 + \frac{1}{\mu_0} \vec{B}^2 \right), \tag{2.9}$$

where the spatial integration extends over the volume V in which the field is considered.

The volume in which the field is considered has to be specified even if the space is infinite. As an example, we will discuss a cubic volume (see Fig. 2.1) without solid walls with linear dimensions L, i. e., we can apply *periodic boundary conditions.*

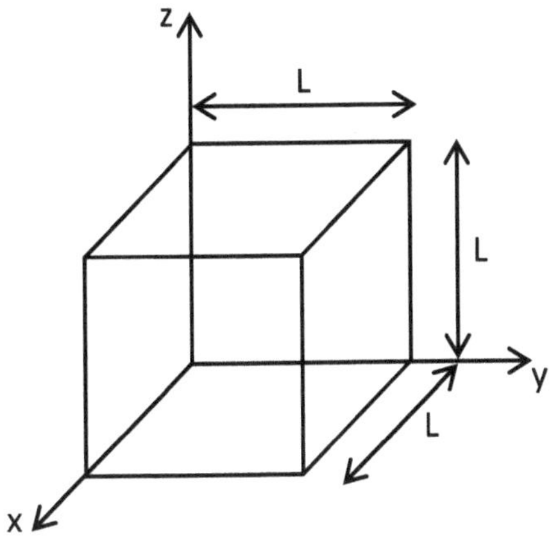

Figure 2.1: Cubic volume of linear dimension L.

For the solution of the wave equation, we can use the well-known expression for plane waves, neglecting first the vector character of $\vec{A}$:

$$A(\vec{r}, t) \sim e^{\pm i \vec{k} \vec{r}} \cdot e^{\mp i \omega t}.$$

We can now consider the periodicity of the field in all three dimensions:

$$A(\vec{r}, t) = A(\vec{r} + \vec{L}, t), \tag{2.10}$$

which restricts the allowed values of the $\vec{k}$-vectors to

$$k_x = \frac{2\pi n_x}{L}, \tag{2.11}$$

$$k_y = \frac{2\pi n_y}{L}, \tag{2.12}$$

$$k_z = \frac{2\pi n_z}{L}, \tag{2.13}$$

where $n_{x,y,z} = \pm 1, \pm 2, \ldots$.

Now we can consider the vector character of $\vec{A}$ within a factor:

$$\sim \vec{e}_{k,\lambda} e^{\pm i\vec{k}\vec{r}}$$

with unit vectors $\vec{e}_{k,\lambda}$, which must be chosen so that the Coulomb gauge ($\nabla \cdot \vec{A} = 0$) is still fulfilled. For each wave vector $\vec{k}$, we have to chose exactly two unit vectors perpendicular to each other and also to $\vec{k}$, with λ taking values 1 and 2. These unit vectors are the polarization vectors of the radiation field (see Fig. 2.2).

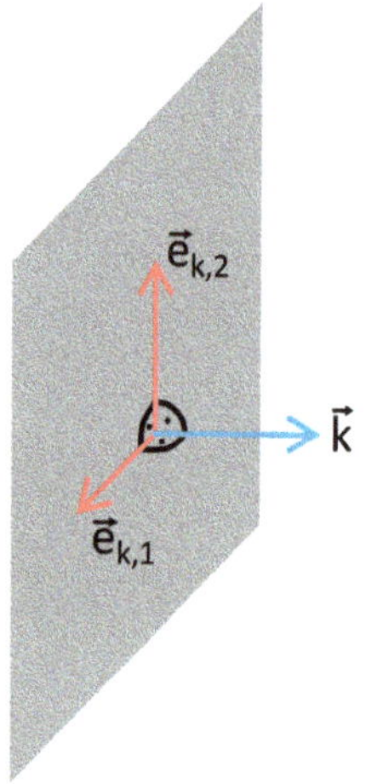

Figure 2.2: Polarization unit vectors and wave vector of the radiation field.

The polarization vectors, together with $\vec{k}$, are chosen to form a right-handed system:

$$\vec{e}_{k,\lambda} \cdot \vec{e}_{k,\lambda'} = \delta_{\lambda,\lambda'} \quad \text{(Kronecker delta)}, \tag{2.14}$$

$$\vec{k} \cdot \vec{e}_{k,\lambda} = 0. \tag{2.15}$$

The vector potential $\vec{A}(\vec{r}, t)$ can therefore be represented as a *sum of modes* $\vec{k}, \lambda$:

$$\vec{A}(\vec{r}, t) = \sum_{\vec{k}} \sum_{\lambda} [A_{\vec{k},\lambda}(t) \cdot \vec{u}_{\vec{k},\lambda}(\vec{r}) + A^*_{\vec{k},\lambda}(t) \cdot \vec{u}^*_{\vec{k},\lambda}(\vec{r})], \tag{2.16}$$

where $\vec{u}_{\vec{k},\lambda}(\vec{r}) = \vec{e}_{\vec{k},\lambda} e^{i\vec{k}\vec{r}}$, and each $A_{\vec{k},\lambda}(t)$ fulfills the differential equation of the harmonic oscillator

$$\left(c^2 k^2 + \frac{\partial^2}{\partial t^2}\right) A_{\vec{k},\lambda}(t) = 0 \tag{2.17}$$

with the dispersion relation for light $\omega_k = c \cdot k$. The Fourier components $A_{\vec{k},\lambda}(t)$ are

$$A_{\vec{k},\lambda}(t) = A_{\vec{k},\lambda} e^{-i\omega_k t}. \tag{2.18}$$

The corresponding expressions for the electric and magnetic fields follow from Eq. (2.6):

$$\vec{E}(\vec{r},t) = -\frac{\partial \vec{A}}{\partial t} = i \sum_{\vec{k},\lambda} \omega_k [A_{\vec{k},\lambda}(t) \cdot \vec{u}_{\vec{k},\lambda}(\vec{r}) - A^*_{\vec{k},\lambda}(t) \cdot \vec{u}^*_{\vec{k},\lambda}(\vec{r})], \tag{2.19}$$

$$\vec{B}(\vec{r},t) = \vec{\nabla} \times \vec{A} = i \sum_{\vec{k},\lambda} \vec{k} \times [A_{\vec{k},\lambda}(t) \cdot \vec{u}_{\vec{k},\lambda}(\vec{r}) - A^*_{\vec{k},\lambda}(t) \cdot \vec{u}^*_{\vec{k},\lambda}(\vec{r})]. \tag{2.20}$$

The expressions for $\vec{E}(\vec{r},t)$ and $\vec{B}(\vec{r},t)$ correspond to Fourier expansions with Fourier components $A_{\vec{k},\lambda}$.

Using the Fourier expansion of $\vec{A}$, we can express the field energy in Eq. (2.9) to

$$\begin{aligned} E &= \varepsilon_0 V \sum_{\vec{k},\lambda} \omega_k^2 (A_{\vec{k},\lambda} A^*_{\vec{k},\lambda} + A^*_{\vec{k},\lambda} A_{\vec{k},\lambda}) \\ &= 2\varepsilon_0 V \sum_{\vec{k},\lambda} |A_{\vec{k},\lambda}|^2 \omega_k^2. \end{aligned} \tag{2.21}$$

It is important to remark here that the latter equation only holds for the classical case.

The goal is now to prepare the quantization of the electromagnetic field. Therefore we introduce new variables. We define

$$A_{\vec{k},\lambda} = \sqrt{\frac{\hbar}{2\varepsilon_0 V \omega_k}} \alpha_{\vec{k},\lambda}, \tag{2.22}$$

and with this definition, it is possible to rewrite the total energy with the new variables:

$$E = \frac{1}{2} \sum_{\vec{k},\lambda} \hbar\omega_k (\alpha_{\vec{k},\lambda} \alpha^*_{\vec{k},\lambda} + \alpha^*_{\vec{k},\lambda} \alpha_{\vec{k},\lambda}). \tag{2.23}$$

Furthermore, we introduce new real variables $q_{\vec{k},\lambda}$ and $p_{\vec{k},\lambda}$:

$$q_{\vec{k},\lambda} \equiv \sqrt{\frac{\hbar}{2\omega_k}} (\alpha_{\vec{k},\lambda} + \alpha^*_{\vec{k},\lambda}), \tag{2.24}$$

$$p_{\vec{k},\lambda} \equiv -i \sqrt{\frac{\hbar\omega_k}{2}} (\alpha_{\vec{k},\lambda} - \alpha^*_{\vec{k},\lambda}), \tag{2.25}$$

and the expression for the energy becomes

$$E = \frac{1}{2} \sum_{\vec{k},\lambda} (p^2_{\vec{k},\lambda} + \omega^2 q^2_{\vec{k},\lambda}). \tag{2.26}$$

We get an expression that is the sum of the energies of an *infinite set of independent harmonic oscillators*. The variables $q_{\vec{k},\lambda}$ and $p_{\vec{k},\lambda}$ play the role of position and its conjugate momentum according to the canonical conjugate variables of Hamilton mechanics.

2.3 Field quantization in open space

Based on the pure classical treatment of the electromagnetic field in the previous section, we will now introduce the quantum mechanical treatment. As seen before, the energy of the electromagnetic field can be expressed as the sum of energies of an *infinite set of independent harmonic oscillators* with the canonical conjugate variables $q_{\vec{k},\lambda}$ and $p_{\vec{k},\lambda}$.

As it is known from basic quantum mechanics, the transition from the classical to the quantum picture is performed by replacing the position $q_{\vec{k},\lambda}$ and its conjugate momentum $p_{\vec{k},\lambda}$ by the corresponding Hermitian operators:

$$q_{\vec{k},\lambda} \to \hat{Q}_{\vec{k},\lambda}, \tag{2.27}$$

$$p_{\vec{k},\lambda} \to \hat{P}_{\vec{k},\lambda}, \tag{2.28}$$

where the operators obey the *commutation relations*

$$[\hat{Q}_{\vec{k},\lambda}, \hat{P}_{\vec{k}',\lambda'}] = i\hbar\, \delta_{\vec{k},\vec{k}'}\, \delta_{\lambda,\lambda'} \tag{2.29}$$

and

$$[\hat{Q}_{\vec{k},\lambda}, \hat{Q}_{\vec{k}',\lambda'}] = [\hat{P}_{\vec{k},\lambda}, \hat{P}_{\vec{k}',\lambda'}] = 0. \tag{2.30}$$

On the basis of these operators, we can define the so-called *annihilation* and *creation operators*:

$$\hat{a}_{\vec{k},\lambda} = \frac{1}{\sqrt{2\hbar\omega_k}}(\omega_k \hat{Q}_{\vec{k},\lambda} + i\hat{P}_{\vec{k},\lambda}), \tag{2.31}$$

$$\hat{a}^{\dagger}_{\vec{k},\lambda} = \frac{1}{\sqrt{2\hbar\omega_k}}(\omega_k \hat{Q}_{\vec{k},\lambda} - i\hat{P}_{\vec{k},\lambda}), \tag{2.32}$$

which fulfill the following *bosonic commutation relations*:

$$[\hat{a}_{\vec{k},\lambda}, \hat{a}^{\dagger}_{\vec{k}',\lambda'}] = \delta_{\vec{k},\vec{k}'}\delta_{\lambda,\lambda'}$$

and

$$[\hat{a}_{\vec{k},\lambda}, \hat{a}_{\vec{k}',\lambda'}] = [\hat{a}^{\dagger}_{\vec{k},\lambda}, \hat{a}^{\dagger}_{\vec{k}',\lambda'}] = 0$$

with the definition

$$[\hat{a}, \hat{a}^\dagger] = \hat{a}\hat{a}^\dagger - \hat{a}^\dagger\hat{a} = 1.$$

With the help of the annihilation and creation operators, we can define the *number operator*

$$\hat{n}_{\vec{k},\lambda} = \hat{a}^\dagger_{\vec{k},\lambda}\hat{a}_{\vec{k},\lambda}. \tag{2.33}$$

It is important to note that $\hat{a}^\dagger$ and $\hat{a}$ are non-Hermitian operators, i. e., they do not represent observables, but they are extremely useful in quantum mechanical calculations.

Taking the creation and annihilation operators, we can express the Hamiltonian of the energy of the electromagnetic field in analogy to the formulation of the classic electromagnetic field (Eq. (2.23)):

$$\begin{aligned}\hat{H} = \frac{1}{2}\sum_{\vec{k},\lambda}\hbar\omega_k(\hat{a}^\dagger_{\vec{k},\lambda}\hat{a}_{\vec{k},\lambda} + \hat{a}_{\vec{k},\lambda}\hat{a}^\dagger_{\vec{k},\lambda}) &= \sum_{\vec{k},\lambda}\hbar\omega_k\left(\hat{a}^\dagger_{\vec{k},\lambda}\hat{a}_{\vec{k},\lambda} + \frac{1}{2}\right)\\ &= \sum_{\vec{k},\lambda}\hbar\omega_k\left(\hat{n}_{\vec{k},\lambda} + \frac{1}{2}\right).\end{aligned} \tag{2.34}$$

The total sum in $\hat{H}$ corresponds to an infinite number of *quantized harmonic oscillators*. The factor $\sum_{\vec{k},\lambda}\frac{1}{2}\hbar\omega_k$ represents the *zero point energy* of all harmonic oscillators.

The vector potential and electric field can be now also expressed by $\hat{a}^\dagger_{\vec{k},\lambda}$ and $\hat{a}_{\vec{k},\lambda}$:

$$\hat{\vec{A}} = \sum_{\vec{k},\lambda}\vec{e}_{\vec{k},\lambda}\sqrt{\frac{\hbar}{2\varepsilon_0 V\omega_k}}\,[\hat{a}_{\vec{k},\lambda}e^{i(\vec{k}\vec{r}-\omega_k t)} + \hat{a}^\dagger_{\vec{k},\lambda}e^{-i(\vec{k}\vec{r}-\omega_k t)}], \tag{2.35}$$

$$\hat{\vec{E}} = i\sum_{\vec{k},\lambda}\vec{e}_{\vec{k},\lambda}\sqrt{\frac{\hbar\omega_k}{2\varepsilon_0 V}}\,[\hat{a}_{\vec{k},\lambda}e^{i(\vec{k}\vec{r}-\omega_k t)} - \hat{a}^\dagger_{\vec{k},\lambda}e^{-i(\vec{k}\vec{r}-\omega_k t)}]. \tag{2.36}$$

It is common to separate the electric field operator $\hat{\vec{E}}$ into the positive and negative frequency parts:

$$\hat{\vec{E}}(\vec{r},t) = \hat{\vec{E}}^+(\vec{r},t) + \hat{\vec{E}}^-(\vec{r},t), \tag{2.37}$$

where $\hat{\vec{E}}^+(\vec{r},t)$ contains the sum of annihilation operators $\hat{a}_{\vec{k},\lambda}$, and $\hat{\vec{E}}^-(\vec{r},t)$ contains the sum of creation operators $\hat{a}^\dagger_{\vec{k},\lambda}$. The connection between the two operators is

$$\hat{\vec{E}}^+ = (\hat{\vec{E}}^-)^\dagger \quad \text{(adjoint operator)}. \tag{2.38}$$

The expression for the electric field (Eq. (2.36)) is of fundamental importance for:
(1) the description of *electric dipole transitions*,
(2) the definition of the *coherence functions*,

(3) especially for a single mode $(\vec{k}, \lambda)$, we have

$$\hat{E}^+ \sim \hat{a} \quad \text{and} \quad \hat{E}^- \sim \hat{a}^\dagger.$$

In the following, we summarize the most fundamental properties of the quantum mechanical oscillator and some basics of the operator algebra of the creation and annihilation operators. For a more detailed discussion, we refer to standard books of quantum mechanics.

– *Hamilton operator of the harmonic oscillator (only one mode $\omega_k = \omega$):*

$$\begin{aligned}\hat{H} &= \frac{1}{2}\hbar\omega(\hat{a}\hat{a}^\dagger + \hat{a}^\dagger\hat{a}) \\ &= \hbar\omega\left(\hat{a}^\dagger\hat{a} + \frac{1}{2}\right) \\ &= \hbar\omega\left(\hat{n} + \frac{1}{2}\right).\end{aligned} \tag{2.39}$$

– *Eigenvalue equation of the quantized harmonic oscillator:*

$$\hat{H}|n\rangle = E_n|n\rangle \tag{2.40}$$

with $|n\rangle$ as eigenstates of $\hat{H}$ and $\hat{n}$.

– *Eigenvalues of the Hamiltonian* (see Fig. 2.3):

$$E_n = \hbar\omega\left(n + \frac{1}{2}\right), \quad n = 0, 1, 2, \dots. \tag{2.41}$$

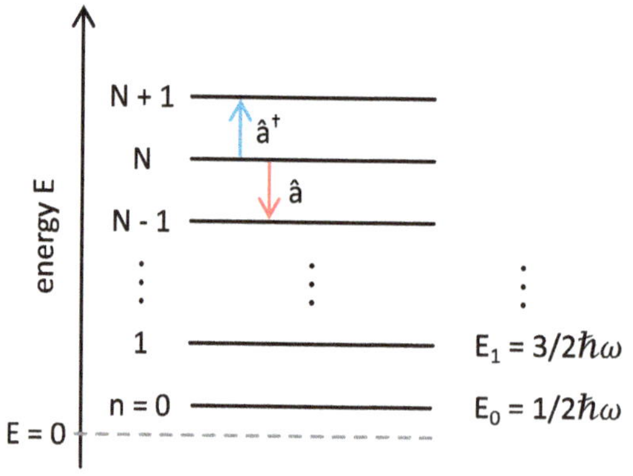

Figure 2.3: Energy spectrum of a single mode $\omega_k = \omega$ of the electromagnetic field (harmonic oscillator). In contrast to the classical harmonic oscillator, there is an additional energy contribution for $n = 0$, which is the zero point energy $E_0 = \frac{1}{2}\hbar\omega$ of the quantized harmonic oscillator.

– *Impact of the annihilation/creation and photon number operators on the eigenstates $|n\rangle$ of the operator $\hat{n}$:*

$$\hat{a}|n\rangle = \sqrt{n}\,|n-1\rangle \qquad \text{annihilation operator,} \tag{2.42}$$

$$\hat{a}^{\dagger}|n\rangle = \sqrt{n+1}\,|n+1\rangle \qquad \text{creation operator,} \tag{2.43}$$

$$\hat{n}|n\rangle = n|n\rangle \qquad \text{eigenvalue equation of } \hat{n}. \tag{2.44}$$

– *Furthermore*:

$$\hat{a}|0\rangle = 0 \qquad \text{(vacuum state cannot be reduced!),} \tag{2.45}$$

$$|n\rangle = \frac{(\hat{a}^{\dagger})^n}{\sqrt{n!}}|0\rangle \qquad \text{(recursive creation of the state } |n\rangle \text{ from } |0\rangle). \tag{2.46}$$

2.4 Field quantization in a cavity

We will now shortly discuss the field quantization in a finite quasi-one-dimensional cavity (quasi-1D) with perfectly reflecting mirrors ($R_1 = R_2 = 1$, see Fig. 2.4). Consequently, the electric field vanishes at the mirrors. Since the field is confined in the cavity, there will be standing instead of running waves. The "quasi-1D" refers to the fact that although a cavity in reality is three-dimensional (3D), here we will only discuss modes with photons propagating in the z-direction. The detailed transverse mode structure in the xy-directions is neglected and depends on the geometry of the respective real cavity. At the end of this section, we will shortly note the necessary modification for describing a three-dimensional cavity.

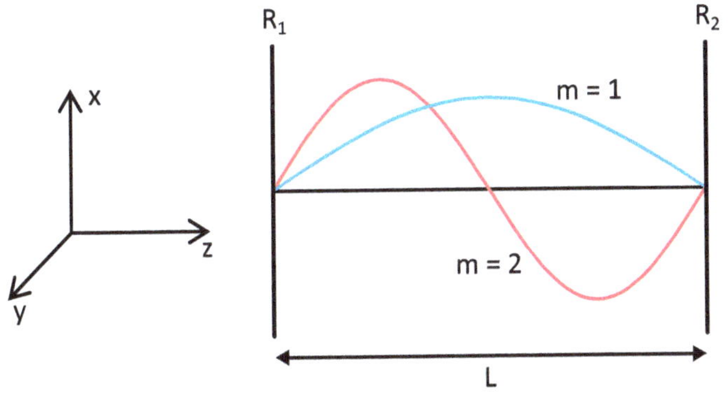

Figure 2.4: Quasi one-dimensional resonator (1D; z-direction) of length L, showing the two lowest allowed standing waves (modes) of the electric field for the case of perfect mirrors with reflectivities $R_1 = R_2 = 1$. V denominates the respective mode volume.

- *Quasi-one-dimensional resonator:* In this case the electric and magnetic field can be simply described by

$$\vec{E}(\vec{r},t) \to \vec{e}E(z,t) = \vec{e}_x E(z,t), \tag{2.47}$$
$$\vec{B}(\vec{r},t) \to \vec{k} \times \vec{e}B(z,t) = \vec{e}_y B(z,t), \tag{2.48}$$

where we assumed that the electric field is polarized along the x-axis. In this notation, $\vec{e}_{(x,y)}$ denominate unit vectors (polarization) by considering their pairwise orthogonality of $\vec{E}$, $\vec{B}$, and $\vec{k}$ (the vectors form a right-handed triad).
- *Boundary conditions:* For perfectly reflecting mirrors, $E(z = 0, t) = E(z = L, t) = 0$.
- *Solution:* The cavity modes have the form $\sim \sin(k_m z)$ with allowed values of wave numbers

$$k_m = m\frac{\pi}{L}, \quad m = 1, 2, 3, \ldots, \tag{2.49}$$

with wavelengths (angular frequencies) of

$$\lambda_m = \frac{2L}{m} \quad (\omega_m = c \cdot k_m). \tag{2.50}$$

- *The electrical field operator of a single mode ω_m in the resonator is*

$$\begin{aligned}\hat{E}_m(z,t) &= \sqrt{\frac{\hbar\omega_m}{\varepsilon_0 V}}\,[\hat{a}_m(t) + \hat{a}_m^\dagger(t)]\sin(k_m z)\\ &= \sqrt{\frac{\hbar\omega_m}{\varepsilon_0 V}}\,[\hat{a}_m e^{-i\omega_m t} + \hat{a}_m^\dagger e^{i\omega_m t}]\sin(k_m z).\end{aligned} \tag{2.51}$$

- *Definition of "Electric field per photon"* (or "electric field of the vacuum"):

$$\mathcal{E}_\omega \equiv \sqrt{\frac{\hbar\omega}{\varepsilon_0 V}}. \tag{2.52}$$

It is interesting to compare this value with the corresponding amplitude factor for a *free running wave* (see last section, Eq. (2.36)) where

$$\mathcal{E}_{f\omega} = \sqrt{\frac{\hbar\omega}{2\varepsilon_0 V}} = \frac{1}{\sqrt{2}}\mathcal{E}_\omega < \mathcal{E}_\omega. \tag{2.53}$$

Due to the resonator structure, the electric field per photon is slightly higher inside the resonator ($\times\sqrt{2}$) than for a free running wave.
- *Three-dimensional cavity:*
The electric field operator of a single mode is given by

$$\hat{\vec{E}}_m(\vec{r},t) = \mathcal{E}_\omega[\hat{a}_m(t) + \hat{a}_m^\dagger(t)]\vec{u}_m(\vec{r}), \tag{2.54}$$

where $\vec{u}_m(\vec{r})$ is the mode function of the associated cavity, and $\vec{u}_m(\vec{r})$ is a complex vector that describes the polarization and the relative field amplitude. Furthermore, $\vec{u}_m(\vec{r})$ fulfills the Maxwell equations and is normalized so that its value is equal to 1 at the antinode of the electric field.

In a solid-state cavity, we have to consider the associated refractive index n of the cavity material to calculate the electric field per photon. This results in

$$\mathcal{E}_\omega \equiv \sqrt{\frac{\hbar\omega}{\varepsilon_0 n^2 V}} \tag{2.55}$$

with the effective mode volume

$$V = \frac{1}{n^2}\iiint_{\vec{r}} \mathrm{d}^3 r \quad n(\vec{r})^2 |\vec{u}_m(\vec{r})|. \tag{2.56}$$

2.5 Quantized correlation functions

In this section, we introduce the quantum mechanical description of the first- and second-order correlation functions. A detailed derivation can be found in quantum optics textbooks, e. g., by Lambropoulos and Petrosyan [108] and Loudon [118].

The quantized normalized first-order correlation function is analogous to the classical definition and gives information on the interference capability of light from two position-time points $(\vec{r}_1, t_1)$ and $(\vec{r}_2, t_2)$. If we consider stationary light fields and plane waves, then the degree of first-order coherence takes the simple form

$$g^{(1)}(\tau) = \frac{\langle \hat{E}^-(t)\hat{E}^+(t+\tau)\rangle}{\langle \hat{E}^-(t)\hat{E}^+(t)\rangle}, \tag{2.57}$$

where $\tau = t_2 - t_1$ (cf. with the classical analogue; see Eq. (1.5)).

The quantized normalized second-order correlation function is analogous to the classical definition and gives information on the joint probability to detect two photons at two position-time points $(\vec{r}_1, t_1)$ and $(\vec{r}_2, t_2)$. If we consider stationary light fields and plane waves, then the degree of second-order coherence takes the simple form (cf. Eq. (1.39))

$$g^{(2)}(\tau) = \frac{\langle \hat{E}^-(t)\hat{E}^-(t+\tau)\hat{E}^+(t+\tau)\hat{E}^+(t)\rangle}{\langle \hat{E}^-(t)\hat{E}^+(t)\rangle^2}. \tag{2.58}$$

With the definition of the intensity operator $\hat{I}(\vec{r}, t) = 2\varepsilon_0 c \hat{E}^-(\vec{r}, t)\hat{E}^+(\vec{r}, t)$, we can rewrite the last expression to

$$g^{(2)}(\tau) = \frac{\langle : \hat{I}(t)\hat{I}(t+\tau) : \rangle}{\langle \hat{I}(t)\rangle^2}. \tag{2.59}$$

The symbol ":" expresses "normal ordering", i. e., the annihilation operators are on the right side of the expression (see Eq. (2.58)).

If we consider the field in only one mode, then this leads to a further simplification of the above expressions for $g^{(1)}(\tau)$ and $g^{(2)}(\tau)$ since the electric field operators reduce to single creation and annihilation operators of this specific mode:

$$\hat{E}^+ \to \hat{a}, \quad \hat{E}^- \to \hat{a}^\dagger.$$

The corresponding first-order correlation function reads

$$g^{(1)}(\tau) = \frac{\langle \hat{a}^\dagger(t)\hat{a}(t+\tau)\rangle}{\langle \hat{a}^\dagger(t)\hat{a}(t)\rangle}, \tag{2.60}$$

and the corresponding second-order correlation function is then given by

$$g^{(2)}(\tau) = \frac{\langle \hat{a}^\dagger(t)\hat{a}^\dagger(t+\tau)\hat{a}(t+\tau)\hat{a}(t)\rangle}{\langle \hat{a}^\dagger(t)\hat{a}(t)\rangle^2}. \tag{2.61}$$

2.6 Photon density of states

The photon density of states is a very useful quantity, which will be needed, e. g., for the discussion of radiative decays of single emitters in vacuum and cavities. We will first discuss the photon density in an assumed cavity, and later we will extend the discussion to the more general case of nonconfined optical systems, i. e., propagation of light pulses in a free space.

If we assume periodic boundary conditions for the field in an open cavity of dimension L, then the following k-values are allowed (see Fig. 2.5):

$$k_x = n_x\frac{\pi}{L}, \quad k_y = n_y\frac{\pi}{L}, \quad k_z = n_z\frac{\pi}{L}, \quad n_x, n_y, n_z = 0, 1, 2, \dots. \tag{2.62}$$

Note that only *one* quantum number can be zero.

The total number of modes, which possesses a wave vector between k and $k + dk$ (for $k \gg \frac{\pi}{L}$) in an octant, can be calculated by the quotient of the volume between two spherical shells with radii k and $k + dk$ and the size of a unit cell $(\pi/L)^3$, which is only occupied by one state.

For the volume between two spherical shells with radii k and $k + dk$, we get

$$\begin{aligned}\frac{4\pi(k+dk)^3}{3} - \frac{4\pi}{3}k^3 &= \frac{4\pi(k^3 + 3k^2\,dk + O(dk^2))}{3} - \frac{4\pi}{3}k^3 \\ &\approx 4\pi k^2\,dk.\end{aligned} \tag{2.63}$$

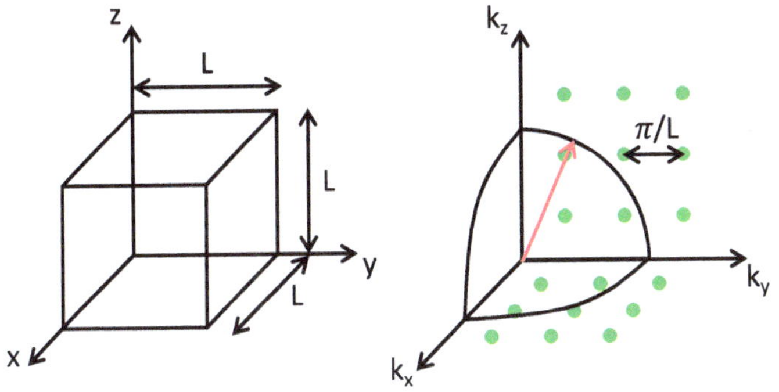

Figure 2.5: Left: Geometry of the open cubic box with linear dimension L; Right: Allowed wave vectors $\vec{k} = (k_x, k_y, k_z)$ for the cubic box. By dividing the size of a unit cell $(\pi/L)^3$ through the volume between two spherical shells with radii k and $k + dk$ (with $k \gg \pi/L$), one gets the number of photonic states in the energy range $[E(k), E(k + dk)]$ (see text).

The unit volume for one k-state is given by $(\pi/L)^3$. Furthermore, we can take into account the two polarization states by a factor of 2 and can consider only one octant by the factor 1/8. The quotient is then given by

$$\rho(k)Vdk = \frac{1}{8} \cdot \frac{4\pi k^2\, dk}{(\frac{\pi}{L})^3} \cdot 2 = V\frac{k^2}{\pi^2}\, dk. \tag{2.64}$$

We find the density of states $\rho(k)$ in k-space by division through the respective total volume $V = L^3$:

$$\rho(k)\, dk = \frac{k^2}{\pi^2}\, dk. \tag{2.65}$$

This holds in general and is independent from the cavity size. Using the dispersion relation of light $\omega = ck$ and $dk = 1/c \cdot d\omega$, this expression can be converted into the angular frequency space:

$$\rho(\omega)\, d\omega = \frac{\omega^2}{\pi^2 c^3}\, d\omega. \tag{2.66}$$

It is important to note that the derived expressions for the density of states $\rho(\omega)\, d\omega$ and $\rho(k)\, dk$ can be also used to convert sums into integrals:

$$\sum_{\vec{k},\lambda} \to \int dk \left(\frac{Vk^2}{\pi^2}\right) \to \int d\omega \left(\frac{V\omega^2}{\pi^2 c^3}\right). \tag{2.67}$$

2.7 Summary

- The Hamilton operator of the quantized electromagnetic field corresponds to an infinite number of quantized harmonic oscillators and can therefore be conveniently described by annihilation and creation operators that fulfill the bosonic commutation relations.
- The electric field operator can be described by a polarization vector, an amplitude factor (electric field per photon), annihilation and creation operators, and the respective mode function (running modes in a free space and standing waves in resonators)
- The photon density of states is a very useful tool in quantum optics and is needed, e. g., for the description of radiative decays of single emitters in vacuum or cavities.

3 Classification of light states and photon statistics

3.1 Introductory remarks

In this chapter, we discuss some physical properties of thermal light, coherent light, and the so-called photon number states of the light. Special emphasis is given to their photon number distributions, fluctuations in their photon numbers, and therefore different photon statistics, which can be quantified by the quantized form of the second-order correlation function. The electromagnetic field is assumed to excite a *single* traveling-wave mode with a given wave vector $\vec{k}$ and a polarization direction specified by the unit vectors $\vec{e}_{k,\lambda}$ with $\lambda = 1, 2$. The transverse mode profile is not important for the discussion here and will thus be neglected. The single-mode theory is therefore limited to the description of experiments that use time-independent light beams. However, many of the nonclassical properties of light can be understood in the framework of single-mode theory [118]. In the next chapter, we will extend the discussion to multimode and continuous-mode theory, which can be used to describe time-dependent light beams, for example, single-photon pulses.

3.2 Thermal light states

In the following, we will derive the photon number distribution $P_{\text{therm}}(n)$ and photon number fluctuations $(\Delta n)^2_{\text{therm}}$ for thermal light from some basic considerations. An ensemble of emitters in the thermal equilibrium with the radiation field, i. e., emission and absorption processes occur, will be considered, and the thermal radiation can be described by the black-body radiation (Planck's law). The occupation probability of a field mode of frequency ω (with energy $E_n = \hbar\omega(n + \frac{1}{2})$) with n photons at a given temperature T is then given by the Boltzmann distribution

$$P_{\text{therm}}(n) = C_0 \mathrm{e}^{-\frac{E(n)}{k_B T}}, \tag{3.1}$$

where the initially undefined constant C_0 can be determined from the normalization condition for P_{thermal}:

$$\begin{aligned}\sum_{n=0}^{\infty} P_{\text{thermal}}(n) &= \sum_{n=0}^{\infty} C_0 \mathrm{e}^{-\frac{E(n)}{k_B T}} = C_0 \sum_{n=0}^{\infty} \mathrm{e}^{-\frac{\hbar\omega(n+\frac{1}{2})}{k_B T}} \\ &= C_0 \mathrm{e}^{-\frac{\hbar\omega}{2k_B T}} \sum_{n=0}^{\infty} \mathrm{e}^{-\frac{n\hbar\omega}{k_B T}} \\ &= C_0 \mathrm{e}^{-\frac{\hbar\omega}{2k_B T}} \sum_{n=0}^{\infty} \mathrm{e}^{-(\frac{\hbar\omega}{k_B T})^n}\end{aligned}$$

https://doi.org/10.1515/9783110703412-003

$$= C_0 e^{-\frac{\hbar\omega}{2k_BT}} \cdot \frac{1}{1 - e^{-\frac{\hbar\omega}{k_BT}}}$$
$$\equiv 1. \tag{3.2}$$

By transformation we get

$$C_0 = \frac{1 - e^{-\frac{\hbar\omega}{k_BT}}}{e^{-\frac{\hbar\omega}{2k_BT}}}. \tag{3.3}$$

This results in a normalized occupation probability for the mode:

$$P_{\text{therm}}(n) = (1 - e^{-\frac{\hbar\omega}{k_BT}})e^{-\frac{n\hbar\omega}{k_BT}}. \tag{3.4}$$

The average photon number $\langle n\rangle_{\text{therm}}$ of the mode at a given temperature results from the weighted sum of $P_{\text{thermal}}(n)$ over all occupation numbers n:

$$\begin{aligned}\langle n\rangle_{\text{therm}} &= \sum_{n=0}^{\infty} n \cdot P_{\text{therm}}(n) \\ &= (1 - e^{-\frac{\hbar\omega}{k_BT}}) \sum_{n=0}^{\infty} n e^{-\frac{n\hbar\omega}{k_BT}}.\end{aligned} \tag{3.5}$$

For the following calculation, we use the abbreviation

$$\xi := e^{-\frac{\hbar\omega}{k_BT}}, \tag{3.6}$$

which results in

$$\begin{aligned}\langle n\rangle_{\text{therm}} &= (1 - \xi) \sum_{n=0}^{\infty} n\xi^n \\ &= (1 - \xi)\xi \frac{\partial}{\partial\xi}\left(\sum_{n=0}^{\infty} \xi^n\right) \\ &= (1 - \xi)\xi \frac{\partial}{\partial\xi}\left(\frac{1}{1 - \xi}\right) \\ &= (1 - \xi)\xi \frac{1}{(1 - \xi)^2} \\ &= \frac{\xi}{1 - \xi} \\ &= \frac{1}{e^{\frac{\hbar\omega}{k_BT}} - 1}.\end{aligned} \tag{3.7}$$

This important result is the Planck thermal excitation function. We can now express the thermal occupation probability $P_{\text{therm}}(n)$ with the average photon number $\langle n\rangle_{\text{therm}}$. We use the abbreviation $\xi = e^{-\frac{\hbar\omega}{k_BT}}$ and neglect the subscript "therm" in $\langle n\rangle$:

$$\langle n\rangle_{\text{therm}} = \frac{\xi}{1-\xi} \quad \leftrightarrow \quad \xi = \frac{\langle n\rangle}{1+\langle n\rangle}. \tag{3.8}$$

Therefore

$$\begin{aligned} P_{\text{therm}}(n) &= (1 - e^{\frac{-\hbar\omega}{k_B T}})(e^{-\frac{\hbar\omega}{k_B T}})^n \\ &= \left(1 - \frac{\langle n\rangle}{\langle n\rangle + 1}\right)\left(\frac{\langle n\rangle}{\langle n\rangle + 1}\right)^n \\ &= \frac{1}{\langle n\rangle + 1} \frac{\langle n\rangle^n}{(\langle n\rangle + 1)^n} \\ &= \frac{\langle n\rangle^n}{(\langle n\rangle + 1)^{n+1}}. \end{aligned} \tag{3.9}$$

Figure 3.1 shows the photon number distributions $P(n; \langle n\rangle)$ for the occupation of a thermal mode of the radiation field for different average photon numbers $\langle n\rangle$. This probability $P_{\text{therm}}(n)$ is the distribution of outcomes obtained for a large number of measurements of the number of photons in the selected mode. We can see that the largest value of $P_{\text{therm}}(n)$ always occurs for $n = 0$ and $P_{\text{therm}}(n)$ decreases monotonically with increasing n.

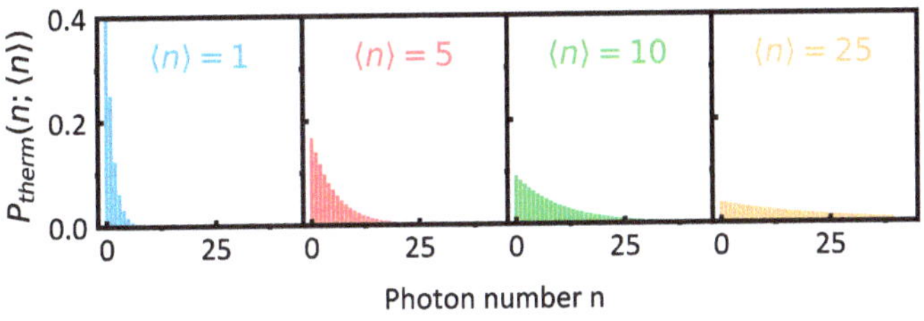

Figure 3.1: Photon number distributions $P(n; \langle n\rangle)$ of the occupation of a thermal mode of the radiation field for different average photon numbers $\langle n\rangle$.

It is now interesting to study the photon number fluctuations, which are statistically described by the variance of the distribution

$$(\Delta n)^2 := \langle n^2\rangle - \langle n\rangle^2. \tag{3.10}$$

A helpful quantity to characterize a distribution is its factorial moment defined as

$$\langle n(n-1)(n-2)\dots(n-k+1)\rangle := \sum_n n(n-1)(n-2)\dots(n-k+1)\cdot P(n),$$

where k is any positive integer. The first and second factorial moments are given as follows:

$$\begin{aligned}
&\text{1. factorial moment} & \langle n\rangle &= \sum_n n\cdot P(n) \mathrel{\hat{=}} \text{average photon number},\\
&\text{2. factorial moment} & \langle n(n-1)\rangle &= \sum_n n(n-1)\cdot P(n)\\
&& &= \sum_n n^2P(n) - \sum_n nP(n)\\
&& &= \langle n^2\rangle - \langle n\rangle.
\end{aligned} \tag{3.11}$$

For thermal radiation, the factorial moment is given by [118]

$$\langle n(n-1)(n-2)\dots(n-k+1)\rangle_{\text{th}} = k!\langle n\rangle^k. \tag{3.12}$$

In particular, for $k = 2$, we have

$$\langle n(n-1)\rangle = 2\langle n\rangle^2. \tag{3.13}$$

By equalizing the expressions in Eqs. (3.11) and (3.13) we get

$$\begin{aligned}
2\langle n\rangle^2 &= \langle n^2\rangle - \langle n\rangle,\\
\langle n^2\rangle_{\text{therm}} &= 2\langle n\rangle^2 + \langle n\rangle.
\end{aligned} \tag{3.14}$$

Putting this into the expression for the variance in Eq. (3.10), we have

$$(\Delta n)^2_{\text{therm}} = \langle n\rangle^2 + \langle n\rangle \quad \text{super-Poissonian fluctuations.} \tag{3.15}$$

Light whose photon number variance exceeds $\langle n\rangle$ exhibits *super-Poissonian* fluctuations, whereas light that exhibits smaller fluctuations than $\langle n\rangle$ is said to exhibit *sub-Poissonian* fluctuations. The relative variance of the thermal light source is given by

$$\left(\frac{(\Delta n)^2}{\langle n\rangle^2}\right)_{\text{therm}} = 1 + \frac{1}{\langle n\rangle} \underset{\langle n\rangle\to\infty}{\to} 1. \tag{3.16}$$

Thermal light is characterized by fluctuations of the order of the average number of photons. This has already been discussed in Chapter 1 for the classical intensity.

For the experimental observation of thermal fluctuations, the time-resolution of the respective detector system has to be better than the coherence time τ_c of the thermal radiation. This is hardly fulfilled for most of the detector systems since the coherence time of thermal light sources is typically in the picosecond (ps)–tens of ps regime. However, the newest generation of single nanowire superconducting detectors (SNSPDs) reach this time regime. In the case of insufficient detector time-resolution, we would measure uncorrelated photons, which follow a Poissonian statistics.

As already remarked in the introduction, the previous discussion holds only for a single-mode contribution. In reality, thermal light consists of a continuum of modes.

We can show that the photon number variance of N_{therm} thermal modes of similar frequency is given by [48]

$$(\Delta n)^2_{\text{therm}} = \frac{\langle n\rangle^2}{N_{\text{therm}}} + \langle n\rangle, \tag{3.17}$$

which reduces to a Poisson distribution with $(\Delta n)^2 = \langle n\rangle$ for a large number of modes N_{therm}. It is not easy the measure a single mode of thermal light, and thus most of the experiments with thermal light exhibit Poissonian statistics. However, the single-mode case can, for example, be fulfilled by a thermal light beam in an optical cavity where all except one of the modes are removed by a filter.

At this point, it is also interesting to remark that there is an uncertainty relation between amplitude and phase of the electric field of a single mode. The well-known energy-time uncertainty relation is given by

$$\Delta E \cdot \Delta t \geq \frac{\hbar}{2}.$$

It is instructive to replace ΔE by $\Delta n \cdot \hbar\omega$ with the standard deviation Δn, and by taking into account $\Delta n := \sqrt{(\Delta n)^2}$ we get

$$\Delta n \cdot (\omega \Delta t) = \Delta n \cdot \Delta\varphi \geq \frac{1}{2}, \tag{3.18}$$

which gives the product of photon-number and phase uncertainties. It represents a trade-off between amplitude and phase value uncertainties of the electric field of a single mode.

3.3 Coherent states (Glauber states)

Coherent light states are generated by laser sources that are operated well above the lasing threshold. Their electric field variation approaches that of a classical wave of fixed amplitude and phase values. In the language of quantum optics, coherent states can be obtained as eigenstates of the annihilation operator $\hat{a}$ (see, e. g., [118]). In the following, we again can assume a single mode expressed by the mode index $i = (\vec{k}, \lambda)$. The eigenvalue equation is given by

$$\hat{a}_i|\alpha_i\rangle = \alpha_i|\alpha_i\rangle, \tag{3.19}$$

$$\langle\alpha_i|\hat{a}_i^\dagger = \alpha_i^*\langle\alpha_i|, \tag{3.20}$$

with

$$\alpha_i = |\alpha_i|e^{i\varphi}.$$

$|\alpha_i\rangle$ is a state of the harmonic oscillator and can therefore be expressed in the basis $\{|n\rangle\}$. In the following, the mode index i will be omitted for better clarity. The coherent state is defined as the following superposition of number states:

$$|\alpha\rangle = \sum_{n=0}^{\infty} \langle n|\alpha\rangle |n\rangle = \mathrm{e}^{-\frac{1}{2}|\alpha|^2} \sum_{n=0}^{\infty} \frac{\alpha^n}{\sqrt{n!}} |n\rangle. \tag{3.21}$$

The probability of finding n photons in the mode is thus given by Poisson statistics:

$$\begin{aligned} P_{\mathrm{coh}}(n) &= |\langle n|\alpha\rangle|^2 = \left| \mathrm{e}^{-\frac{1}{2}|\alpha|^2} \frac{\alpha^n}{\sqrt{n!}} \right|^2 \\ &= \mathrm{e}^{-|\alpha|^2} \frac{|\alpha|^{2n}}{n!}. \end{aligned} \tag{3.22}$$

The expectation value of the photon number operator $\langle \hat{n} \rangle = \langle n \rangle_{\mathrm{coherent}}$, i. e., the mean photon number in the considered mode is given by

$$\langle n \rangle_{\mathrm{coh}} = \langle \alpha|\hat{n}|\alpha\rangle = \langle \alpha|\hat{a}^\dagger \hat{a}|\alpha\rangle = |\alpha|^2. \tag{3.23}$$

Now we can express the probability of finding n photons in the mode $P_{\mathrm{coh}}(n)$ with the expectation value for $\langle n \rangle_{\mathrm{coh}}$:

$$P_{\mathrm{coh}}(n, \langle n \rangle) = \mathrm{e}^{-\langle n \rangle} \frac{\langle n \rangle^n}{n!}. \tag{3.24}$$

Figure 3.2 shows the photon number distributions $P_{\mathrm{coh}}(n; \langle n \rangle)$ of the occupation of a coherent mode of the radiation field for different average photon numbers $\langle n \rangle$. This probability $P_{\mathrm{coh}}(n)$ is the distribution of outcomes obtained for a large number of measurements of the number of photons in the selected mode. We can see that the largest value of $P_{\mathrm{coh}}(n)$ always occurs for $n = \langle n \rangle_{\mathrm{coh}}$. This is in strong contrast with $P_{\mathrm{therm}}(n)$, the highest value of which is always achieved for $n = 0$.

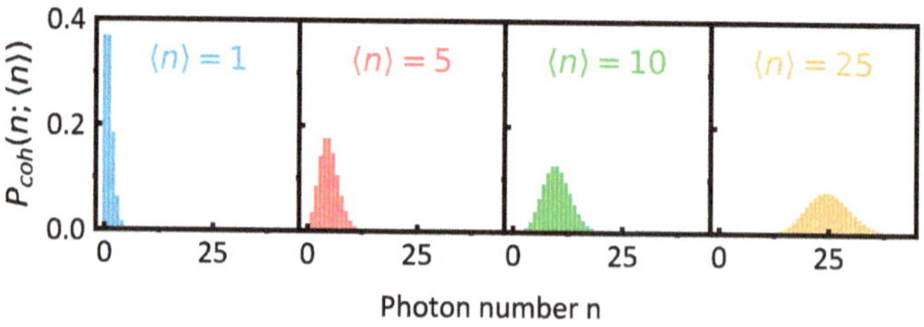

Figure 3.2: Photon number distributions $P(n; \langle n \rangle)$ of the occupation of a coherent mode of the radiation field for different average photon numbers $\langle n \rangle$.

Now we will determine the variance $(\Delta n)^2_{\text{coh}}$ of the distribution of the coherent field. In general, the variance is defined by $(\Delta n)^2 = \langle n^2\rangle - \langle n\rangle^2$. Using the eigenvalue properties of the number operator, we can calculate the variance:

$$\begin{aligned}(\Delta n)^2_{\text{coh}} &= \langle \hat{n}^2\rangle - \langle \hat{n}\rangle^2 \\ &= \langle \alpha|\hat{n}^2|\alpha\rangle - \langle \alpha|\hat{n}|\alpha\rangle^2 \\ &= \langle \alpha|\hat{n}^2|\alpha\rangle - |\alpha|^4.\end{aligned}$$

For the first expression, we get

$$\begin{aligned}\langle \alpha|\hat{n}^2|\alpha\rangle &= \langle \alpha|\hat{a}^\dagger\hat{a}\hat{a}^\dagger\hat{a}|\alpha\rangle \\ &= \langle \alpha|\hat{a}^\dagger(\hat{a}^\dagger\hat{a}+1)\hat{a}|\alpha\rangle \\ &= \langle \alpha|\hat{a}^\dagger\hat{a}^\dagger\hat{a}\hat{a}|\alpha\rangle + \langle \alpha|\hat{a}^\dagger\hat{a}|\alpha\rangle \\ &= |\alpha|^4 + |\alpha|^2.\end{aligned}$$

Note that the property $\langle \alpha|\hat{a}^\dagger = (\hat{a}|\alpha\rangle)^* = \alpha^*\langle \alpha|$ has been used for the calculation above. We can use this expression to get the final result:

$$(\Delta n)^2_{\text{coh}} = |\alpha|^4 + |\alpha|^2 - |\alpha|^4 = |\alpha|^2 = \langle n\rangle_{\text{coh}}, \tag{3.25}$$

i. e., light exhibits Poissonian fluctuations. The standard deviation is given by

$$\Delta n := \sqrt{(\Delta n)^2} = \sqrt{\langle n\rangle_{\text{coh}}}. \tag{3.26}$$

Finally, the *relative variance* is given by the expression

$$\left(\frac{(\Delta n)^2}{\langle n\rangle^2}\right)_{\text{coherent}} = \frac{1}{\langle n\rangle} \underset{n\to\infty}{\longrightarrow} 0. \tag{3.27}$$

The relative photon number variance vanishes with increasing values of the mean photon number $\langle n\rangle$.

3.4 Photon number states (Fock states)

The purest states of light are photon number states (see Fig. 3.3). These states $\{|n_i\rangle\}$ are eigenstates of the photon number operator $\hat{n}_i = \hat{a}_i^\dagger\hat{a}_i$. The simplest photon number state, the single-photon state, can be generated by single quantum emitters, e. g., an atom, an ion, a molecule, or a semiconductor quantum dot. The generation of higher photon number states (>1) is more difficult and currently a major research task in the field

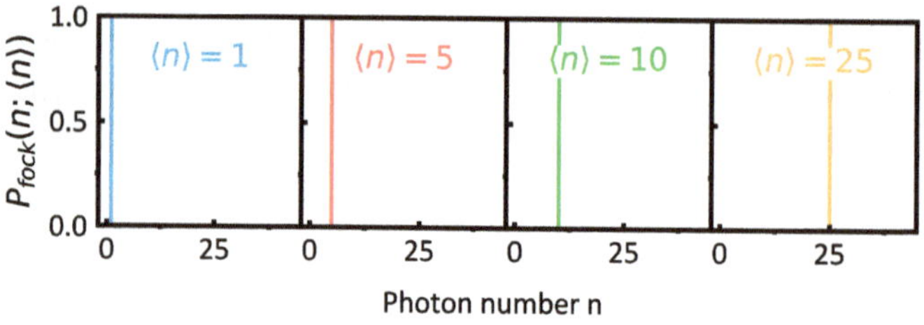

Figure 3.3: Photon number distributions $P(n; \langle n\rangle)$ of the occupation of a photon number state (Fock state) of the radiation field for different average photon numbers $\langle n\rangle$.

of quantum optics science. The eigenvalue equation of the photon number operator is given by (cf. Eqs. (2.42)–(2.44))

$$\hat{n}_i|n_i\rangle = \hat{a}_i^\dagger \hat{a}_i|n_i\rangle = n_i|n_i\rangle, \tag{3.28}$$

where n_i is the occupation number of the respective mode $i = (\vec{k}, \lambda)$. As outlined above, the states $\{|n_i\rangle\}$ represent a complete set of eigenstates of a single mode and satisfy the orthonormality condition

$$\langle n_i|n_j\rangle = \delta_{i,j}.$$

The occupation probability of a photon number states is given by

$$P_{\text{Fock}}(n) = \begin{cases} 1, & n = n_i, \\ 0, & n \neq n_i. \end{cases} \tag{3.29}$$

According to the deterministic nature of the Fock states, the average photon number (expectation value) $\langle n\rangle_{\text{Fock}}$ corresponds exactly to the actual occupation number n_i:

$$\langle n\rangle_{\text{Fock}} = \langle n_i|\hat{n}_i|n_i\rangle = n_i. \tag{3.30}$$

As a consequence, both the variance and relative variance are zero:

$$(\Delta n)^2_{\text{Fock}} = 0 \quad \text{sub-Poissonian fluctuations}, \tag{3.31}$$

$$\left(\frac{(\Delta n)^2}{\langle n\rangle^2}\right)_{\text{Fock}} = 0. \tag{3.32}$$

3.5 Distinction between classical and nonclassical lights

In the following, we will discuss the question how we can distinguish between quantum light (photon number states) and classical light (thermal and coherent light). To get insight, different measurement techniques will be considered, i. e., measuring the first- and second-order correlation functions (cf. Section 2.5).

a) Consider first the most simple setup by putting a single detector into the beam path to monitor the field of a single mode, i. e., one detector measures the incoming light field (see Fig. 3.4). The corresponding first-order correlation function for $\tau = 0$ reads

$$g^{(1)} = \frac{\langle \hat{a}^\dagger(t)\hat{a}(t)\rangle}{\langle \hat{a}^\dagger(t)\hat{a}(t)\rangle} = \frac{\langle \hat{n}\rangle}{\langle \hat{n}\rangle} = 1. \tag{3.33}$$

Obviously, no information of the type of light field can be gained, since $g^{(1)}(0) = 1$ for classical and quantum lights.

Figure 3.4: Single detector in the beam path of a single-mode light field.

b) Next, let us consider an interferometric setup where the field is first split and then recombined on the detector as shown in Fig. 3.5. The corresponding first-order correlation function for the path difference $\delta S = S_2 - S_1$ is

$$g^{(1)}(\delta S) = \frac{\langle \hat{a}^\dagger(t)\hat{a}(t + \frac{\delta S}{c})\rangle}{\langle \hat{a}^\dagger(t)\hat{a}(t)\rangle} = \frac{\langle \hat{a}^\dagger(t)\hat{a}(t)\rangle}{\langle \hat{a}^\dagger(t)\hat{a}(t)\rangle} \mathrm{e}^{-ik\delta S} = \mathrm{e}^{-ik\delta S},$$

where we have used

$$\hat{a}(t) = \hat{a}\mathrm{e}^{-i\omega t} \quad \Rightarrow \quad \hat{a}\left(t + \frac{\delta S}{c}\right) = \underbrace{\hat{a}\mathrm{e}^{-i\omega t}}_{\hat{a}(t)} \mathrm{e}^{-ik\delta S} \quad \text{with } k = \frac{\omega}{c}.$$

This result is a purely classical result and gives no information about the quantum state of light.

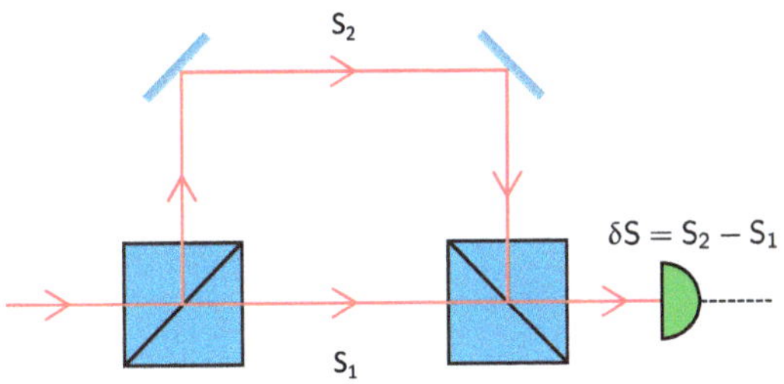

Figure 3.5: A single-photon interference setup.

c) Finally, a Hanbury Brown–Twiss setup (see Fig. 3.6) for measuring the second-order correlation function is considered (cf. Sections 1.6 and 5.2.5):

$$g^{(2)}(t_1, t_2) = \frac{\langle \hat{a}^\dagger(t_1)\hat{a}^\dagger(t_2)\hat{a}(t_2)\hat{a}(t_1)\rangle}{\langle \hat{a}^\dagger(t_1)\hat{a}(t_1)\rangle\langle \hat{a}^\dagger(t_2)\hat{a}(t_2)\rangle}.$$

In the case of a stationary field, $g^{(2)}$ only depends on the time difference $\tau = t_2 - t_1$:

$$g^{(2)}(\tau) = \frac{\langle \hat{a}^\dagger(0)\hat{a}^\dagger(\tau)\hat{a}(\tau)\hat{a}(0)\rangle}{\langle \hat{a}^\dagger\hat{a}\rangle^2}.$$

Now $g^{(2)}(\tau)$ depends on the type of light field, which becomes obvious by explicitly calculating $g^{(2)}(\tau)$ for different light fields at $\tau = 0$. In the following, we consider coherent light, photon number states (Fock states), and thermal light.

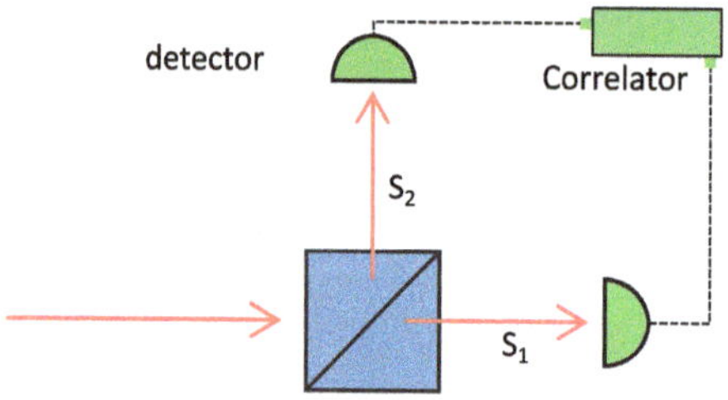

Figure 3.6: Hanbury Brown–Twiss setup.

1. *Coherent light* $|\alpha\rangle$:

$$g^{(2)}(0) = \frac{\langle\alpha|\hat{a}^\dagger\hat{a}^\dagger\hat{a}\hat{a}|\alpha\rangle}{\langle\alpha|\hat{a}^\dagger\hat{a}|\alpha\rangle^2}$$
$$\text{with} \quad \hat{a}|\alpha = \alpha|\alpha\rangle \quad \text{and} \quad \langle\alpha|\hat{a}^\dagger = \langle\alpha|\alpha^*$$
$$= \frac{\alpha\alpha\alpha^*\alpha^*}{(\alpha\alpha^*)^2} = \frac{|\alpha|^2}{|\alpha|^2} = 1.$$

It is important to note that the second-order correlation function for coherent light is also 1 for all delay times ($g^{(2)}(\tau) = 1$).

2. *Fock states* $|n\rangle$

$$g^{(2)}(0) = \frac{\langle n|\hat{a}^\dagger\hat{a}^\dagger\hat{a}\hat{a}|n\rangle}{\langle n|\hat{a}^\dagger\hat{a}|n\rangle^2}$$
$$\text{with} \quad [\hat{a}, \hat{a}^\dagger] = 1 \to \hat{a}\hat{a}^\dagger - \hat{a}^\dagger\hat{a} = 1,$$
$$g^{(2)}(0) = \frac{\langle n|\hat{a}^\dagger\hat{a}\hat{a}^\dagger\hat{a} - \hat{a}^\dagger\hat{a}|n\rangle}{n^2} \tag{3.34}$$
$$= \frac{n^2}{n^2} - \frac{n}{n^2} = 1 - \frac{1}{n} \quad (g^{(2)} < 1 \Leftrightarrow \text{photon antibunching}).$$

For a single-photon state ($n = 1$), we get $g^{(2)}(0) = 0$, i. e., we observe photon antibunching. This is a truly nonclassical result since $g^{(2)}(0) \geq 1$ for all classical light fields (see points 1 and 2).

3. *Thermal light*

$$g^{(2)}(0) \overset{(3.34)}{=} \frac{\langle(\hat{a}^\dagger\hat{a})^2\rangle - \langle\hat{a}^\dagger\hat{a}\rangle}{\langle\hat{a}^\dagger\hat{a}\rangle^2} = \frac{\langle(\hat{a}^\dagger\hat{a})^2\rangle - \overbrace{\langle\hat{a}^\dagger\hat{a}\rangle^2 + \langle\hat{a}^\dagger\hat{a}\rangle^2}^{\text{zero added!}} - \langle\hat{a}^\dagger\hat{a}\rangle}{\langle\hat{a}^\dagger\hat{a}\rangle^2},$$

$\langle\hat{a}^\dagger\hat{a}\rangle = \langle n\rangle$ = average photon number,
$\langle(\hat{a}^\dagger\hat{a})^2\rangle - \langle\hat{a}^\dagger\hat{a}\rangle^2 = (\Delta n)^2$ = variance,

$$g^{(2)}(0) = 1 + \frac{(\Delta n)^2 - \langle n\rangle}{\langle n\rangle^2},$$

for thermal light, $(\Delta n)^2 = \langle n\rangle^2 + \langle n\rangle$,

$$\rightarrow g^{(2)}(0) = 2.$$

This result indicates that photons show the tendency to appear simultaneous at both detectors, i. e., the effect of photon bunching is observed. It is important to point out that for both thermal light and photon number states, $g^{(2)}(\tau)$ approaches the value 1 as $\tau \rightarrow \infty$.

3.6 Vacuum field fluctuations

Vacuum field fluctuations play an important role in quantum photonic technologies, e. g., to enhance or suppress radiative transitions of quantum emitters in cavities. To get insight into this pure quantum phenomenon, we will consider the electromagnetic field of a single mode in a one-dimensional resonator with exactly n photons (Fock state).

First, we calculate the expectation value of the electric field:

$$\langle E\rangle = \langle n|\hat{E}(z,t)|n\rangle = \varepsilon_\omega \langle n|\hat{a}(t) + \hat{a}^\dagger(t)|n\rangle \sin(kz) \tag{3.35}$$

with $\varepsilon_\omega = \sqrt{\frac{\hbar\omega}{\varepsilon_0 V}}$ as a "field per photon" (see Eq. (2.52)). Using the operator properties for the photon number states

$$\langle n|\hat{a}|n\rangle = \sqrt{n}\langle n|n-1\rangle = 0, \tag{3.36}$$

$$\langle n|\hat{a}^\dagger|n\rangle = \sqrt{n+1}\langle n|n+1\rangle = 0, \tag{3.37}$$

we get the expectation value of the electric field inside the resonator:

$$\langle E\rangle = \langle n|\hat{E}(z,t)|n\rangle = 0, \tag{3.38}$$

which means that the expectation value of the electric field is zero independent of the photon number inside the mode.

Second, we calculate the variance of the electric field. With the previous result for the expectation value, we get

$$(\Delta E)^2 = \langle \hat{E}^2 \rangle - \langle \hat{E} \rangle^2 = \langle \hat{E}^2 \rangle. \tag{3.39}$$

Using the commutator relation $[\hat{a}, \hat{a}^\dagger] = 1$, we get the expectation value

$$\begin{aligned}
\langle \hat{E}^2 \rangle = \langle n|\hat{E}^2|n \rangle &= \varepsilon_\omega^2 \sin^2(kz) \langle n|\hat{a}^\dagger \hat{a}^\dagger e^{i2\omega t} + \hat{a}\hat{a}^\dagger + \hat{a}^\dagger \hat{a} + \hat{a}\hat{a} e^{-i2\omega t}|n \rangle \\
&= \varepsilon_\omega^2 \sin^2(kz) \langle n|2\hat{a}^\dagger \hat{a} + 1|n \rangle \\
&= 2\varepsilon_\omega^2 \sin^2(kz) \left(n + \frac{1}{2} \right).
\end{aligned}$$

This results in the standard deviation

$$\begin{aligned}
\Delta E &= \sqrt{(\Delta E)^2} \\
&= \sqrt{2} \varepsilon_\omega |\sin(kz)| \sqrt{n + \frac{1}{2}}.
\end{aligned} \tag{3.40}$$

Note that even for the vacuum state $|n\rangle = |0\rangle$ (no photons in the cavity), there is a nonzero uncertainty in the electric field:

$$\Delta E = \varepsilon_\omega |\sin(kz)| \neq 0. \tag{3.41}$$

The nonzero value of the expectation value of the electric field implies that the quantized electric field exhibits fluctuations even when no photons are present. The size of the fluctuations are determined by the prefactor ε_ω, which goes as $\sim \sqrt{1/V}$ with V the mode volume of the resonator mode. This shows that the field fluctuations increase with decreasing mode volume. If an emitter is placed in the center of a microcavity, then it experiences enhanced electric field fluctuations according to Eq. (3.41), resulting in an enhanced spontaneous emission rate, and explains the strong interest in low mode volume microcavities. This is known as the Purcell effect and will be discussed in detail in Chapter 6.

3.7 Summary

- The photon number distribution $P(n; \langle n \rangle)$ of thermal light in a single mode is described by a Bose–Einstein photon probability function, where $P(n; \langle n \rangle)$ falls of monotonically with increasing n.
- Thermal light in a single mode is characterized by fluctuations of the order of the average number of photons and shows super-Poisson fluctuations. However, in most

practical situations, thermal light consists of a continuum of modes, and in this case the photon number variance is described by Poisson statistics.

- The photon number distribution of coherent light is described by a Poisson distribution, where the largest value always occur at the expectation value of the distribution.
- The purest states of light are photon number states (Fock states). The average photon number always corresponds to the actual photon occupation number, and thus the variance is zero.
- By measuring the second-order correlation function we can distinguish between quantum light (photon number states) and classical light (thermal and coherent light).
- Even in the vacuum state (no photons in a cavity), there is a nonzero uncertainty in the electric field inside a cavity. This pure quantum mechanical effect is called vacuum field fluctuations, which play an important role for the light emission of quantum emitters in cavities.
- There is a trade-off between the photon number Δn and phase uncertainties $\Delta\varphi$ of the electric field of a single mode, which can be expressed by an uncertainty relation $\Delta n \cdot \Delta\varphi \geq 1/2$.

4 Quantum optics with photon number states

4.1 Introductory remarks

Photonic quantum technologies, such as quantum communication via quantum repeaters, quantum simulation, and linear optics-based and measurement-based quantum computing, rely on highly sophisticated single-photon sources, which are able to deliver on-demand photon number states, highly efficient single-photon detectors, and linear optical gate operations based on single photons interfering at optical beamsplitters. Therefore it is mandatory to understand the quantum behavior of single and interfering photons at beamsplitters. We will start our discussion with the reflection/transmission of a single-photon state at a beamsplitter and two-photon interference experiments of single photons on a beamsplitter, which gives us insight into the famous Hong–Ou–Mandel effect, an exclusive quantum mechanical effect. To understand single-photon experiments, we have to extend the description of photons by introducing a wavepacket approach, where the spectrum and temporal shape of the photon are determined by its generation process. The treatments of the theory in this chapter include some shortcuts and omissions of more advanced derivations by Loudon [118].

4.2 Single- and two-photon interference at the beamsplitter

The beamsplitter belongs to the most important components in quantum optical experiments and quantum optical-based technologies. Figure 4.1 displays a representation of a beamsplitter with input and output channels for the respective field operators containing the annihilation operators. The discussion starts by recapitulating the basic classical beamsplitter relations based on the conservation of the classical energy flow (having defined $\mathcal{R}$ and $\mathcal{T}$ as the reflection and transmission coefficients of the symmetric beamsplitter):

$$|\mathcal{R}|^2 + |\mathcal{T}|^2 = 1, \quad \mathcal{R}\mathcal{T}^* + \mathcal{T}\mathcal{R}^* = 0. \tag{4.1}$$

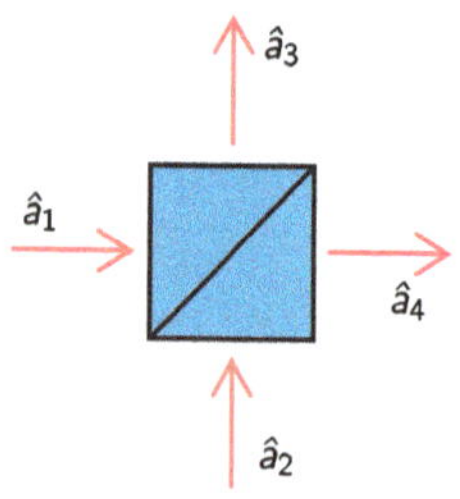

Figure 4.1: Beamsplitter with the notations for the annihilation operators associated with the input and output fields.

https://doi.org/10.1515/9783110703412-004

The boundary conditions are the same for classical fields and for the quantum-mechanical field operators. Therefore the quantized field operators $\hat{a}_i$ and $\hat{a}_i^\dagger$ convert in the same way as classical fields. For a symmetric beamsplitter, we have

$$\hat{a}_3 = \mathcal{R}\hat{a}_1 + \mathcal{T}\hat{a}_2 \quad \text{and} \quad \hat{a}_4 = \mathcal{T}\hat{a}_1 + \mathcal{R}\hat{a}_2, \tag{4.2}$$

$$\hat{a}_1 = \mathcal{R}^*\hat{a}_3 + \mathcal{T}^*\hat{a}_4 \quad \text{and} \quad \hat{a}_2 = \mathcal{T}^*\hat{a}_3 + \mathcal{R}^*\hat{a}_4. \tag{4.3}$$

The relations for the creation operators $\hat{a}_i^\dagger$ are given by the Hermitian conjugates of Eqs. (4.2) and (4.3). Furthermore, we assume that the input fields in channels 1 and 2 are independent:

$$[\hat{a}_1, \hat{a}_1^\dagger] = [\hat{a}_2, \hat{a}_2^\dagger] = 1, \tag{4.4}$$

$$[\hat{a}_1, \hat{a}_2^\dagger] = [\hat{a}_2, \hat{a}_1^\dagger] = 0 \quad \text{(bosonic commutator relations)}, \tag{4.5}$$

and by Eqs. (4.2) and (4.3) we get the corresponding commutator relations for the output fields:

$$[\hat{a}_3, \hat{a}_3^\dagger] = [\mathcal{R}\hat{a}_1 + \mathcal{T}\hat{a}_2, \mathcal{R}^*\hat{a}_1^\dagger + \mathcal{T}^*\hat{a}_2^\dagger] = |\mathcal{R}|^2 + |\mathcal{T}|^2 = 1, \tag{4.6}$$

$$[\hat{a}_3, \hat{a}_4^\dagger] = [\mathcal{R}\hat{a}_1 + \mathcal{T}\hat{a}_2, \mathcal{T}^*\hat{a}_1^\dagger + \mathcal{R}^*\hat{a}_2^\dagger] = \mathcal{R}\mathcal{T}^* + \mathcal{T}\mathcal{R}^* = 0, \quad \text{and} \tag{4.7}$$

$$[\hat{a}_4, \hat{a}_4^\dagger] = 1. \tag{4.8}$$

Obviously, the output mode operators also satisfy the independent bosonic commutation relations.

Now let us first consider the simplest situation, i. e., a single-photon input in one arm and vacuum in the other input arm of the beamsplitter (see Fig. 4.2(a)). The input state can be formally written as $|1\rangle_1|0\rangle_2 = \hat{a}_1^\dagger|0\rangle_1|0\rangle_2 = \hat{a}_1^\dagger|0\rangle$, where $|0\rangle$ represents the joint vacuum state of the beamsplitter arms. By using the Hermitian conjugate of Eq. (4.3) for $\hat{a}_1^\dagger$ the input state is converted to the corresponding output state:

$$|1\rangle_1|0\rangle_2 = \hat{a}_1^\dagger|0\rangle_1|0\rangle_2 = (\mathcal{R}\hat{a}_3^\dagger + \mathcal{T}\hat{a}_4^\dagger)|0\rangle = \mathcal{R}|1\rangle_3|0\rangle_4 + \mathcal{T}|0\rangle_3|1\rangle_4, \tag{4.9}$$

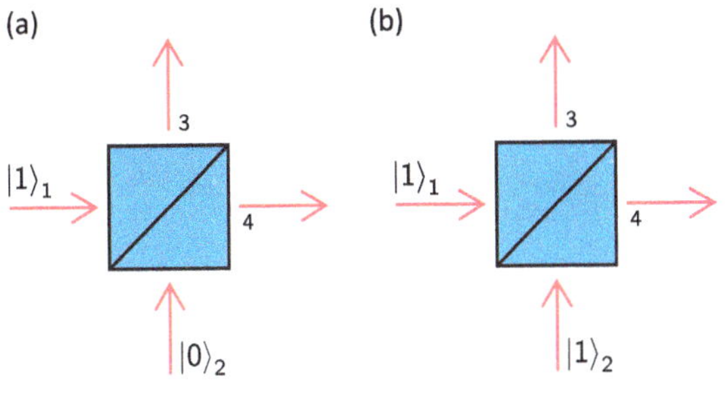

Figure 4.2: (a) Single-photon input on a beamsplitter. (b) Two-photon interference experiment on a beamsplitter. The two photons impinge from two different input arms on the beamsplitter.

where we have used the ladder property of the creation operator ($\mathcal{R}a_3^\dagger|0\rangle_3|0\rangle_4 = \mathcal{R}|1\rangle_3|0\rangle_4$). The input state is converted into a linear superposition of two possible output states. Both are products of states for different output arms: these represent one photon in one arm and none in the other. The resulting state is said to be *entangled* since it cannot be written as a pure product state.

Now we will discuss the result on the photonic state of a certain measurement on one of the beamsplitter arms. The state of the system after a measurement is given by the projection of the state before the measurement onto the state determined by the measurement. For example, if we perform a measurement at the output arm 3 and the result is "no photon at arm 3", i. e., $\rightarrow |0\rangle_3$, then the resulting photonic state after the measurement is given by

$$\begin{aligned} N_3\langle 0|\{\mathcal{R}|1\rangle_3|0\rangle_4 + \mathcal{T}|0\rangle_3|1\rangle_4\} &= N\{\mathcal{R}\underbrace{{}_3\langle 0|1\rangle_3}_{\substack{=0\\ \text{orthogonal}}}|0\rangle_4 + \mathcal{T}\underbrace{{}_3\langle 0|0\rangle_3}_{=1}|1\rangle_4\} \\ &= N\mathcal{T}|1\rangle_4 \\ &= |1\rangle_4, \end{aligned} \tag{4.10}$$

where N is a normalization constant equal to $1/\mathcal{T}$. The outcome means that if there is no photon measured in arm 3 (vacuum state), then the single photon is in arm 4.

Next, we discuss the process with two photons impinging on the beamsplitter from two different input arms of the beamsplitter (see Fig. 4.2(b)). This represents a two-photon interference process on a beamsplitter, and it is one of the most fundamental processes used in photonic quantum technologies. The seminal experiment has been first performed by Hong, Ou, and Mandel, and the observed effect is named after them [81]. First, we will discuss the outcome of such an experiment using so-called indistinguishable photons for input in arms 1 and 2. Indistinguishability means that the two photons are identical in all their properties, and we can neglect a more elaborated description with wavepackets, which will be introduced in the next chapter.

The input state can be formally written as $|1,1\rangle_{1,2} = \hat{a}_1^\dagger\hat{a}_2^\dagger|0,0\rangle_{1,2}$ with $\hat{a}_1^\dagger$ and $\hat{a}_2^\dagger$ the photon creation operators in arms 1 and 2, respectively. The input state is converted into the corresponding output state by using the beamsplitter matrix of the symmetric balanced 50:50 beamsplitter (input and output now in a more compact vector formalism):

$$\begin{pmatrix} \hat{a}_1^\dagger \\ \hat{a}_2^\dagger \end{pmatrix} \Rightarrow \frac{1}{\sqrt{2}}\begin{pmatrix} 1 & 1 \\ 1 & -1 \end{pmatrix}\begin{pmatrix} \hat{a}_3^\dagger \\ \hat{a}_4^\dagger \end{pmatrix} = \frac{1}{\sqrt{2}}\begin{pmatrix} \hat{a}_3^\dagger + \hat{a}_4^\dagger \\ \hat{a}_3^\dagger - \hat{a}_4^\dagger \end{pmatrix}. \tag{4.11}$$

Therefore the conversion is $\hat{a}_1^\dagger \Rightarrow \frac{1}{\sqrt{2}}(\hat{a}_3^\dagger + \hat{a}_4^\dagger)$ and $\hat{a}_2^\dagger \Rightarrow \frac{1}{\sqrt{2}}(\hat{a}_3^\dagger - \hat{a}_4^\dagger)$. Now we can calculate the expected output state:

$$|1,1\rangle_{1,2} = \hat{a}_1^\dagger\hat{a}_2^\dagger|0,0\rangle_{1,2} \Rightarrow \frac{1}{2}(\hat{a}_3^\dagger + \hat{a}_4^\dagger)(\hat{a}_3^\dagger - \hat{a}_4^\dagger)|0,0\rangle_{3,4}, \tag{4.12}$$

and further considering the commutator relation and the ladder property of the creation operator, we get

$$\begin{aligned}\frac{1}{2}(\hat{a}_3^\dagger + \hat{a}_4^\dagger)(\hat{a}_3^\dagger - \hat{a}_4^\dagger)|0,0\rangle_{3,4} &= \frac{1}{2}((\hat{a}_3^\dagger)^2 - \hat{a}_3^\dagger\hat{a}_4^\dagger + \hat{a}_4^\dagger\hat{a}_3^\dagger - (\hat{a}_4^\dagger)^2)|0,0\rangle_{3,4} \\ &= \frac{1}{2}((\hat{a}_3^\dagger)^2 - (\hat{a}_4^\dagger)^2)|0,0\rangle_{3,4} \\ &= \frac{\sqrt{2}}{2}(|2,0\rangle_{3,4} - |0,2\rangle_{3,4}) \\ &= \frac{1}{\sqrt{2}}(|2,0\rangle_{3,4} - |0,2\rangle_{3,4}). \qquad (4.13)\end{aligned}$$

This result represents the famous Hong–Ou–Mandel (HOM) effect for identical bosonic particles. This causes the generation of the two-particle path entangled state which is also called NOON state and is indicated as "2002-state" for two photons ($N = 2$). This implies that both photons can only be detected together in either one of the two exit arms (3 or 4) of the beamsplitter.

As we will see, real photons from quantum emitters are not necessarily indistinguishable, and the degree of distinguishability degrades the visibility of the HOM effect, meaning that sometimes in each output port c and d, a photon can be detected simultaneously. Thus, from a practical point of view, the dependency of the two-photon interference on the photon indistinguishability is nowadays used to quantify the indistinguishability of photons from single-photon sources (see Chapters 5 and 11).

4.3 Time-dependent light fields and wavepacket description

In the previous sections, we mainly discussed the quantum properties of light within a single-mode theory that uses time-independent light beams. In many practical situations, experiments and applications are usually performed with time-dependent light beams, i. e., with optical pulses. In this case, many modes are excited, and it is necessary to quantize the light field in free space with a set of modes characterized by a continuous wave vector. Moreover, photons stemming from a single quantum emitter in the solid-state are not necessarily indistinguishable due to fluctuations in the environment (see Chapters 5 and 11). Both properties can be accommodated by an appropriate wavepacket description of the single-photon pulse.

First, it is worth noting that the pulse shape of an optical pulse is ultimately connected to its corresponding spectrum by a Fourier transformation (see Fig. 4.3). Second, the continuous-mode annihilation and creation operators, denoted by $\hat{a}(\omega)$ and $\hat{a}^\dagger(\omega)$, are introduced. The relations to their discrete-mode counterparts (mode index k) are $\hat{a}_k \rightarrow \sqrt{\Delta\omega}\hat{a}(\omega)$ and $\hat{a}_k^\dagger \rightarrow \sqrt{\Delta\omega}\hat{a}^\dagger(\omega)$ with the one-dimensional mode spacing $\Delta\omega =$

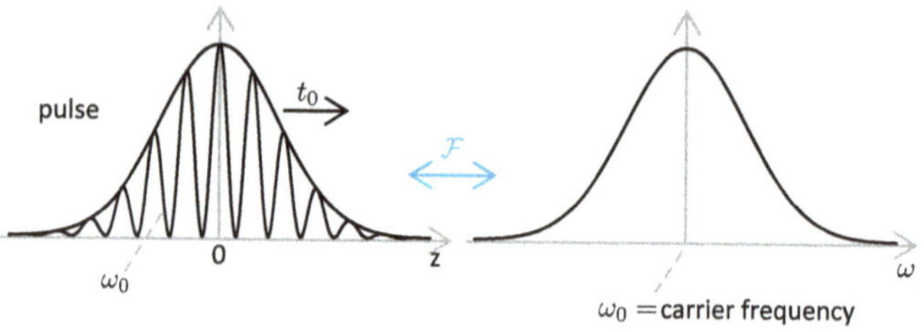

Figure 4.3: Schematic picture of an optical pulse in the time and frequency space.

$2\pi c/L$, which goes to zero as the length goes to infinity [118]. Thus the continuous-mode commutation relation is

$$[\hat{a}(\omega), \hat{a}^\dagger(\omega')] = \delta(\omega - \omega'), \tag{4.14}$$

$$[\hat{a}(t), \hat{a}^\dagger(t')] = \delta(t - t'), \tag{4.15}$$

where the operators in the frequency and time domain are connected by a Fourier transform:

$$\hat{a}(t) = \frac{1}{\sqrt{2\pi}} \int_{-\infty}^{\infty} d\omega\, \hat{a}(\omega) e^{-i\omega t}. \tag{4.16}$$

The integration range over ω strictly extends from 0 to ∞ as the frequency is a positive number. However, since the bandwidth of the optical pulse is typically much narrower than its central frequency ω, the integration can be extended over the full axis from $-\infty$ to ∞ without significant error.

A single photon stemming from a radiative transition of a single quantum emitter can be described by a wavepacket. The temporal shape and spectrum of the photon is defined by the physical process within the source. The details of this process and the corresponding wavepacket description for semiconductor quantum dots will be discussed in Chapter 7. For simplicity, here we introduce a description based on a Gaussian wavepacket (model wavepacket) to show the influence of a possible diversity of photons on beamsplitter operations. The normalized Gaussian wavepacket is defined by

$$\xi(\omega) = \frac{1}{(2\pi\Delta^2)^{\frac{1}{4}}} e^{-i(\omega_0 - \omega)t_0 - \frac{(\omega_0 - \omega)^2}{4\Delta^2}}, \tag{4.17}$$

where ω_0 is the central frequency of the pulse spectrum, and t_0 is the time at which the peak of the pulse passes the origin at $z = 0$. The variance of the spectrum $|\xi(\omega)|^2$ is given by Δ^2.

The Fourier transform of Eq. (4.17) gives

$$\xi(t) = \left(\frac{2\Delta^2}{\pi}\right)^{\frac{1}{4}} e^{-i\omega_0 t - \Delta^2 (t_0 - t)^2}. \tag{4.18}$$

Further, the following normalization conditions are fulfilled:

$$\int \mathrm{d}\omega |\xi(\omega)|^2 = \int \mathrm{d}t |\xi(t)|^2 = 1. \tag{4.19}$$

We can now define a photon-wavepacket creation operator [118]:

$$\hat{a}_\xi^\dagger = \int \mathrm{d}\omega \xi(\omega) \hat{a}^\dagger(\omega) = \int \mathrm{d}t \xi(t) \hat{a}^\dagger(t) \tag{4.20}$$

with $[\hat{a}_\xi, \hat{a}_\xi^\dagger] = 1$.

Continuous-mode number states can then be constructed with the photon wavepacket creation operator in a similar way as described for their discrete counterparts in Eq. (2.45):

$$|n_\xi\rangle = \frac{1}{\sqrt{n!}} (\hat{a}_\xi^\dagger)^n |0\rangle; \quad |0\rangle \text{ is the continuous mode vacuum state.} \tag{4.21}$$

It is important to note that single-discrete-mode number states and continuous-mode-number states have identical degrees of first- and second-order coherence [118].

Now we will discuss the description of photon pair states. Photon pair states can be generated, e. g., by cascaded photon emission in quantum dots or by parametric down-conversion processes. Their description depends on the physical process of their generation. The description has to be extended from a single-photon wavepacket to the more general case of a two-photon wavepacket, where possible correlations between the photons have to be considered. In the following, we will consider three different photon pairs states:

a) Two independent photons in the same continuous-mode field and same wavepacket. In this case the number state ($n = 2$) is defined by Eq. (4.21):

$$|2_\xi\rangle = \frac{1}{\sqrt{2}} (\hat{a}_\xi^\dagger)^2 |0\rangle, \tag{4.22}$$

where the subscript ξ symbolizes the wavepacket function $\xi(\omega)$ or rather $\xi(t)$.

b) Two correlated photons in the same continuous-mode field with different wavepackets ($\omega \neq \omega'$).

Here we have to introduce a more generalized two-photon pair creation operator [118]:

$$|(2_1)_\beta\rangle = P_{\beta 11}^\dagger |0\rangle, \tag{4.23}$$

where the label 1 denotes the same continuous-mode field, the label β denotes the two-photon wavepacket, and

$$P_{\beta 11}^\dagger = \frac{1}{\sqrt{2}} \int \mathrm{d}\omega \int \mathrm{d}\omega' \beta(\omega, \omega') \hat{a}_1^\dagger(\omega) \hat{a}_1^\dagger(\omega') \tag{4.24}$$

represents the photon-pair creation operator, where $\beta(\omega,\omega')$ describes the actual two-photon wavepacket. The joint two-photon wavepacket is normalized according to

$$\int d\omega \int d\omega' |\beta(\omega,\omega')|^2 = 1. \tag{4.25}$$

The photon-pair creation operator in the time domain is given by

$$P^{\dagger}_{\beta 11} = \frac{1}{\sqrt{2}} \int dt \int dt' \beta(t,t') \hat{a}_1^{\dagger}(t) \hat{a}_1^{\dagger}(t') \tag{4.26}$$

with the normalization

$$\int dt \int dt' |\beta(t,t')|^2 = 1 \tag{4.27}$$

and the corresponding Fourier transform of the two-photon spectrum given by

$$\beta(\omega,\omega') \overset{\mathcal{F}}{\longleftrightarrow} \beta(t,t') = \frac{1}{2\pi} \int_{-\infty}^{\infty} d\omega \int_{-\infty}^{\infty} d\omega' \beta(\omega,\omega') e^{-i\omega t - i\omega' t'}. \tag{4.28}$$

It is now important to note that the orders of frequencies or times in the β functions of the two photon-pair creation operators in Eqs. (4.24) and (4.26) refer to the same mode field a. As a consequence, the β functions satisfy the symmetry relations

$$\beta(\omega',\omega) = \beta(\omega,\omega') \quad \text{and} \quad \beta(t',t) = \beta(t,t'). \tag{4.29}$$

In the most general case, possible correlations between the photons have to be considered, e. g., if the photons stem from a radiative cascade of a three-level system (e. g., biexciton–exciton cascade in a quantum dot). This will be discussed in more detail later in the book (see Chapter 7). If the photons are not entangled or correlated, then their wavepacket factorizes to

$$\beta(t,t') = \xi(t)\xi(t'). \tag{4.30}$$

c) Two correlated photons in different continuous-mode fields (denoted by 1 and 2) with different wavepackets ($\omega \neq \omega'$).
We will discuss this case in the framework of the Hong–Ou–Mandel effect, where two photons enter the beamsplitter via two different mode fields (e. g. input ports 1 and 2 of a beamsplitter). This situation will be discussed in more detail below. Here the two-photon state is described by

$$|(1_1, 1_2)_\beta\rangle = P^{\dagger}_{\beta 12}|0\rangle, \tag{4.31}$$

where the photon-pair creation operator is now [118]

$$P^{\dagger}_{\beta 12} = \int d\omega \int d\omega' \beta(\omega, \omega') \hat{a}^{\dagger}_1(\omega) \hat{a}^{\dagger}_2(\omega') \tag{4.32}$$

or written in the time domain as

$$P^{\dagger}_{\beta 12} = \int dt \int dt' \beta(t, t') \hat{a}^{\dagger}_1(t) \hat{a}^{\dagger}_2(t'); \tag{4.33}$$

$\beta(\omega, \omega')$ and $\beta(t, t')$ refer now to the two different mode fields 1 and 2 and therefore do not fulfill the symmetry relation of Eq. (4.29), i. e., $\beta(\omega, \omega') \neq \beta(\omega', \omega)$ and $\beta(t, t') \neq \beta(t', t)$.

In the following, we will discuss cases b) and c) as input beams of a symmetric beamsplitter.

4.4 Hong–Ou–Mandel effect with single-photon wavepackets

We will now extend the discussion of photons impinging on a beamsplitterby introducing the previously discussed wavepacket approach, for which it is necessary to consider the spectral and temporal properties of the generated photons. In the following, we will discuss two different beamsplitter operations with photon-pair states. In the first case, the two photons excite a single mode in one input arm of the beamsplitter (case b of the previous chapter; see also Fig. 4.4(a)), whereas in the second case the photon pair state excites two different input arms of the beamsplitter (case c of the previous chapter; see also Fig. 4.4(b)). In the first case, we will see that the transmission of the photons through the beamsplitter exhibits a behavior similar to classical particles and the shape of two-photon wavepacket does not play any role for the outcome, whereas in the second case the photons show nonclassical behavior (Hong–Ou–Mandel effect), and the wavepackets of the photons determine the outcome of the experiment.

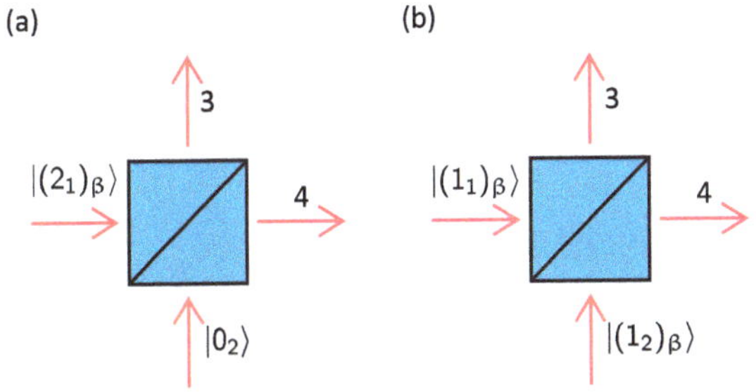

Figure 4.4: (a) Two photons excite a single mode of a beamsplitter. (b) A photon pair state excites two different input arms of a beamsplitter.

Let us start the discussion with two photons entering input arm 1 (see Fig. 4.4(a)). The two-photon state is then defined by Eq. (4.23):

$$|(2_1)_\beta\rangle = \frac{1}{\sqrt{2}} \int dt \int dt' \beta(t,t') \hat{a}_1^\dagger(t) \hat{a}_1^\dagger(t')|0\rangle. \tag{4.34}$$

In this case the symmetry relation $\beta(t,t') = \beta(t',t)$ (Eq. (4.29)) holds. Using the beamsplitter relations for the creation operators according to Eq. (4.3), we get

$$\hat{a}_1^\dagger(t) = \mathcal{R}\hat{a}_3^\dagger(t) + \mathcal{T}\hat{a}_4^\dagger(t) \quad \text{and} \quad \hat{a}_2^\dagger(t) = \mathcal{T}\hat{a}_3^\dagger(t) + \mathcal{R}\hat{a}_4^\dagger(t). \tag{4.35}$$

Substituting the input operators $\hat{a}_1^\dagger(t)$ and $\hat{a}_1^\dagger(t')$ into Eq. (4.34), we get

$$\begin{aligned} |(2_1)_\beta\rangle = \frac{1}{\sqrt{2}} \int dt \int dt' \beta(t,t') \times [&\mathcal{R}^2 \hat{a}_3^\dagger(t)\hat{a}_3^\dagger(t') + \mathcal{R}\mathcal{T}\hat{a}_3^\dagger(t)\hat{a}_4^\dagger(t') \\ &+ \mathcal{T}\mathcal{R}\hat{a}_4^\dagger(t)\hat{a}_3^\dagger(t') + \mathcal{T}^2\hat{a}_4^\dagger(t)\hat{a}_4^\dagger(t')]|0\rangle. \end{aligned} \tag{4.36}$$

Using the symmetry relation (Eq. (4.29)), we can rewrite the central term $\mathcal{R}\mathcal{T}\hat{a}_3^\dagger(t)\hat{a}_4^\dagger(t') + \mathcal{R}\mathcal{T}\hat{a}_3^\dagger(t')\hat{a}_4^\dagger(t) = 2\mathcal{R}\mathcal{T}\hat{a}_3^\dagger(t)\hat{a}_4^\dagger(t')$ and express the output state by using Eq. (4.23) as

$$|(2_1)_\beta\rangle = \mathcal{R}^2|(2_3)_\beta\rangle + \sqrt{2}\mathcal{R}\mathcal{T}|(1_3,1_4)_\beta\rangle + \mathcal{T}^2|(2_4)_\beta\rangle. \tag{4.37}$$

This is an entangled state, and we can now determine the probabilities for finding the photons in the different output ports of the beamsplitter:

$$\begin{aligned} P(2_3,0_4) &= |\mathcal{R}|^4 \quad \text{for a 50:50 beamsplitter this results} \to 0.25, \\ P(1_3,1_4) &= 2|\mathcal{R}|^2|\mathcal{T}|^2 \qquad\qquad\qquad\qquad\qquad\quad \to 0.5, \\ P(0_3,2_4) &= |\mathcal{T}|^4 \qquad\qquad\qquad\qquad\qquad\qquad\quad\; \to 0.25. \end{aligned}$$

Obviously, the probabilities are independent of the shape of the photon wavepacket and reproduce those for classical particles.

Let us now consider two photons entering into two different input ports (labels 1 and 2) of the beamsplitter (see Fig. 4.4(b)). The corresponding two-photon input state is given by

$$|(1_1,1_2)_\beta\rangle = \int dt \int dt' \beta(t,t') \hat{a}_1^\dagger(t)\hat{a}_2^\dagger(t')|0\rangle. \tag{4.38}$$

The times t, t' in the β-function are now associated with the beamsplitter arms 1 and 2, respectively. Consequently, $\beta(t,t')$ no longer satisfies the symmetry relation $\beta(t,t') = \beta(t',t)$ (Eq. (4.29)). Using the beamsplitter relations for the creation operators (Eqn. (4.3)) and substituting the input operator $\hat{a}_1^\dagger(t)$ and $\hat{a}_2^\dagger(t')$, we get

$$|(1_1,1_2)_\beta\rangle = \int dt \int dt' \beta(t,t') \times [\mathcal{R}\mathcal{T}\hat{a}_3^\dagger(t)\hat{a}_3^\dagger(t') + \mathcal{R}^2\hat{a}_3^\dagger(t)\hat{a}_4^\dagger(t')$$

$$+\mathcal{T}^2\hat{a}_4^\dagger(t)\hat{a}_3^\dagger(t') + \mathcal{T}\mathcal{R}\hat{a}_4^\dagger(t)\hat{a}_4^\dagger(t')]|0\rangle \tag{4.39}$$

with

$$\begin{aligned}
&\mathcal{T}^2\beta(t,t')\hat{a}_4^\dagger(t)\hat{a}_3^\dagger(t') \quad [\hat{a}_4^\dagger,\hat{a}_3^\dagger]=0\\
&=\mathcal{T}^2\beta(t,t')\hat{a}_3^\dagger(t')\hat{a}_4^\dagger(t)\\
&=\mathcal{T}^2\beta(t',t)\hat{a}_3^\dagger(t)\hat{a}_4^\dagger(t') \quad t,t' \text{ interchanged, different modes}
\end{aligned}$$

and with

$$\begin{aligned}
&\frac{1}{2}\mathcal{R}\mathcal{T}\beta(t,t')\hat{a}_3^\dagger(t)\hat{a}_3^\dagger(t')\\
&=\frac{1}{2}\mathcal{R}\mathcal{T}\beta(t',t)\hat{a}_3^\dagger(t)\hat{a}_3^\dagger(t') \quad t,t' \text{ interchanged, same mode, } \beta \text{ invariant.}
\end{aligned}$$

Therefore

$$|(1_1,1_2)_\beta\rangle = \int \mathrm{d}t \int \mathrm{d}t' \left\{\frac{1}{2}\mathcal{R}\mathcal{T}[\beta(t,t')+\beta(t',t)][\hat{a}_3^\dagger(t)\hat{a}_3^\dagger(t') + \hat{a}_4^\dagger(t)\hat{a}_4^\dagger(t')] \right. \tag{4.40}$$

$$\left. + [\mathcal{R}^2\beta(t,t') + \mathcal{T}^2\beta(t',t)]\hat{a}_3^\dagger(t)\hat{a}_4^\dagger(t')\right\}|0\rangle. \tag{4.41}$$

Considering different normalizations in the definitions for the two-photon creation operators in Eqs. (4.24) and (4.32), we can again determine the probabilities for finding the photons in the different output ports of the beamsplitter:

$$P(2_3,0_4) = \frac{1}{2}|\mathcal{R}|^2|\mathcal{T}|^2 \int \mathrm{d}t \int \mathrm{d}t' |\beta(t,t')+\beta(t',t)|^2, \tag{4.42}$$

$$P(1_3,1_4) = \int \mathrm{d}t \int \mathrm{d}t' |\mathcal{R}^2\beta(t,t') + \mathcal{T}^2\beta(t',t)|^2. \tag{4.43}$$

We can further simplify the expression by considering the invariance of the double integral under interchange of t' and t to

$$\begin{aligned}
\int \mathrm{d}t \int \mathrm{d}t' |\beta(t,t')+\beta(t',t)|^2 &= \int \mathrm{d}t \int \mathrm{d}t' (\beta(t,t')+\beta(t',t))(\beta^*(t,t')+\beta^*(t',t))\\
&= \int \mathrm{d}t \int \mathrm{d}t' [\underbrace{\beta(t,t')\beta^*(t,t')}_{\substack{=1\\ \text{normalization}}} + \beta(t,t')\beta^*(t',t)\\
&\quad + \beta(t',t)\beta^*(t,t') + \underbrace{\beta(t',t)\beta^*(t',t)}_{\substack{=1\\ \text{normalization}}}]\\
&= 2\cdot\left(1+\int \mathrm{d}t \int \mathrm{d}t' \beta^*(t,t')\beta(t',t)\right).
\end{aligned}$$

This results in

$$P(2_3,0_4) = P(0_3,2_4) = |\mathcal{R}|^2|\mathcal{T}|^2(1+|\mathcal{J}|^2) \tag{4.44}$$

with the overlap integral of the two input states, $|J|^2 = \int dt \int dt' \beta^*(t,t')\beta(t',t)$.

Furthermore, by considering the normalizations $\int dt \int dt' |\beta(t,t')|^2 = 1$, $|\mathcal{R}|^2 + |\mathcal{T}|^2 = 1$, and $\mathcal{R}\mathcal{T}^* + \mathcal{T}\mathcal{R}^* = 0$ we get

$$P(1_3, 1_4) = 1 - 2|\mathcal{R}|^2|\mathcal{T}|^2(1 + |J|^2). \tag{4.45}$$

It is important to note that in contrast to the situation where the two photons impinge on the same input port of the beamsplitter (see Fig. 4.4(a)), the output probabilities now depend on the form of the photon wavepackets expressed by the result of the integral J. It is also interesting to compare this result with the expectation for classical particles that reflect and transmit independently on the beamsplitter with the probabilities $|\mathcal{R}|^2$ and $|\mathcal{T}|^2$, respectively:

$$P(2_3, 0_4)_{\text{class}} = P(0_3, 2_4)_{\text{class}} = |\mathcal{R}|^2|\mathcal{T}|^2 \quad \text{and} \quad P(1_3, 1_4)_{\text{class}} = |\mathcal{R}|^4 + |\mathcal{T}|^4 = 1 - 2|\mathcal{R}|^2|\mathcal{T}|^2. \tag{4.46}$$

By comparing the classical and quantum mechanical results we see that it is more likely in the quantum mechanical case that the photons emerge from the same output arm and less likely from different arms.

Now we will discuss an important special case where the two photons are not entangled (cf. Eq. (4.30)). Therefore the two-photon wavepacket factorizes $\beta(t,t') = \xi_1(t)\xi_2(t')$, and the overlap integral becomes

$$|J|^2 = \left| \int dt \xi_1^*(t)\xi_2(t) \right|^2. \tag{4.47}$$

If we assume a Gaussian shape (see Eq. (4.18)) for the two independent wavepackets $\xi_1(t)$ and $\xi_2(t)$, then the integral can be expressed as

$$J = e^{-\frac{1}{2}\Delta^2(t_{01}-t_{02})^2}, \tag{4.48}$$

where $\omega_1 = \omega_2 = \omega_0$, the variances $\Delta_1^2 = \Delta_2^2 = \Delta^2$, but with different arrival times $t_{01} \neq t_{02}$ at the beamsplitter. For equal arrival times of the two photons at the beamsplitter, i. e., $t_{01} = t_{02}$, we get

$$P(2_3, 0_4) = P(0_3, 2_4) = 2|\mathcal{R}|^2|\mathcal{T}|^2 \quad \text{and} \tag{4.49}$$

$$P(1_3, 1_4) = 1 - 4|\mathcal{R}|^2|\mathcal{T}|^2. \tag{4.50}$$

This means that the probability of the two photons leaving the same beamsplitter arm is twice the classical value given in Eq. (4.46). If we now choose a balanced beamsplitter (50:50), i. e., $|\mathcal{R}|^2 = |\mathcal{T}|^2 = \frac{1}{2}$, then the probabilities are

$$P(2_3, 0_4) = P(0_3, 2_4) = 0.5, \tag{4.51}$$

$$P(1_3, 1_4) = 0. \tag{4.52}$$

This is the famous result of the Hong–Ou–Mandel experiment in quantum optics, i. e., the two photons can never be detected simultaneously in both output arms 3 and 4. This only holds for indistinguishable photons as chosen here ($\omega_1 = \omega_2 = \omega_0$, $\Delta_1^2 = \Delta_2^2 = \Delta^2$, and $t_{01} = t_{02}$). This particular case has been already discussed in the previous chapter (see Eq. (4.13)).

4.5 Summary

- The quantized field operators $\hat{a}_i$ and $\hat{a}_i^\dagger$ convert in the same way as classical fields at the beamsplitter.
- The state of a quantum mechanical system after a measurement is given by the projection of the state before the measurement onto the state determined by the measurement.
- For the description of single-photon light pulses, an appropriate wavepacket description of the single-photon pulse is necessary.
- In the case of photon pair states (e. g., entangled photons), the description has to be extended from a single-photon wavepacket to the more general case of a two-photon wavepacket.
- Indistinguishability of two photons means that the two photons are identical in all their properties.
- When two indistinguishable single photons enter a 50:50 beamsplitter on two different input arms, both photons can only be detected together in either one of the two exit arms of the beamsplitter. This is the famous Hong–Ou–Mandel (HOM) effect.
- Photons stemming from quantum emitters in solids are not necessarily indistinguishable due to fluctuations in their environment. As a consequence, the degree of distinguishability degrades the visibility of the HOM effect.

5 Single photons from single emitters: properties and their experimental characterization

5.1 Introductory remarks

In this chapter, we will discuss the fundamental properties that characterize a realistic nonclassical light emitter. We will start with the characterization techniques commonly used in spectroscopy and quantum optics laboratories where quantum light emitters are investigated and developed. For the basic description of the light emission process, the nonclassical light emitter can be thought as a two-level system, comprising a ground and one excited state. Once carriers confined in the system are found in the excited state, they can radiatively recombine reaching the system ground state with corresponding emission of light. For simplicity, in the following sections, we will explain the techniques considering optical excitation of the quantum emitters. With few abstraction efforts, the described experiments can be understood including an electrical source that injects the carriers via a current flow rather than an optical source of photons as excitation medium.[1] Toward the end of the chapter, we will consider a more realistic picture of the nonclassical emitters, where more than two levels will be included. Methods for providing optical and electrical excitation of single emitters can be found in Chapter 10.

5.2 Properties of single emitters

In this section, we discuss some of the most common and widely utilized spectroscopic techniques. The described measurement methods are very important to reach a deep comprehension of the characteristics of any emitter and can provide fundamental insight into its physics and behavior. The title of each section contains the emitter's property and respective techniques that can be employed for its characterization.

5.2.1 Electronic states: photoluminescence (PL), micro-PL spectroscopy, and photoluminescence excitation

Spectroscopy indicates the study of the optical material response under probing with light. In solid-state systems, the electrical carriers can be brought from low-energy states to excited ones employing optical (or electrical) excitation. Once excited, the carriers relax down to the lowest excited state and finally reach the ground state. This last step

1 It is worth mentioning that electrical excitation has been demonstrated only for a subset of single quantum emitters. In addition, optically excited sources still represent the state-of-the-art in terms of emitted photon properties.

https://doi.org/10.1515/9783110703412-005

can take place radiatively, i. e., accompanied by the emission of light, which is called luminescence. Its observation under optical pumping has been and is currently highly employed when dealing with solid-state systems, from bulk materials to nanostructures in semiconductors or 2D materials, since it can provide insights into the emitter's electronic states. In *photoluminescence*-spectroscopy (PL-spectroscopy), carriers are photo-excited in the material when illuminated with a light source (i. e., thermal lamp, LED, laser, etc.): the energy carried by the impinging photons can promote the carriers from the low-energy states to excited ones.

For this to happen, the impinging photons need to be absorbed by the sample under investigation and their energy transferred to the carriers. As in the exemplary case depicted in Fig. 5.1, the arrival of a photon enables the transition from the ground state $|g\rangle$ to higher excited states (here simplified as one high excited level $|i\rangle$). Fast relaxation processes can take place resulting in the transition (often nonradiative) $|i\rangle \rightarrow |e\rangle$ to the lowest excited state. The last transition between the levels $|e\rangle \rightarrow |g\rangle$ can happen radiatively, resulting in the emission of a photon and the system to be found in the ground state. This simplified excitation, relaxation, and radiative recombination cycle is repeated several times in the experiments, having the light emitted by the system statistically collected and analyzed. This analysis is commonly performed by means of a spectrometer (for example, a spectrograph or monochromator depending on the experimental configuration). Typically, an important information is given by the observation of the light intensity as a function of the wavelength (or energy, frequency, etc.). Indeed, observing the so-called emission *spectrum* provides fundamental knowledge regarding the physics happening in the investigated sample: as an example, considering a semiconductor nanostructure, its size or the number of charges confined therein can influence the emission wavelength and the number of transitions observed in the emission spectrum. The excitation process can be similar to the previously discussed, as seen in

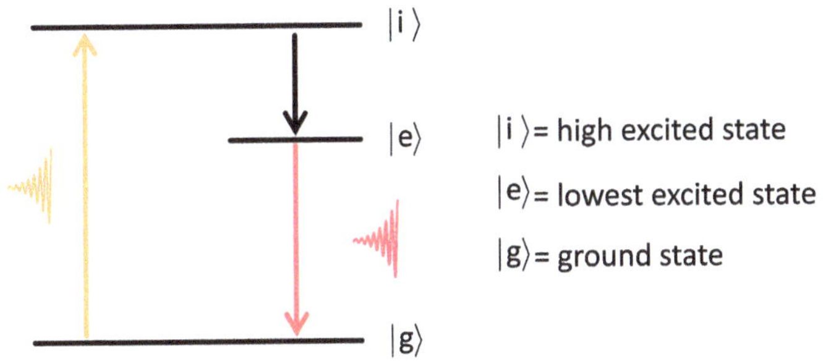

Figure 5.1: Schematic excitation, relaxation, and radiative recombination among three levels. The incident light (yellow arrow) allows the excitation from the ground state $|g\rangle$ to the excited state $|i\rangle$. Then a nonradiative relaxation takes place having the system transitioning from the higher excited state $|i\rangle$ to the lowest excited state $|e\rangle$. The last transition $|e\rangle \rightarrow |g\rangle$ can happen radiatively, i. e., accompanied by the emission of a photon (red arrow). The system is now back in the ground state, ready for another excitation, relaxation, and radiative recombination cycle.

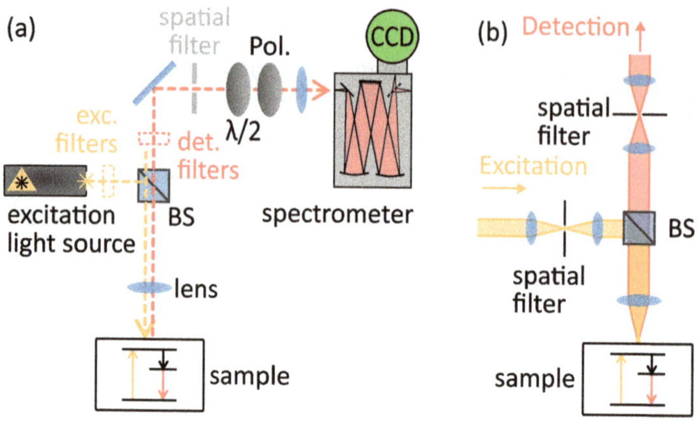

Figure 5.2: (a) Schematic of the basic building blocks required for photoluminescence experiments. The excitation light source (a laser, a thermal lamp, LED, etc.) can be further tailored to experimental needs employing excitation filters (i. e., laser clean-up, short/long pass filters, etc.). The shown beamsplitter (BS), used for guiding the excitation light on the sample and for collecting the emitted light, can have various splitting ratios depending on the intended experiments. In some cases, it can also be replaced by a dichroic beamsplitter to discriminate in wavelength, excitation, and luminescence. Additionally, unwanted light components (either from the excitation or from the sample) can be filtered before reaching the detector (detection filters). Optical lenses or objectives are employed to focus the excitation light down to specific areas of the sample and to collect the largest possible fraction of the emitted luminescence. The detection can be performed using a spectrometer equipped with a charge-coupled device (CCD): a diffraction grating is used to spatially separate the spectral components that reach the CCD. A careful calibration allows for the observation of the light intensity over the wavelength (*spectrum*). When the polarization properties of the light need to be investigated, polarization optics, as a half-waveplate and a polarizer, can be implemented prior the spectrometer (an additional quarter-waveplate can be first added to investigate circularly polarized components). In case the diffraction grating is polarization sensitive, the polarizer should be aligned to maximize the transmitted signal. (b) Schematic of a confocal optical microscopy: spatial filters, as pinholes or single mode fibers, are set in conjugate image planes. Therefore excitation and detection paths are aligned to ensure their simultaneous focusing on the sample. Few modifications in the collection path, i. e., inserting a half-waveplate and a polarizer, can be implemented to investigate the polarization properties of the luminescence. Confocal and collinear setups are employed when resonant optical excitation is needed (see Section 10.2.3).

Fig. 5.1, and explained in more detail in Chapter 10, where more than one single radiative transition are considered. A schematic of an exemplary setup for performing the aforementioned studies can be found in Fig. 5.2.

Excitation light is guided toward the sample via a beamsplitter (BS) and can be focused on the sample by means of optical elements as lenses or objectives. These further collect the photoluminescence signal, which is transmitted through the BS and sent toward the spectrometer. Ideally, losses of photons emitted by the sample should be minimized, and therefore beamsplitters with asymmetric splitting ratios are more favorable. As an example, in the experimental scheme depicted in Fig. 5.2(a), a BS with high transmission and low reflection would be the ideal choice to direct the largest fraction of luminescence toward the spectrometer: although it is important to collect as much luminescence as possible, losses of the excitation light are not that critical since high pumping intensities are typically available.[2] It is often the case that excitation and emitted light have very different wavelengths (from few tens to few hundreds of nanometers): in this case, the experimental configuration can be adapted, and the beamsplitter can be replaced by a dichroic element. Still referring to Fig. 5.2(a), a dichroic BS would allow us to reflect the excitation wavelength while transmitting the photoluminescence signal. This would have a twofold advantage: on the one hand, it will maximize the amount of sample's emitted light that can be directed to the spectrometer. On the other hand, it will help avoiding unwanted excitation light (resulting from scattering or reflection from the sample or employed optical elements) to reach the spectrometer. Indeed, the excitation light can constitute an unwanted background in the detection process. Furthermore, in single quantum emitter spectroscopy, the excitation intensity largely exceeds the emitted luminescence. Therefore it becomes imperative to suppress unwanted light components before they reach the spectrometer; otherwise, they may constitute a background even larger than the investigated signal. In the experimental implementations where the use of dichroic elements is not possible or not desired, filtering of unwanted excitation light can be performed before reaching the detection elements, as shown in Fig. 5.2(a). This filtration stage can also be employed to avoid further undesired spectral components to reach the detector (see also Fig. 5.6).[3]

Differently from PL-spectroscopy, microphotoluminescence (μ-PL) spectroscopy is often used to investigate much smaller areas of the sample: the prefix *micro* indeed refers to the dimension of the excitation/collection area, typically around few μm^2. The possibility of investigating micron-sized areas becomes important when an ensemble of micro- or nanostructures are embedded in the sample under study. A small investigation area can give the possibility to address isolated emitters rather than collecting a luminescence signal, which is the average from a broad ensemble. The utilized lens or objective further defines the excitation spot size and the area from which the light is collected, at a first approximation given by the Airy disk diameter as $1.22\lambda/\mathrm{NA}$ (where λ is the light wavelength, and $\mathrm{NA} = n \sin\theta$ is the optics numerical aperture, with n the refractive index of the material surrounding the lens, and θ the half-angle of the maximum

2 The design of the setup has to be carefully performed taking into account all experimental conditions: if losses on the excitation line cannot be sustained (for example, if the employed excitation source does not have enough power), then trade-off can be made increasing the losses on the collected luminescence, which means using a beamsplitter with lower transmission and higher reflection.

3 For simplicity, the emission process in Fig. 5.1 is schematically depicted as if only light with a very specific wavelength is emitted by the sample. It is worth mentioning that in actual experimental conditions, various radiative processes can take place while PL is investigated. This may then require the filtering of light, which is not subject of investigation from the light that needs to be studied.

cone set by the light entering or leaving the optics).[4] Typical beam sizes in micro-PL experiments performed at near infrared wavelength result in diameters of circa 1–2 µm (as, for example, with an objective of NA = 0.7 at ≈ 900 nm). In single quantum emitter spectroscopy, where the emitter itself has a dimension much smaller than the excitation spot size (even when diffraction limited), an unwanted signal may come from the surrounding area that is still photo-excited. Spatial filters, as, for example, pinholes or optical fibers (down to single-mode ones), can be employed to ensure collecting light from a smaller sample area. A confocal microscopy geometry further ensures that light from one selected focal plane can be detected. Figure 5.2(b) shows an exemplary setup for confocal experiments: excitation and detection optics are aligned to ensure that both light paths are simultaneously focused. Although in the figure a pinhole is sketched as an exemplary spatial filter, practical implementations often employ single-mode fibers acting as pinholes. Such a confocal geometry finds application in the development of setups for the resonant excitation of single quantum emitters (see Section 10.2.3): the use of two orthogonal polarizers, one in the excitation and one in the detection path, allows for suppressing the excitation laser light, preventing it to reach the detector. In this configuration, where excitation and detection have the same wavelength, implementing a collinear (and confocal) geometry helps increasing the laser light rejection, reaching experimentally proven values above 10^7 (see early works as [69] and [104]).

Together with the wavelength of the emitted light, further insight on the investigated sample can be gained measuring the polarization characteristics of luminescence. To do so, polarization optics is typically added in the light collection path transforming the polarization information into intensity information. For example, a half-waveplate and a polarizer are a useful combination to investigate linear polarization components in the emitted light: the half-waveplate can be used to rotate the linear polarization of the incoming luminescence until one component is parallel to the following polarizer and therefore can be transmitted toward the detector (see Fig. 5.2(a)). All others polarization components of the light are suppressed and will only be visible at different angular configurations of the waveplate.[5] It is worth noting that the most common spectrometers employ reflection gratings to separate the spectral components of the incoming light. These gratings are polarization sensitive, meaning that their efficiency varies with the incident light polarization.

Typically, during the investigation of new and unknown samples, these described techniques are both employed. Indeed, whereas PL-spectroscopy can give a fast and clear idea on the emission wavelength of an "inhomogeneously broadened" emitters

4 The definition via the Airy disk is not the only used parameter to define the spot size. In some cases, the $1/e^2$ or the full width at half maximum of the intensity is used. Furthermore, Gaussian optics can also be utilized to obtain the minimum beam waist.

5 Circularly polarized components can be detected adding a quarter waveplate in front of the half-waveplate. This allows for transforming any polarization state into a linear one.

ensemble, μ-PL can be utilized to investigate the properties of single emitters as a first step toward more complex quantum optical experiments. As a final remark, it is worth noting that these types of measurements are normally performed integrating over a time scale, which is typically longer than emitter's lifetime: other techniques need to be employed to resolve their temporal dynamics (see Section 5.2.3). This also means that in PL and μ-PL experiments, exciting the sample via a continuous wave (CW) or pulsed light source only alters the overall measured intensity, and typically the spectra show similar wavelength-dependent features.

The last photoluminescence technique here discussed is the so-called *photoluminescence excitation* (PLE). This technique represents a very valuable tool in solid-state spectroscopy, in particular, when the electronic states of a single emitter need to be investigated. With this experimental approach, the excitation wavelength of a laser is scanned while the PL signal of the investigated transition is recorded. Every time the excitation laser (typically, a CW source with a narrow spectrum to ensure high spectral resolution in the PLE) hits a resonance that is coupled to the luminescent transition, a PL signal is observed (see Fig. 5.3(a)). The emission spectrum does not change its frequency behavior, but its intensity results to be proportional to the excitation laser absorption as well as the quantum efficiency of the emission process. For each excited state, or phonon resonances (i. e., longitudinal optical) that are coupled to the transition under study, after excitation, the photoexcited carriers relax in the lowest state giving rise to luminescence. Thanks to PLE, it is possible to obtain information on the electronic states and recombination dynamics, valuable knowledge in solid-state spectroscopy. This technique has also a practical advantage, since it can also be performed on samples found on ab-

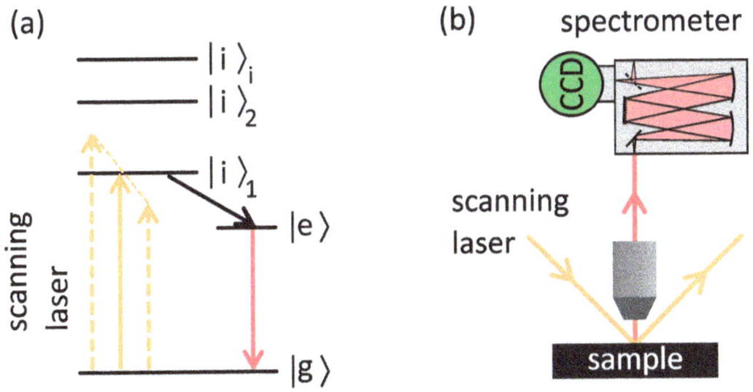

Figure 5.3: (a) Sketch of a multilevel system probed by PLE. The excitation laser is exemplary scanned in frequency over the first high excited state. When hitting the resonance, it is absorbed (depicted as a solid line in comparison to dashed ones), and then the system relaxes down to the lowest excited state, followed by radiative emission. (b) Simplified experimental setup. The scanning laser is depicted off axis with respect to the collection direction to minimize the unwanted scattered laser reaching the spectrometer. Similarly to photoluminescence, other filtering methods can be employed even using a collinear excitation/detection setup. The scanning laser power can also be monitored to provide a normalization value for the PLE signal.

sorptive substrates, for which other approaches as transmission measurements are not employable [47].

5.2.2 Transition linewidth: Fabry–Perot interferometry

It is now clear that photoluminescence (or microphotoluminescence) spectroscopy represents a fundamental tool in the investigation of real emitters: the analysis of the light intensity as a function of the emission wavelength is of utmost importance for a deep understanding of underlying physical properties of the sample under study. On the other hand, typically employed spectrometers have a spectral resolution of a few tens of picometers: this means that all spectral features below the aforementioned value will not be resolved and will only appear as one narrow line in the acquired spectrum.[6] Nevertheless, in several cases the capability of acquiring high resolution spectra can provide further important information on the physics of the investigated system. In some research fields, multiple coupled spectrometers are used to increase the final spectral resolution. This improvement comes at the expense of signal intensity since photons will experience multiple losses when propagating through multiple monochromators. Alternatively, an appealing approach to perform high-resolution spectroscopy is based on the use of *Fabry–Perot interferometers* (FPIs).[7]

On a simple picture, an FPI consists of a cavity formed by two reflective surfaces, either two spaced mirrors or two surfaces of a solid plate (also called the Fabry–Perot etalon). The incoming light reaches the first surface and gets partially reflected and partially transmitted: amplitude transmittance and reflectance for the incoming wave (propagating from left to right in Fig. 5.4(a)) are labeled as t_1 and r_1. The propagating component can then reach the second surface being also partially reflected and transmitted (with amplitude transmittance and reflectance for the second surface set as t_2 and r_2). Multiple reflections result in several counter propagating waves, which can constructively or destructively interfere. As result, light will propagate only at specific, equally spaced, frequencies which depend on the length of the path followed by the incident light. In the particular case where no losses are present and the amplitude re-

6 Typically, when the light illuminates one single pixel of a CCD, the nearby pixels may also provide a signal: therefore a resolution-limited spectrum can still appear as a narrow peak on two to three pixels.

7 In the literature, there is sometimes confusion regarding the correct spelling of the name, either Fabry–Perot or Fabry–Pérot. So which one is correct? With or without an accent? The confusion on the surname of Jean-Baptiste Alfred Perot apparently comes from the fact that although he is registered at the civil registry as Perot, he himself used the surname Pérot in some scientific publications. So it is often referred to as Perot when talking about the person but as Pérot when discussing his discoveries. Since in many books the most common nomenclature is Fabry–Perot, we decided to employ this version in our work.

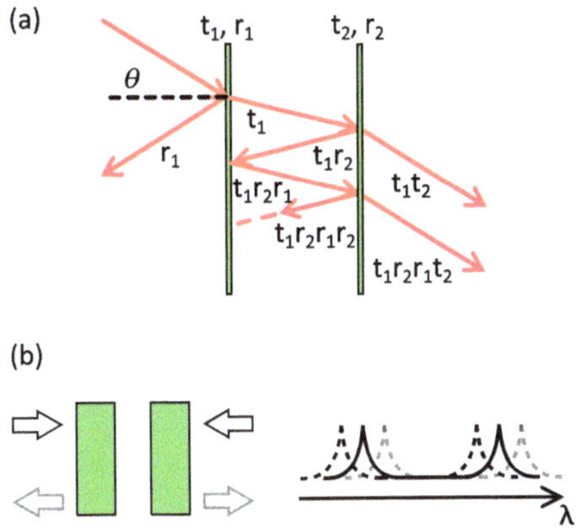

Figure 5.4: (a) Sketch of reflected and transmitted beams at the two surfaces 1 and 2 for a defined angle θ. In the figure the amplitude transmission and reflection coefficients r and t are reported. (b) Sketch of an etalon where the transmission peaks (for $\theta = 0$) move under an applied external perturbation, which effectively changes the etalon length. For air-spaced etalons, as in the figure, the relative distance can be modified. For solid etalons, strain or temperature can effectively change the etalon length.

flectance/transmittance at the two mirrors are equal,[8] the intensity transmittance can be written as

$$T = \frac{1}{1 + (2F/\pi)^2 \sin^2(\varphi)}, \tag{5.1}$$

where φ is the phase between adjacent beams, and F is the so-called *finesse*. This can be defined as $F = \pi\sqrt{R}/(1 - R)$ with R the intensity reflectance of the considered reflector ($R = |r_1 r_2|$). In case the two parallel surfaces are spaced by d, the phase can be defined as

$$\varphi = \frac{2\pi dn}{\lambda_0}\cos(\theta), \tag{5.2}$$

where n is the refractive index of the material between the reflective surfaces, and λ_0 is the wavelength (in vacuum) of the incident light reaching the FPI with an angle θ. In the most common case of perpendicular incidence ($\theta = 0$), the phase becomes $\varphi = 2\pi dn/\lambda_0 = dnk_0$. From Eq. (5.1) we see that the intensity transmittance is maximal when the phase φ is an integer multiple of π. The transmission spectrum is therefore constituted by a periodic repetition of peaks as in Fig. 5.5 with periodicity set by a phase difference of π, and the peaks become narrower with increasing the finesse.

8 In the general case where the amplitude transmittance and reflectance for the two mirrors are different, the intensity transmittance takes the more general form $T = \frac{T_{max}}{1+(2F/\pi)^2 \sin^2(\varphi)}$, where $T_{max} = \frac{|t_1 t_2|^2}{(1-|r_1 r_2|)^2}$ [163].

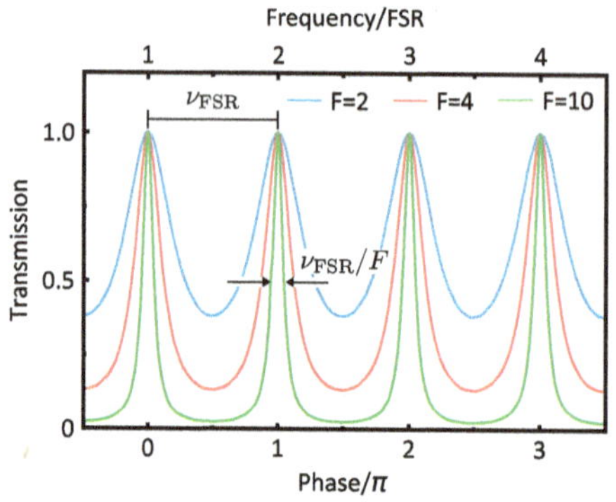

Figure 5.5: Calculated transmission from Eq. (5.1). Three different *finesse* values F have been considered. The x-axis reports the transmission versus phase (as multiple of π, bottom) or frequency (as multiple of the free spectral range, top).

The phase in Eq. (5.2) can also be written in terms of frequency, meaning that the conditions of maximum transmittance (for normal incidence) are reached when

$$\varphi = \pi = dnk_0 = dk = d\frac{\omega_{FSR}}{c} \rightarrow \omega_{FSR} = \pi\frac{c}{d}, \quad \nu_{FSR} = \frac{c}{2d}. \tag{5.3}$$

The aforementioned is referred to as a *free spectral range*, which represents the spectral separation between two nearby transmission peaks: in case of large finesse ($F \gg 1$), the width of the peaks becomes as ν_{FSR}/F. Equation (5.1) can be rewritten in terms of frequency as

$$T = \frac{1}{1 + (2F/\pi)^2 \sin^2(\pi\nu/\nu_{FSR})}. \tag{5.4}$$

Finesse and *free spectral range* constitute the two main characterizing parameters (at a specific wavelength) for a Fabry–Perot interferometer. From the equations above it is possible to understand the importance of a Fabry–Perot interferometer in high-resolution spectroscopy: actively controlling the separation between the reflective surfaces, for example, by tuning the relative position of one mirror (via piezoactuators) or changing the etalon properties by applying mechanical strain or varying its temperature, would allow us to scan the transmission features in wavelength (see Fig. 5.4(b)). If the FPI finesse is high enough that the transmission peaks are narrower than the spectral features to be analyzed, then this type of interferometer would allow only a spectrally narrow part of the incident light to be transmitted. Tuning of the transmission, once calibrated in wavelength, would result in the reconstruction of the spectrum of the incident light with high spectral resolution. This means that the Fabry–Perot interferometer needs to be carefully tailored in terms of *finesse* and *free spectral range* to the intended applications: the *finesse* needs to be high enough to achieve the intended spectral resolution, whereas the *free spectral range* needs to be sufficiently large to have

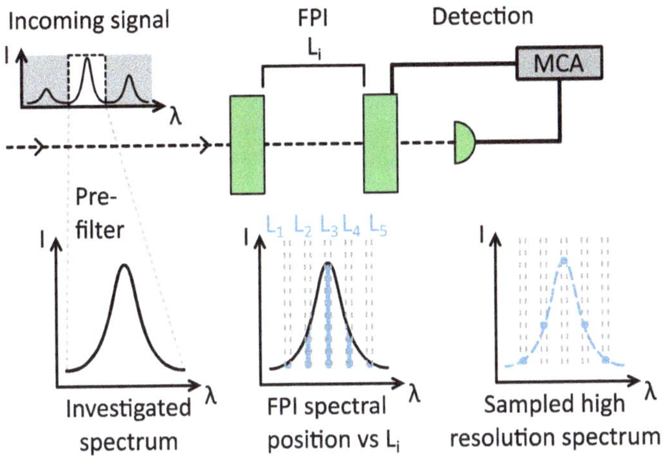

Figure 5.6: (top) The incoming signal is first filtered to allow only the investigated spectrum to be transmitted through the Fabry–Perot interferometer. The prefilter spectral bandwidth should be smaller than the FPI free-spectral range, so that only one peak is scanned over the investigated spectrum. The fraction of the signal transmitted through the FPI is sent to a detector (reaching a single-photon level in case of single quantum emitters). A multichannel analyzer (MCA) relates the detected intensity (or detected events) to the length of the FPI (which can be controlled via temperature or applied voltage in case of piezoactuators controlling the FPI length). The calibration allows for assigning the FPI length to a specific wavelength. (bottom) Exemplary investigated spectrum: when sent through the FPI, only a certain fraction is transmitted (depending on the wavelength/FPI length L_i). The integrated intensity per bin (defined by the FPI length) allows for sampling the high-resolution spectrum.

only one FPI peak scanning over the investigated spectral feature. To experimentally employ the FPI for high-resolution spectroscopy, a setup as depicted in Fig. 5.6 can be employed. First, the potentially complex signal under investigation is prefiltered (via a monochromator or other elements) to isolate the spectral feature to be further analyzed in high resolution (see in Fig. 5.6 the incoming prefiltered signal to become the investigated spectrum through the FPI). This first filtration stage should have a transmission bandwidth $\Delta\omega_{\text{monochr.}}$ similar to (or smaller than) the free spectral range ω_{FSR} to avoid collecting the investigated signal with more than one FPI transmission peak. The incoming signal is then sent through the interferometer. The transmitted light reaches a detector (in case of single quantum emitter spectroscopy, a detector sensitive up to the single-photon level is needed). Changing the FPI length L_i results in the spectral shift of the FPI transmission window, enabling other parts of the spectrum to be transmitted and detected. Once calibrated, i. e., having assigned a specific wavelength for each FPI length L_i, a multichannel analyzer allows us to relate the FPI spectral position with the respective detected intensity, reconstructing the high-resolution spectrum.

As a final remark, although here the FPI has been discussed as an important tool for high-resolution spectroscopy, the availability of a tunable narrow-band filter becomes appealing for the spectral filtering of an incoming signal. Once again, a careful design of the *finesse* and *free spectral range* becomes necessary to select the desired filter spectral width. Then the FPI can be tuned to match the intended spectral feature and, in some cases, stabilized to avoid frequency drifts of the transmission peaks. This type of filters are widely used in different kinds of experiments.

5.2.3 Transition lifetime: time-resolved spectroscopy

Temporal dynamics of the light emission process is another important property together with the emission spectrum. Measuring the decay time of a two-level system population may give important insights on the recombination dynamics. Various charge states recombine with a different probability, and hence this reflects on the measured decay time. As an additional example, the presence of fine-structure splitting in semiconductor quantum dots can lead to the observation of oscillations superimposed to the commonly observed exponential behavior of the exciton decay time (in case of resonant excitation of the QD, as explained in Section 10.2.3): this measurement can then be employed to discriminate between exciton or charged exciton states (if the temporal resolution is high enough to resolve these oscillations) [178]. Additionally, the presence of nonradiative decay channels can be inferred via the observation of a decay time shorter than the commonly measured one. Cavity quantum electrodynamic effects can also be observed for as they modify the emitter's lifetime: the presence of Purcell effect is reflected in a measured shortening of the emitter's decay time in comparison to the case of an emitter in a free space, i. e., not interacting with the cavity (see Chapter 6, Section 6.4). These are few exemplary reasons pointing out the importance of carefully analyze the light emission temporal dynamics.

The basic principle behind time-resolved spectroscopy techniques can be understood as follows: first, the sample under investigation is excited, and then the excitation signal is turned off, and the temporal evolution of the luminescence intensity is observed. For this type of measurements, it becomes evident that a pulsed source needs to be employed to excite the emitters and that the intensity detection has to be precisely synchronized with the excitation pulse. This is often performed by time-correlating the transistor–transistor logic (TTL) signal of the laser with the emitter's intensity (see Fig. 5.7(a)): this is detected with a photodiode having a fast time resolution and an appropriate sensitivity, often close to or at the single-photon level. In this last case, this type of measurements is named time-correlated single-photon counting (TCSPC). In this kind of experiments, the generation of the laser pulse can be seen as a synchronization start-trigger, whereas the click of the single photon on the detector can be utilized as a stop signal. The periodic nature of this kind of measurement together with its precise temporal trigger allows for summing up multiple measuring cycles and then improving the signal-to-noise ratio. As a result, the emission intensity behavior over time is obtained.

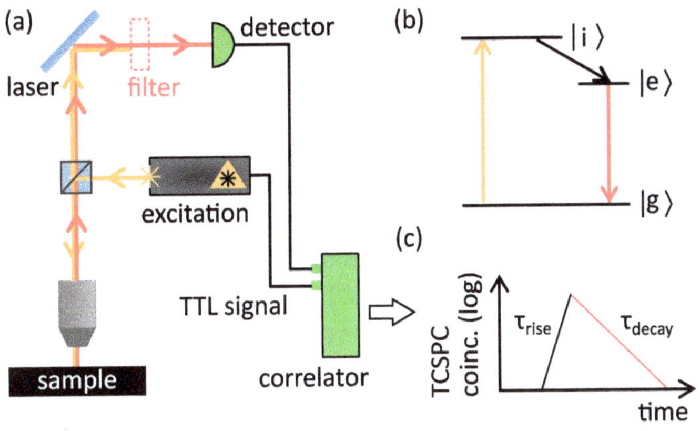

Figure 5.7: (a) Sketch of an exemplary time-correlated single-photon counting experiment used to measure the temporal dynamics of the emitter's intensity. The excitation pulse is sent on the sample using a beamsplitter and then filtered before reaching the single-photon detector. The observed counts on the detector are time-correlated with the pulsed excitation signal providing the histogram in (c). (b) Schematic level system with a high excited state $|i\rangle$, a lowest excited state $|e\rangle$, and the ground state $|g\rangle$. The excitation brings the system to $|i\rangle$, followed by a relaxation to $|e\rangle$. Finally, the radiative transition to $|g\rangle$ generates the photons measured in the TCSPC. (c) Exemplary coincidence histogram over time (ordinate axis in logarithmic scale): the excitation, followed by the relaxation to $|e\rangle$, is responsible for the finite rise time τ_{rise} (mainly due to the transition $|i\rangle \rightarrow |e\rangle$, which is usually longer than the excitation process, despite the fact that it is often shorter than the transition $|e\rangle \rightarrow |g\rangle$). The radiative transition from the lowest excited to the ground state generates the photons measured in the time-correlated measurement; the single exponential (linear in a semilogarithmic plot) decrease follows the emitter's excited state (de)population over time.

At this point, the capability of discriminating the excitation signal from the emitted light becomes important: firstly, because in normal scenarios, the excitation signal is stronger than the emitted one, which means that the excitation pulse can saturate the detector (if not damage it) preventing the detection of the investigated luminescence. Secondly, being able to detect the emission signal also during the excitation process provides important additional information on the emitter's dynamics. This becomes particularly significant when the emitter is not considered as a simple two-level system but as a realistic multilevel element. Considering the exemplary level structure in Fig. 5.7(b), the excitation process brings the system to a higher excited state, followed by a relaxation to the lowest excited state. There the transition to the ground state is accompanied by the photon emission, which is measured in the time-correlated measurement. The excitation and following relaxation (radiative or nonradiative) to $|e\rangle$ results in a TCSPC coincidences increase with a rise time τ_{rise} mostly dominated by the transition $|i\rangle \rightarrow |e\rangle$: this can be understood considering that the photon emission from the transition $|e\rangle \rightarrow |g\rangle$ is proportional to the (time-dependent) population of the excited state $|e\rangle$. This rise in the detected intensity is then followed by the decaying part of the curve, which arises from the signal stemming from the transition $|e\rangle \rightarrow |g\rangle$ when no excitation is present (see Fig. 5.7(c)). From a practical point of view, it is important to make sure that the emitter is only excited ones per cycle. This means that the excitation pulse Δt_{Laser} should be much shorter than the two-level system decay time τ_{decay}. This ensures that before the excited-to-ground state transition happens, no excitation pulse is present. This condition $\Delta t_{\text{Laser}} \ll \tau_{\text{decay}}$ becomes particularly relevant when transitions with short decays (for example, in the presence of strong Purcell enhancement; see Secs. 6.4 and 13.3) are investigated. Furthermore, the detection setup time response, i. e., detector and correlation electronics time resolution, should follow $\Delta t_{\text{detection}} \ll \tau_{\text{rise}}, \tau_{\text{decay}}$ to clearly reconstruct the TCSPC histogram. Even for high time resolutions, good praxis in time-resolved measurements requires the acquisition of the setup time response to allow for the correct deconvolution of the acquired data. Indeed, the setup time response will reflect in the measured single photon counting histogram (Fig. 5.7(c)). This means that before proceeding with the data analysis, it is important to deconvolve the system response from the measured data. The utilized setup can be characterized by performing a TCSPC measurement of a pulsed signal with a time duration much shorter than the instrument resolution, for example, a short laser pulse (i. e., the signal is ideally negligibly short versus the detector and correlation electronics time resolution). Once this measurement is available, it can be used to deconvolve the measured TCSPC.[9]

It is important to remark that only in the ideal case where nonradiative decay channels are absent, the measured decay time represents the radiative transition lifetime. On the contrary, the presence of fast nonradiative decay channels (τ_{nonrad}) can shorten the observed decay time following $1/\tau_{\text{decay}} = 1/\tau_{\text{rad}} + 1/\tau_{\text{nonrad}}$. In case of semiconductor nanostructures, local defects can give rise to nonradiative decay channels that result in the observation of shorter decay times with respect to the expected radiative lifetime. From an experimental point of view, a good approach to reasonably exclude the presence of these decay channels is to compare the observed decay with a statistically relevant number of similar emitters. Being nonradiative channels related to the local presence of structural defects, emitters located at different points are less likely to have a similar imperfection at a comparable distance.[10]

---

9 Even in cases where the temporal dynamics of the investigated emitter is much longer than the system response, the deconvolution discussed above can still be necessary, e. g., when the detector response function is deviating from a trivial shape. A good experimental praxis requires that the system response is always measured when performing TCSPC: in particular, when operating with free-space single-photon detectors, their time resolution may change upon different illumination of the active area. This means that only a system response measured under the same conditions of the sample TCSPC can provide an accurate deconvolution.

10 Time-correlated single-photon counting is one of the most used techniques to study the population transients that follows the excitation of the emitter. This technique is very popular since it requires the same devices, i. e., single-photon detectors and correlators, which are also necessary for studying the photon coherence or the single-photon nature of the emission. An alternative technique is represented by the use of streak cameras for the implementation of time-resolved measurements. These high-speed

It is worth mentioning that although for the excited-to-ground state transition of a two-level system, the intensity behavior over time is expected to follow a single exponential decay, in actual solid-state samples the measured behavior can be more complex. Additional mechanisms, like the presence of charge traps nearby the emitter, can introduce additional slower decays dynamics (in the TCSPC results), which are superimposed to the expected single exponential signal. The presence of a double exponential in the observed decay can therefore give an insight of the emitter's surrounding (see also Sec. 8.4). These considerations further highlight the importance of measuring the time response of the spontaneous decay of an emitter.

As a final remark, as it will also be discussed in Section 5.2.5, when time-to-amplitude converters (TAC) are employed, the statistics collected with these time-resolved measurements is only correct under the assumption of low detected count rates; in other words, $1/R \gg \tau$ with R as the signal count rate and τ as the time separation between the start event (here the TTL of the laser) and the stop (the single-photon detector click). When this condition is not met, the acquired statistics will not correspond to the desired decay time. It is important to say that even if for decades such a kind of start-and-stop measurements have been employed (for a more detailed explanation, see Section 5.2.5), currently the improvements in correlation electronics and computational power allow for utilizing more powerful time-tagging techniques. Whereas start-and-stop measurements were performed using time-to-amplitude converters, time tagging allows for recording the absolute detection time of a photon assigning it a digital time stamp on a scale that runs continuously. Time taggers have an internal precise clock, which can be synchronized with an experimental reference, as the utilized excitation laser transistor–transistor logic signal. This approach is particularly advantageous since it circumvents the nonutilized time arising from the detector dead time. Time tagging allows for achieving a better signal-to-noise ratio for the same integration time, together with a much higher flexibility in the way of analyzing and correlating the data (in particular, when more than one detector needs to be employed) with the small price to pay related to the higher required computational power.

5.2.4 Brightness: photon counting

In the last years the development of photonic-based quantum technology drove the search of high-performances sources of nonclassical light. To be considered *ideal* for these purposes, a quantum emitter needs to be able to generate a high flux of single, identical photons in an "on-demand way": this last property means that one photon is deterministically emitted once a certain trigger has been pulled. Therefore the photon

---

detectors can detect the intensity variation over time and have been used even in correlation experiments [226].

emission process needs to be 100 % efficient in terms of excitation of the emitter as well as in generation and collection of the photons. Ideally, the photon stream should be as sketched in Fig. 5.8.

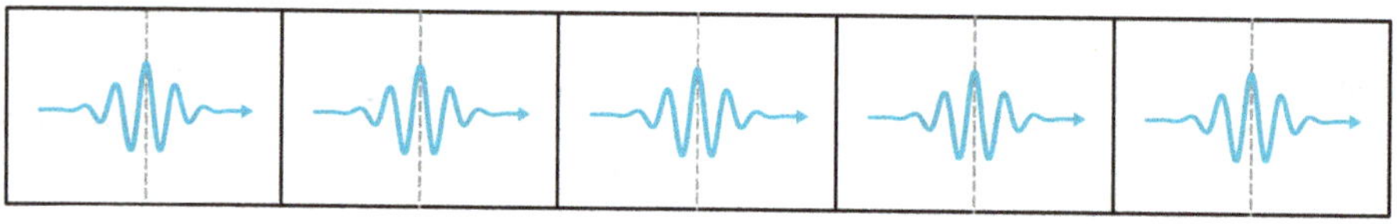

Figure 5.8: Sketch of an ideal nonclassical light emission process: the source should generate a stream of single, identical, and timewise equally spaced photons. For every excitation repetition, one single photon is emitted. For clarity, five excitation cycles are shown. In this and some of the following figures, a Gaussian photon wavepacket is considered as a general example. Nonetheless, when the photon emission stems from a two-level system, an exponentially decaying photon profile is considered.

Therefore the excitation performed with a high repetition rate pulsed source should generate a stream of single, identical photons equally spaced in time. In the following, we will discuss all fundamental properties, together with the experimental techniques to characterize them and the potential limitations to the ideal case.

The first property to be discussed is often referred to as *brightness* [59, 126, 194]. In layman's terms, it constitutes the number of photons emitted per excitation pulse (collected at a defined optical element). It takes values [0, 1], and in combination with the frequency of excitation (in pulsed measurements, the excitation source repetition rate), it gives

$$\text{Number}/s = \text{Bs}\,[0,1] \times f_{\text{rep,exc}}\left[\frac{1}{s}\right], \tag{5.5}$$

which constitutes the number of photons generated by the source, which are available for experiments at the considered optical element.

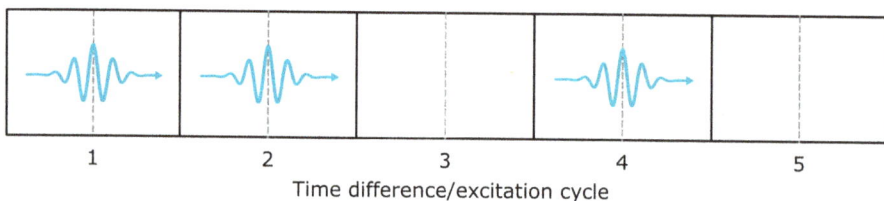

Figure 5.9: Pictorial view of a source with brightness <100 %: photons are not always present after excitation, so the number of photons is lower than the number of excitations.

More quantitatively, the brightness can be defined as

$$\mathrm{Bs} = p \times \eta, \tag{5.6}$$

where p is the state occupation factor of the first excited state (i. e., the population probability of the excited state $|e\rangle$), and η is the overall photon extraction efficiency. In the most general case, an excited two-level system can be considered as sketched in Fig. 5.10. Once the excited two-level atom is placed into an optical cavity, the recombination to the ground state leads to the emission of a photon, which can happen in the cavity mode or in the continuum of the free-space modes (which do not include the cavity mode). The former corresponds to a decay rate in the cavity mode named Γ_c, whereas the latter to a decay rate in modes other than the cavity Γ. Operating in the *weak coupling* regime (see Section 6.4), the Purcell effect brings an acceleration of the spontaneous emission in the cavity mode over the others, making it a more probable decay channel. Therefore we can define the fraction of the emission in the cavity mode with respect to all modes via the factor β defined as

$$\beta = \frac{\Gamma_c}{\Gamma_c + \Gamma} = \frac{\Gamma_c/\Gamma_{\text{bulk}}}{(\Gamma_c + \Gamma)/\Gamma_{\text{bulk}}} = \frac{F_P}{F_P + \Gamma/\Gamma_{\text{bulk}}}. \tag{5.7}$$

In the last equivalence, after dividing both terms by the spontaneous emission rate in bulk Γ_{bulk}, we defined the Purcell factor $F_P = \Gamma_c/\Gamma_{\text{bulk}}$, i. e., the ratio of the decay rate in the cavity mode to the spontaneous emission rate for the emitter in bulk.[11] From the last equivalence in Eq. (5.7) we can see that two strategies can be employed to increase the factor β: either via Purcell enhancement of the transition decay in the cavity mode either by reducing the decay rate in the modes other than the cavity one (Γ). Although Purcell enhancement can be achieved exploiting cavity quantum electrodynamics (cQED) in the weak coupling regime (see Chapter 6), reducing Γ can only require waveguiding-based approaches [59]. It is often the case that when employing cQED, although the decay rate in the cavity mode gets Purcell enhanced, the decay in the other modes maintains the same value as in bulk, meaning that β can take the form

$$\beta = \frac{F_P}{F_P + 1}. \tag{5.8}$$

In the second discussed approach, i. e., considering that no Purcell effect takes place ($F_P = 1$), Eq. (5.7) can take the form

$$\beta = \frac{1}{1 + \Gamma/\Gamma_{\text{bulk}}} = \frac{\Gamma_{\text{bulk}}}{\Gamma_{\text{bulk}} + \Gamma}, \tag{5.9}$$

11 Some studies include in the denominator also the non-radiative decay rate Γ_{nr}, having therefore $\beta = \Gamma_c/(\Gamma_c + \Gamma + \Gamma_{\text{nr}})$. Being non-radiative decay channels often negligible for semiconductor emitters operating at cryogenic temperatures, Γ_{nr} is hereafter neglected.

meaning that reducing the decay rate Γ into modes other than one specific of the employed waveguiding element (which for simplicity can be seen as the directional mode labeled before as *c*) also enables reaching β values close to unity. Avoiding the use of cQED simplifies the realization of photonic structures, since precise spectral or spatial matching with one cavity mode is not required (see Chapter 13). On the other hand, a reduction of the spontaneous decay time has the advantage of reducing the impact of dephasing and spectral diffusion mechanisms (as explained in Sections 5.2.6 and 5.2.7), and it allows exciting the quantum light source with higher rate (i. e., the time difference between excitation cycles as depicted in Fig. 5.8 and Eq. (5.5) can be reduced, being the temporal profile of the photons shorter), resulting in a higher number of available photons for the intended experiments [136].

The photon found in the cavity mode can then leave the cavity with a certain efficiency η_{out}, meaning that the photon will escape through a certain defined direction over all other possible paths. This out-coupling efficiency needs to be designed according to the experimental conditions (as the collection numerical aperture), and several methods are employed for this purpose [59]. Considering these arguments, the source brightness Bs can be defined as:

$$\text{Bs}_{\text{unpolarized}} = p \times \beta \times \eta_{\text{out}}. \tag{5.10}$$

In a general scenario, where the source is emitting unpolarized photons, it becomes interesting to define the brightness for a certain polarization. In experimental quantum optics, having the photon in a defined polarization state is often necessary and usually obtained using a polarizer. The eventual photon loss[12] is accounted with the term η_{pol}, defining the brightness for the polarized source as

$$\text{Bs}_{\text{polarized}} = p \times \beta \times \eta_{\text{out}} \times \eta_{\text{pol}}. \tag{5.11}$$

Recent studies focused on the realization of cavities that exhibit a well-defined polarization [217], hence enabling maximizing the term η_{pol}. In quantum photonics, photons should have not only a well-defined polarization but also a well-defined spatial mode: together with designing the out-coupling of the cavity to ensure light collection (as discussed for the term η_{out}) having a Gaussian mode profile plays an important role when it comes to coupling into single-mode fibers. Losses due to fiber coupling are accounted for via the term η_{FC}, resulting in

$$\text{Bs}_{\text{fiber coupled}} = p \times \beta \times \eta_{\text{out}} \times \eta_{\text{pol}} \times \eta_{\text{FC}}. \tag{5.12}$$

This last, rather laborious, expression of the source brightness has an inherent interest for experimental realizations, since it considers the actual nonclassical light source

12 The term η_{pol} is not only accounting for losses related to a non-perfect optical element but also the photons which are filtered by the polarized itself, i. e. the light orthogonal to the chosen polarization direction.

brightness at the output of a single mode fiber. This represents the amount of photons available for the intended experiment. Indeed, when performing actual quantum optical experiments, the achieved count rate can be a discriminant on their feasibility. Hence number of clicks observed on the detector represents an important experimental parameter. The observed count rate on the detector[13] will be given considering the setup losses η_{setup} and the detection efficiency η_{det}:

$$\text{Detector counts/s} = \text{Bs}_{\text{fiber coupled}} \times \eta_{\text{setup}} \times \eta_{\text{det}} \times f_{\text{rep,exc}}, \tag{5.13}$$

being also proportional to the excitation rate (in Hz). With this information, together with the precisely calibrated measurement of the efficiencies (setup, detector, etc.), it is possible to experimentally evaluate the source brightness. Evidently, even for constant brightness, a higher setup transmission or detection efficiencies translate in higher detected count rates.

Different factors can have a detrimental impact on the brightness. First, we can consider a nonunity occupation factor p, which can be due to decoherence, to a nonperfectly effective excitation process (also eventual blinking can lower the overall brightness; see Chapter 8), or to loss of carriers via nonradiative processes before they could recombine with emission of light. A nonunitary out-coupling efficiency can be attributed to photon coupled to a channel that cannot be efficiently collected (for example, if the photons leave the resonators at angles larger than the experimental collection numerical aperture). Finally, the photon could also be emitted in a mode (with rate Γ) that is different from the one that can be experimentally collected. The aforesaid elements can be summarized in the sketch shown in Fig. 5.10.

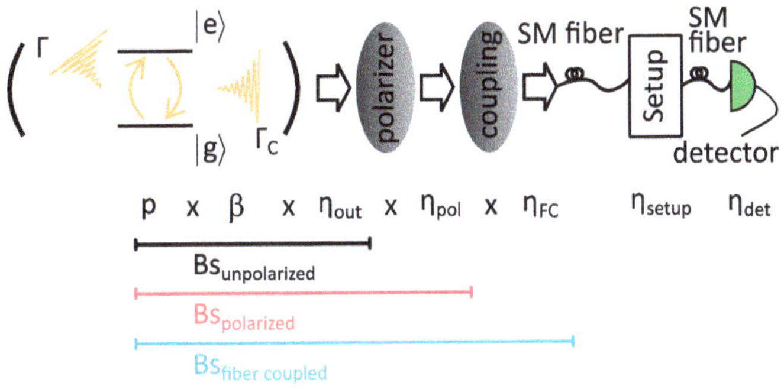

Figure 5.10: Sketch of terms in Eqs. (5.6)–(5.12). A two-level system relaxes from the excited to the ground state emitting photons, which can be in the specific Γ_c or in other radiation Γ modes. Being the photons in the sketch generated by a two-level system, an exponential profile has been drawn.

13 The actual measured detector count rate includes three terms $R + R_{\text{dc}} + R_{\text{bck}}$: the detected counts from the light source, namely R, the detector dark counts R_{dc} and eventual unwanted background counts R_{bck}. It is important to consider only R in the estimation of the source brightness.

A high brightness is highly desirable since it leads to high achievable experimental complexity. On the other hand, as explained, photon losses can also arise from the experimental setup itself (absorption in optics and materials, limited coupling efficiency between elements). From the experiment point of view, a low brightness or low setup efficiency would both affect the *end-to-end efficiency*, i. e., the overall source and setup efficiency from photon generation to light filtering and experimental implementation. Therefore the final countrate on the utilized detectors as in Eq. (5.13) represents a parameter as important as the source brightness.

Multiphoton brightness

Everything discussed so far includes a comprehensive discussion of the brightness for a single-photon source. Nonetheless, state-of-the-art performances of semiconductor quantum light sources allow designing and execution of experiments where two or more photons are employed. With particular attention to two case studies, we have to take into account the following considerations in the evaluation of the *usable* brightness:

- *Two-photon source*: one example well discussed in this book is based on the cascaded emission of biexciton (XX) and exciton (X) photons (see Section 7.3). In this case the brightness defined earlier enters into play for both photons:

$$\mathrm{Bs}_{XX,X} \propto p \times \eta_{XX} \times \eta_X := p \times \eta^2_{XX,X} := \mathrm{Bs}_{XX} \times \mathrm{Bs}_X, \tag{5.14}$$

  still representing the brightness in a two-photon emission process deriving from one source and therefore having one population probability (the one of the biexciton in case of the radiative cascade). The first equivalence can be expressed in case the photon extraction efficiency is the same for both photons (often the case, even if care must be taken for advanced cavity geometries where two separate modes are employed for extracting the two photons; see Chapter 14). The second equivalence can be written when the population probability is equal to one.
- *Multiple single-photon sources*: in experimental implementations where multiple photons are generated by distinct single photon sources (SPSs), the brightness scales with the number of n-photon coincidences as Bs^n. For an exemplary case of five SPS with a brightness of 0.8, a 5-fold coincidence would happen with a probability of less than 0.33. This underlines the importance of achieving high brightness in scalable multiphoton experiments employing distinct sources.

Brightness correction for imperfect single-photon purity

So far, the brightness discussion assumed an ideal source of single photons, therefore having no more than one photon per excitation cycle as in Figs. 5.8 and 5.9. Nevertheless, as discussed in the following section, the photon stream generated by the SPS could have more than one photon in certain time slots. To avoid an overestimation of the brightness, the detected count rate can be corrected for a nonzero second-order correlation function $g^{(2)}(0)$ following the formula [147]

$$\text{Single-photon counts} = \text{Detector counts} \times \sqrt{1 - g^{(2)}(0)} \tag{5.15}$$

under the assumption that the ideal stream of single photons is statistically mixed with a small background constituted by photons following a Poissonian statistics.

5.2.5 Single-photon purity: photon autocorrelation measurements

When talking about *pure* single-photon emission, this typically refers (in solid-state physics) to the high probability of having exactly one photon into the considered time slot (see Fig. 5.11). The emission of single photons can be quantified via the second-order correlation function $g^{(2)}(\tau)$.[14] If $g^{(2)}(\tau = 0) > 0$, then more than one photon can be present in some of the time slots: in particular, for photon number states, $g^{(2)}(0) = 1 - \frac{1}{n}$, where n is the number of photons (see Chapter 3). The availability of pure single photons is considered key to ensure security in quantum information and to avoid impacting the achievable fidelity in multiphoton experiments.

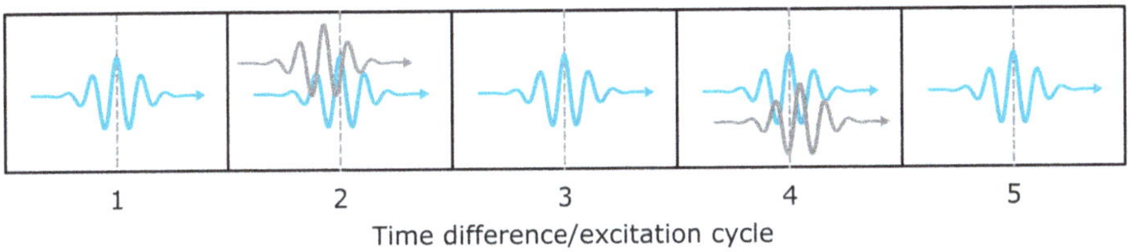

Figure 5.11: Nonideal single-photon stream: in a single time slot (exemplary in slots 2 and 4), more than one photon can be found.

To verify the single-photon nature of the emission, it is therefore necessary to measure the second-order correlation function $g^{(2)}(\tau)$. This can be done via the so-called Hanbury Brown–Twiss (HBT) interferometer (see Fig. 5.12(a)), which was originally implemented to record spatial coherence of classical signals in astronomy [21].

The setup itself is rather simple, i. e., the photon stream is sent on a 50:50 beamsplitter where two detectors are placed at the two output ports (see Fig. 5.12(a)). Correlating the recorded clicks (in the limit of a strongly attenuated signal or a low detection efficiency for start and stop measurements; see the following discussion) allows for

14 Some groups refer to single-photon purity as defined using the second-order correlation function via $1 - g^{(2)}(0)$.

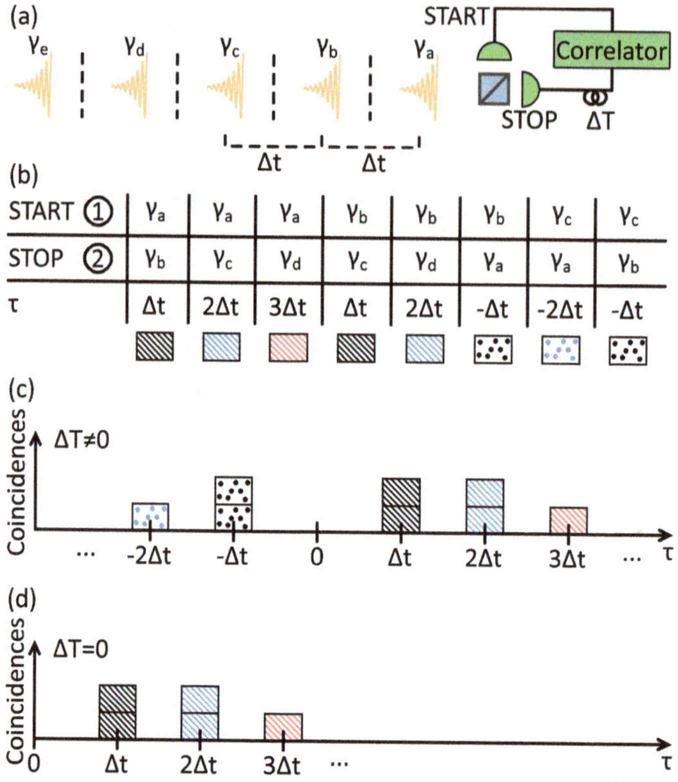

Figure 5.12: (a) Schematic of the Hanbury Brown–Twiss interferometer fed by a stream of single photons. Two single-photon detectors are placed at the two output ports of a 50:50 beamsplitter, and their clicks are time-correlated, hence enabling the extrapolation of the second-order correlation function. For simplicity, the study case of on-demand single photon emission is sketched, having an equal time separation Δt between time-adjacent photons. (b) The table showing exemplary cases of photon detection and respective time separation τ. The color coding allows for comparing the detection events and their respective impact on the coincidences. The temporal position of the zero delay peak (i. e., simultaneous detection on the two detectors) is normally shifted inserting a delay ΔT in one detection path. (c) Coincidence histogram for $\Delta T \neq 0$: thanks to the time shifting of the zero-delay peak, negative coincidence times can be observed. (d) Coincidence histogram for $\Delta T = 0$: only positive delays are shown. In both cases (c) and (d), pure single photons are considered. In case of a photon stream as in Fig. 5.11, additional coincidence events would appear for a delay time equal to 0.

recording a coincidence histogram, which becomes proportional to the second-order correlation function of the measured signal. Typically, a single-photon detection event is referred as a "click" on the detector. For HBT measurements, two single-photon counting modules (SPCM), as the one employed for TCSPC measurements, are utilized. These types of devices are indeed capable of detecting photon fluxes as low as one single photon, employing different kinds of mechanisms from avalanche processes[15] to changes in the superconductivity properties of the material.

Classically, the second-order correlation function can be written as (see Section 1.5)

$$g^{(2)}_{\text{class}}(\tau) = \frac{\langle \bar{I}(t)\bar{I}(t+\tau)\rangle}{\langle \bar{I}(t)\rangle^2} = \frac{\langle E^*(t)E^*(t+\tau)E(t+\tau)E(t)\rangle}{\langle E^*(t)E(t)\rangle^2}, \tag{5.16}$$

where $I(t)$ and $E(t)$ are the intensity and electric field at time t for the impinging light beam. In the measurements, the time average $\langle \dots \rangle$ is obtained by integrating over long time periods. Therefore the term $\langle \bar{I}(t)\bar{I}(t+\tau)\rangle$ in Eq. (5.16) indicates that the function $g^{(2)}(\tau)$ considers correlations between the light beam intensities at times t and $t+\tau$.

Employing quantum mechanical formalism, as discussed in Section 2.5, Eq. (5.16) takes the form

$$g^{(2)}_{\text{QM}}(\tau) = \frac{\langle \hat{E}^-(t)\hat{E}^-(t+\tau)\hat{E}^+(t+\tau)\hat{E}^+(t)\rangle}{\langle \hat{E}^-(t)\hat{E}^+(t)\rangle^2} \tag{5.17}$$

$$\overset{\tau\to 0}{=} \frac{\langle (\hat{a})^\dagger(\hat{a})^\dagger(\hat{a})(\hat{a})\rangle}{\langle (\hat{a})^\dagger(\hat{a})\rangle^2} = \frac{\langle n(n-1)\rangle}{\langle n\rangle^2}, \tag{5.18}$$

where we assume that for the second equivalence, the field is in a single mode. The lack of coincidences at $\tau = 0$ is a signature of presence of single photons. Despite a hand-waving understanding of this finding is that a single photon can either go toward one detector or the other (i. e., either reflected or transmitted on the beamsplitter), this picture is not correct. Indeed, a single photon is a quantum object, and Eq. (5.18) highlights the difference between the classical and the quantum description. The single photon indeed follows both paths toward the detectors, but it is measured by only one of them (i. e., a single click on one single-photon detector) collapsing the wavefunction: a single photon being indivisible (it is the single quantum of the radiation field), it can only be detected once at a certain time t. This decreases (by one quantum) the number of photons available for detection at time $t+\tau$. Employing classical field theory, this effect is not taken into account, so lacking a correct description of nonclassical light states. With this in mind, it becomes evident that to verify the single-photon nature of the radiation stream under investigation, the $g^{(2)}(\tau \to 0)$ is of interest: indeed, this value of the

15 The process of photon detection into an avalanche photodiode is similar to a standard semiconductor detector, with an important difference: the carriers generated by the light absorption are accelerated by an applied voltage, and they can produce more pairs by impacting on the atoms constituting the semiconductor lattice. This avalanche process produces an output voltage, which is easily detectable (rather than trying to directly measure one electron-hole pair generated by one single photon absorption). In the text, we refer to single-photon counting modules, which are a special kind of avalanche detectors: whereas for the latter the output voltage is directly proportional to the incident light intensity, the former produce output voltage pulses with a fixed shape and amplitude. The stronger the incident light intensity, the higher the number of generated pulses.

second-order correlation function indicates if a second photon has been detected simultaneously to the detection of the first one (otherwise said as the conditional probability of measuring a photon at a certain time $t + \tau$ after detecting a first photon at t). For n photons being in a Fock state $|n\rangle$, Eq. (5.18) takes a value <1 (for $n > 1$), which is a nonclassical finding (see Section 3.5). It is now clear that recording simultaneous clicks on both detectors is not possible when measuring pure single photons.

A first approach for performing Hanbury Brown and Twiss experiments is based on the so-called start-and-stop measurement: one single-photon detector acts as a *start* signal, while the time passed until a click is recorded on the second *stop* detector is measured (Fig. 5.12(a)). This time difference feeds a time-to-amplitude converter (TAC) whose output is sent to a multichannel analyzer (MCA). This allows for reconstructing a histogram of the coincidence events $n(\tau)$ (see Fig. 5.12(c)): this represents the probability of detecting a *stop* event at time difference τ, having detected a *start* signal at $\tau = 0$, jointly having no stop clicks before the time τ, i. e.,

$$n(\tau) = (G^{(2)}(\tau) + R_{\mathrm{dc}}) \times \left(1 - \int_0^{\tau} n(\tau')\,\mathrm{d}\tau'\right). \tag{5.19}$$

In this equation the unnormalized second-order correlation function has been introduced, whereas R_{dc} represents the dark counts detected during the measurement, in contrast to the signal count rate R. Equation (5.19) shows that the coincidence histogram recorded in an HBT measurement differs from the second-order correlation function, but it approaches $G^{(2)}(\tau)$ if the detector dark counts are much smaller than the signal count rate, together with the assumption that the arrival time of the photons (defined as $1/R$) is in average much larger than the time difference τ (this last assumption becomes an important experimental operation limit, as explained in the following and in Fig. 5.13(b)), i. e., the second term in Eq. (5.19) is ≈ 1. Performing measurements under the above conditions, means that $n(\tau) \approx G^{(2)}(\tau)$, and the normalization of the correlation function can be obtained in comparison with a Poissonian light source: as discussed in Chapter 3, for a coherent light source, $g^{(2)}(\tau) = 1$ is expected for each τ. Considering such a light source with a comparable average count rate on the detectors, the expected counts for a continuous wave measurement (cw) would be

$$C_{\mathrm{Poisson}}^{\mathrm{cw}} = R_{\mathrm{start}} R_{\mathrm{stop}} \Delta t_{\mathrm{MCA}} t_{\mathrm{int}}, \tag{5.20}$$

where R_{start} and R_{stop} are the count rates on the two detectors *start* and *stop*, respectively, Δt_{MCA} represents the employed multichannel analyzer time-bin, and t_{int} is the integration time. This leads to the normalization

$$g^{(2)}(\tau) = \frac{G^{(2)}(\tau)}{C_{\mathrm{Poisson}}^{\mathrm{cw}}}. \tag{5.21}$$

In case of pulsed operation, the coincidence counts for the reference Poissonian source would become

$$C_{\text{Poisson}}^{\text{pulsed}} = R_{\text{start}} R_{\text{stop}} \Delta t_{\text{laser}} t_{\text{int}}, \tag{5.22}$$

where this time the repetition period of the source Δt_{laser} comes into play. Therefore $C_{\text{Poisson}}^{\text{pulsed}}$ can be used to normalize each integrated peak in the correlation measurement.

The measurement of $g^{(2)}(\tau)$ under pulsed excitation further demonstrates the deterministic single-photon emission or, in the firstly used terminology, the signature of the turnstile operation [123]. This can be understood by looking at the expected correlation histogram in case of emission of single photons under a pulsed excitation scheme (compare, for instance, the ideal case of Fig. 5.8). For each excitation cycle, the emitter will generate one single photon. This will propagate up to the beamsplitter and will be detected by either detector; let us assume here detector 1 (Fig. 5.12(a)). A second photon will be generated during the following excitation cycle and reach the beamsplitter. In the case it is detected by detector 2, the correlation will produce an event for a time separation Δt, the time difference between two excitation pulses (i. e., the inverse of the repetition rate of the excitation process). When the second photon is not detected (or it clicks on detector 1 instead), the third photon generated by the following excitation cycle can still produce a click on detector 2, hence adding one coincidence event on the histogram for a time difference of two cycles, i. e., $2\Delta t$ (some more examples are shown in Fig. 5.12(b)). The described process produces the histogram in Fig. 5.12(c): in actual experiments the shape of the peaks (oversimplified in the figure) depends on the temporal profile of the photons (convolved with the system response function), i. e., on the temporal dynamics as explained in Section 5.2.3. The events recorded for a negative time difference are due to an inverse order in the detection events (i. e., 2 instead of 1 as above). Extending these arguments to all photons in the measured stream results in a correlation histogram as that in Fig. 5.13, where the shape of the peaks is also depicted in a more realistic manner.

With this description in mind, we can understand why for start and stop measurements, we have to experimentally operate in the limit of a strongly attenuated signal or a low detection efficiency: if all photons are always detected, then observing a click at shorter time separation will be much more probable. This will modify the temporal statistics of the histogram in Fig. 5.13(a), observing more counts at the first repetition and decreasing then the observed coincidences for increasing time separation, as in Fig. 5.13(b). This can be solved, even for very bright sources and high-efficiency setups, employing time-tagging techniques: as for the case of time-correlated single-photon counting, these techniques allow for flagging the time of detection events with respect to a precise clock and then reconstructing afterward the correlation for different time separations between the two detectors. Using time-tagging electronics (or in start and stop measurement in the aforementioned limit), the coincidence histogram for an ideal single photon source under pulsed excitation will be as in Figs. 5.12(c) and 5.13(a):

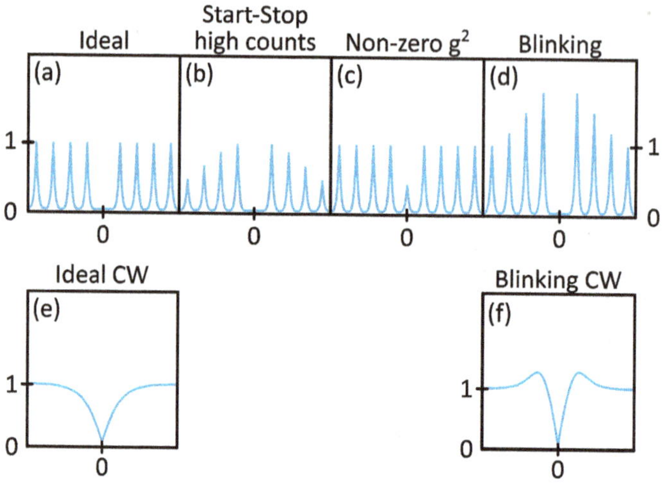

Figure 5.13: (a) Sketch of a correlation histogram expected for a Hanbury Brown and Twiss experiment with an ideal on-demand source of single photons. (b) In start-and-stop measurements, for high photon fluxes ($1/R_{start}, 1/R_{stop} < \tau$) or high setup efficiencies, the statistics from the ideal case changes as depicted. (c) Expected correlation histogram for a source having $g^{(2)}(0) \neq 0$, still with $g^{(2)}(0) < 0.5$. (d) Correlation histogram in case of pure single photons but in the presence of blinking in the emission statistics. (e) Expected correlation histogram for a source of single photons operated in continuous wave. (f) Presence of blinking for a CW-operated source of single photons.

no events are recorded for zero time separation (i. e., simultaneous clicks on both detectors), having peaks evenly spaced by the time separation Δt set by the repetition rate of the excitation source. Once again in the ideal case, these peaks have the same area and the same shape, and they can be normalized to the *Poissonian level*, i. e., the coincidence level expected for a coherent light source (as discussed before). This level represents the lack of temporal correlations as for the case of a Poissonian light source. Even when dealing with Fock states, increasing the delay between photon emission events reaches statistical independence.

In realistic experiments the histogram can be different from that just described for several reasons. In case of a nonperfect single-photon purity as that depicted in Fig. 5.11, having few cases where more than one photon is present in one time slot will result in the observation of few events recorded at zero time delay. This would result in the appearance of an additional peak, ideally with an area lower than the other peaks (see Fig. 5.13(c)). Setting the area of nonzero peaks as 1 (or their height as 1 if all areas are identical), if the central peak area is smaller than 0.5, then the photon stream would be constituted by mostly single photons (Eq. (5.18) provides $g^{(2)}_{QM}(\tau = 0) = 0.5$ for $n = 2$).

Then even in case of a photon stream constituted by single photons, the population dynamics of the emitter's excited states can also be reflected in a different time distribution of the observed coincidences: in particular, for solid-state nanostructures, the emitter may switch from a bright state to a dark state, hence resulting in a reduced overall brightness. This phenomenon, called *blinking*, can make the observation of light more or less probable in certain time scales (see Chapter 8). This can complicate the determination of the Poissonian level, which is then reached only for a sufficiently long time separation where the blinking is no longer affecting the photon emission dynamics (see Fig. 5.13(d)). These arguments made clear that a correct normalization of the measurements needs to be performed to avoid misjudging the actual $g^{(2)}(0)$ and hence the purity of the single-photon emission (see [222] and references therein). The arguments here described still apply to measurements where the excitation is not pulsed but rather continuous, for example, when using a continuous wave (CW) laser. In this case, it is no longer possible to define a time slot where a photon can be found, but the emitter is continuously excited, having the emission of a photon directly followed by the subsequent excitation of the excited state population and respective new photon emission. Still, the emission of single photons is reflected in the observation of no coincidences for zero time delay, as in Fig. 5.13(e). Increasing the time separation between events measured on detectors 1 and 2 would still result in the observation of coincidences reaching the Poissonian level, whereas the near-zero profile of the histogram is still defined by the spontaneous decay behavior and the excitation rate of the emitter (and hence the emitted photon temporal profile).[16] Also for CW measurements, the effect of blinking can appear in the histogram as schematically depicted in Fig. 5.13(f).

Semiconductor quantum dots proved their capability of generating triggered single photons with unprecedented purity ($g^{(2)}(0) < 10^{-4}$), comparable to values obtained with single, isolated atoms. Nevertheless, for this purpose, an appropriate excitation scheme needs to be employed (for record values, a two-photon excitation resonantly addressing the biexciton state was utilized; see Sec. 10.2.3.4). When nonideal pumping schemes are employed, the single-photon purity may degrade due to background emission from the transition surrounding or due to eventual reexcitation of the emitter during the same excitation cycle [45], as described in Sec. 7.3.4. Finally, it is worth mentioning that charge refilling from eventual neighboring traps in the vicinity of a quantum dot (Fig. 8.13(a)) may cause a change in the temporal shape (which can become biexponential, as mentioned in Sec. 5.2.3 and discussed in Sec. 8.4) of the intensity autocorrelation histogram observed around zero time delays: the actual shape of the central correlation peak varies as a function of the trap relaxation time even though, with sufficiently high time resolution, no coincidences can be observed at zero time delay, still highlighting the single-photon nature of the emission process (for more detail, see Fig. 8.13).

16 It is worth mentioning that the shape of the function $g^{(2)}(\tau)$ in CW excitation does not only depend from the emitter's recombination rate τ_{rad} (later also defined as T_1). Indeed, the carrier injection rate P enters in the second-order correlation function as $g^{(2)}_{CW}(\tau) = 1 - \exp\{-(1/(\tau_{rad} + P)) \cdot \tau\}$. This means that for increasing carrier injection rate, the antibunching curve shown schematically in Fig. 5.13(e,f) would get narrower. This would mean, experimentally, that for a high carrier injection rate, it is necessary to operate with high temporal resolution detectors to observe the actual antibunching at zero time delays.

5.2.6 Photon coherence: Michelson interferometry

In strict similarity as for the classical case, *coherence* describes the phase stability of the light. When dealing with single photons, the coherence and impact of dephasing is quantified as follows:

$$\frac{1}{T_2} = \frac{1}{2T_1} + \frac{1}{T_2^*}, \tag{5.23}$$

where T_2 indicates the coherence time, T_1 defines the investigated transition lifetime, and T_2^* is the pure dephasing time, i. e., the loss of coherence without recombination. The values of T_1 and T_2 characterize the photon wavepacket. If we consider the case of solid-state systems, like semiconductor quantum dots, then the electrons e^- and holes h^+ forming the excitonic states can interact with the surroundings. This interaction may result in a loss of coherence during radiative emission, leading to a change of the photon phase. This is schematically depicted in Fig. 5.14, where the broken lines indicate the change of the photon phase. When the environmental fluctuations happen faster than radiative transitions, the term T_2^* comes into play. These fluctuations will have an impact on the photon linewidth as a homogeneous broadening term, resulting in a Lorentzian shape larger than that expected from the lifetime alone (see the following). It is worth mentioning at this point that T_2^* includes mechanisms leading not only to homogeneous broadening, but also to inhomogeneous broadening, as explained in Section 5.2.7.

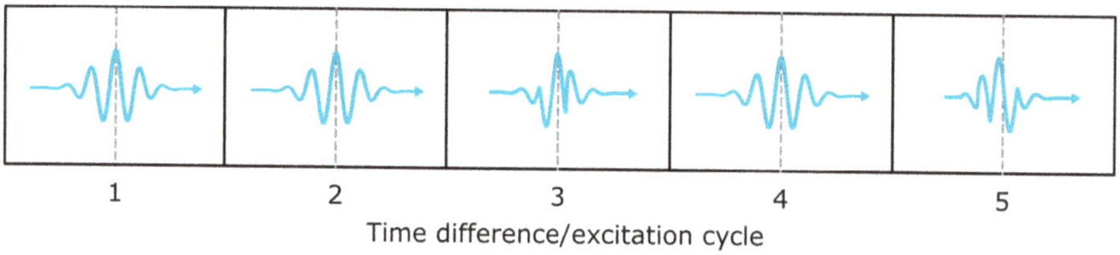

Figure 5.14: Loss of photon coherence is here depicted as broken lines indicating the change of photon phase (time slots 3 and 5). In this case, $T_2 < 2T_1$ applies.

In the ideal case where decoherence mechanisms are not present and non-radiative recombination is absent ($T_1 = \tau_{rad}$), or in any case negligible, Eq. (5.23) becomes

$$\frac{T_2}{2T_1} = 1. \tag{5.24}$$

The aforementioned equivalence is known as the *Fourier-transform limit*. The Fourier-limited photon lineshape is Lorentzian with a linewidth defined as $\Delta\nu_{FL} = 1/2\pi T_1$. In the presence of decoherence, depending on the timescale of the mechanisms in place (see the next section and Chapter 8), the lineshape will deviate from the Fourier-limited linewidth, having a lineshape that can follow a Voigt profile (with more Lorentzian or Gaussian component depending on the involved decoherence processes). From the quantum optics point of view, reaching the Fourier limit is highly desirable. Indeed, having $T_2 = 2T_1$ is a fundamental requirement for the generation of indistinguishable photons. Once the photons generated by the light source are identical, i. e., have the same spectral bandwidth, polarization, pulse width, carrier frequency, mode, and temporal profile, they can be defined as *indistinguishable*. As it will be discussed in the following (Sec. 5.2.7), indistinguishable photons will be able to quantum interfere, which is a basis for more advanced quantum operations (see Sec. 4.4).

From the previous discussion it becomes evident that the photon coherence time and hence the photon linewidth are important experimental parameters. The coherence time can be obtained measuring the first-order correlation function. As defined in detail in Section 1.3, Eq. (1.5) gives

$$g^{(1)}(\tau) = \frac{\langle E^*(t)E(t+\tau)\rangle}{\langle E^*(t)E(t)\rangle}. \tag{5.25}$$

Similarly to Eq. (1.33), the *Wiener–Khintchine theorem* relates the photon coherence to its normalized spectrum, and therefore

$$F(\nu) = \int_{-\infty}^{\infty} g^{(1)}(\tau)\exp(-i2\pi\nu\tau)\,d\tau. \tag{5.26}$$

The function $g^{(1)}$ can then be measured by Fourier spectroscopy carried out via *Michelson interferometry*.

For this scope, the light to be measured is sent to a nonpolarizing 50:50 beamsplitter (see Fig. 5.15(a)): the reflected and transmitted paths impinge on plane mirrors (or retroreflectors, more commonly used for practical reasons) and reach back the beamsplitter (BS). The two paths are aligned so that the reflected light can interfere on the BS and then travel to the detector. One arm of the Michelson interferometer can then be moved so that it varies the path length of one interferometer arm over the other. In typical experiments employing semiconductor quantum dots, this arm length needs to be controlled with a precision of few tens of nanometers to provide reliable measurements of the coherence time. Once the path length is varied, the interference conditions (either constructive or destructive) change periodically, and this modification can be observed as intensity oscillations in the detected signal (see Fig. 5.15(b)). From the measured data we can extract the visibility defined as (see Sec. 1.4)

$$V = \frac{I_{\max} - I_{\min}}{I_{\max} + I_{\min}} = |g^{(1)}(\tau)|. \tag{5.27}$$

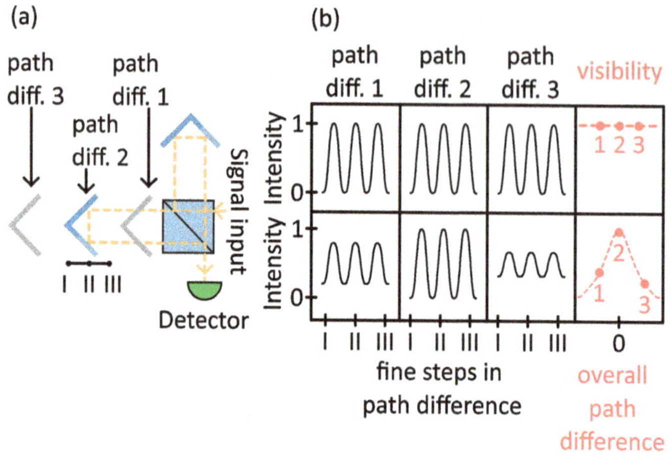

Figure 5.15: (a) Sketch of a Michelson interferometer: the incident light is sent onto a 50:50 (nonpolarizing) beamsplitter, where it is reflected or transmitted. At the end of the two paths, retroreflectors are placed. Whereas the length of one interferometer arm is fixed, the other can be actively changed, being one mirror movable. The light recombines on the beamsplitter and is detected at one output. The movable retroreflector modifies the delay difference acquired between the two paths, resulting in a variation of the interference conditions at the output. Three exemplary cases are shown (experimentally, more path differences than those depicted are used): path difference 2 refers to the same length in both arms. The Roman numbers I, II, and III indicate the fine steps in the path difference resulting in the observation of intensity oscillations as in (b). These fine steps are then repeated for multiple path length differences (in the figure, three are sketched). (b) Intensity oscillations expected for different path length differences for a fully coherent signal (top) and a partially coherent (bottom) light source. Each quadrant is constructed continuously moving the mirror in fine steps around the central position set by the chosen path difference. From the oscillations it is possible to extract the visibility for the investigated path length difference (last quadrant on the right). The results obtained for a fully coherent light source, as for example a laser, can be employed for determining the deviation of the visibility from the ideal case due to setup imperfections.

The intensity maximum I_{max} and minimum I_{min} can be determined by fitting the intensity measurements with a sinusoidal function (see below). For fully coherent light, $|g^{(1)}(\tau)| = 1$ for all τ, whereas $0 < |g^{(1)}(\tau)| \leq 1$ when the light is only partially coherent. These two cases are exemplified in Fig. 5.15(b), top and bottom panels, respectively. To properly measure the coherence of the investigated photons, the visibility needs to be evaluated for different delay values, i. e., for different path length differences in the interferometer, then reconstructing the $g^{(1)}(\tau)$. Experimentally, to account for setup imperfections in the measured visibility, a measurement of the first-order correlation function can be performed investigating highly coherent photons (like those generated by a highly coherent laser). In this case the expected $g^{(1)}(\tau) = 1$ for all accessible path differences (typically smaller than the coherence time) will only deviate due to setup imperfections. This calibration measurement can then be employed to correct the actual observed visibilities of the investigated photons. The drop in visibility due to an increase in path length difference (hence delay acquired by the photon propagating in the variable arm versus the other) provides then an estimation of the photon coherence time. Indeed, the *coherence time* $\tau_c = T_2$ follows

$$\tau_c = T_2 = \int_{-\infty}^{+\infty} |g^{(1)}(\tau)|^2 \, d\tau. \tag{5.28}$$

Experimentally, the oscillatory intensity curves obtained for each path length difference (see Fig. 5.15(b)) are fitted in a sinusoidal fashion to provide the visibility value at the specific position. This allows reconstructing the behavior of the visibility versus the path length difference, which can be then fitted with an exponential fit function, which contains information on the homogeneous (T_{hom}) and inhomogeneous (T_{inhom}) coherence times (see Eq. (5.31)). The first-order coherence function can be then written in the form (see Sec. 1.3)

$$g^{(1)}(\tau) = \exp\left(-\frac{\pi}{2}\left(\frac{\tau}{T_{\text{inhom}}}\right)^2 - \frac{|\tau|}{T_{\text{hom}}}\right). \tag{5.29}$$

These two coherence times can be then employed to evaluate the respective linewidths as discussed in the following section. Equation (5.28) takes the form

$$\tau_c = T_2 = T_{\text{inhom}} \exp\left(\frac{T_{\text{inhom}}}{\sqrt{\pi} T_{\text{hom}}}\right)^2 \operatorname{erfc}\left(\frac{T_{\text{inhom}}}{\pi T_{\text{hom}}}\right), \tag{5.30}$$

where erfc is the complementary error function.

5.2.7 Spectral diffusion: photon-correlation Fourier spectroscopy

In the previous section, we considered the detrimental effect on the photon coherence and indistinguishability induced by fluctuations faster than the transition lifetime. Additional to these, environmental fluctuations that are slower than the considered transition lifetime can also impact the photon indistinguishability. These so-called *spectral diffusion* or *spectral wandering* mechanisms, being slower than the considered lifetime, still contribute to the term T_2^* in Eq. (5.23) but have a different impact on the photon linewidth. These mechanisms contribute as inhomogeneous broadening, resulting then in a Gaussian-shaped linewidth contribution. In a simplified picture, once a (even Fourier-limited) photon is emitted, the environmental conditions of the emitter's surrounding may change, affecting the property of the two-level system. Once a second photon is emitted, the carrier frequency can be different because of the impact of a varying electric or magnetic fields. These fields can be induced by nearby fluctuating charges or by oscillations of the nuclear spins (the Overhauser field), and they may vary the energy of the two-level system under investigation (see Fig. 5.16 and Sec. 8.3 for a semiconductor quantum dot as case study).

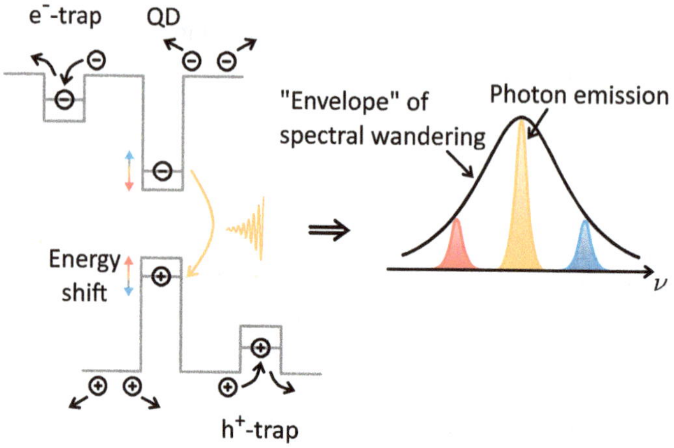

Figure 5.16: Sketch of fluctuating charge environment around a quantum dot as an exemplary studied system (see also Sec. 8.3). These fluctuations influence the energy of the emitted photons (yellow, blue, and red emission lines). This results in the emission of photons with different frequencies. Their envelope is measured as a Gaussian broadened spectrum in high-resolution photoluminescence when integrated over a time longer than the typical fluctuation timescales.

At the end, the carrier frequency of every emitted photon may then differ from the others, limiting the photon indistinguishability and hence the achievable two-photon interference (as explained in Sec. 5.2.8 and schematically sketched in Fig. 5.20). The wandering of the emitted photon linewidth for a varying environment is sketched in Fig. 5.16, whereas its potential effect on the photon wavelength is depicted in Fig. 5.17.

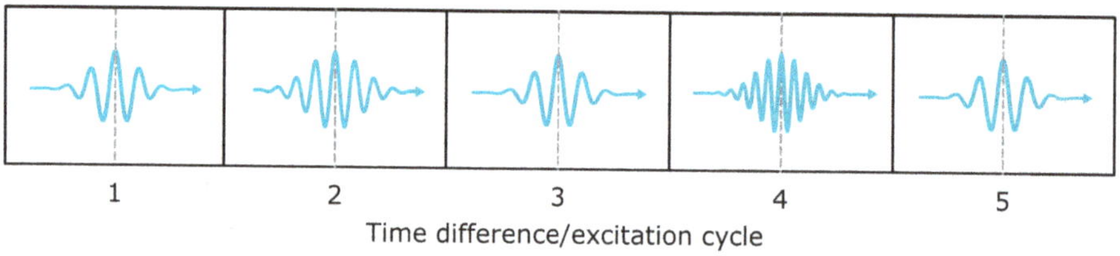

Figure 5.17: The presence of spectral wandering alters the emitted photon wavelength over time, as depicted by the change in the photon frequency in time slots 2 and 4.

The effect of this spectral wandering is observed not only as a degradation of the photon indistinguishability, but its presence can be inferred from high-resolution linewidth measurements (as that discussed in Sec. 5.2.2). Indeed, a time-integrated PL spectrum (acquired on timescales much longer than the lifetime and spectral wandering timescales) would result in a linewidth larger than that expected from the Fourier transformation of the transition lifetime: when nonradiative recombinations are absent, the measured decay time provides the radiative lifetime. Hence the linewidth for Fourier-limited photons (full width at half maximum) can be inferred as $\Delta\nu_{FL} = 1/\pi T_2 = 1/2\pi T_1$ ($T_1 = \tau_{rad}$). As explained in detail in the following, dephasing and spectral diffusion mechanisms can take place at various timescales. This means that even a perfectly Lorentzian, Fourier-limited emission spectrum will deviate from this ideal lineshape while spectral diffusion starts to take place. After all dephasing and spectral diffusion mechanisms happened, the time-integrated emission spectrum will not change any further, and it can be referred to as the *stationary spectrum* (envelope in Fig. 5.16 (right)).

In the most general case, the observed linewidth would be comprised by Lorentzian and Gaussian components because of homogeneous and inhomogeneous broadening mechanisms: fitting with a Voigt profile would then provide information on the linewidths, and respective coherence times of the Gaussian and Lorentzian contributions. In summary, recalling Eq. (5.23),

$$\frac{1}{T_2} = \underbrace{\frac{1}{2T_1}}_{\text{homogeneous}} + \underbrace{\frac{1}{T_2^*}}_{\substack{\text{homogeneous}\\+\\\text{inhomogeneous}}}, \tag{5.31}$$

we can note the various contributions to the observed linewidth. Deviating from the homogeneously broadened linewidth due to the transition lifetime, dephasing mechanisms provide an homogeneous contribution, whereas spectral diffusion contributes with an inhomogeneous term (see Chapter 8). Furthermore, the homogeneous and inhomogeneous coherence time constants introduced in the previous section are connected to the respective linewidth (FWHM) contributions as

$$\Delta\nu_{hom} = \frac{1}{\pi T_{hom}}, \tag{5.32}$$

$$\Delta\nu_{inhom} = \frac{\sqrt{2\ln(2)}}{\sqrt{\pi}T_{inhom}}, \tag{5.33}$$

which enter in the overall Voigt linewidth that can be approximated as [207]:

$$\Delta\nu_{tot} = 0.535\Delta\nu_{hom} + \sqrt{0.217\Delta\nu_{hom}^2 + \Delta\nu_{inhom}^2}. \tag{5.34}$$

Of course, a linewidth measurement can provide limited information regarding spectral wandering or dephasing mechanisms in general, so more sophisticated techniques are required (two examples are described in the upcoming section): in this way the timescale of the dephasing and diffusion mechanisms and their origin can be identified and addressed (see [20, 213, 215] and references therein for a more detailed discussion).

Depending on the nature of the investigated emitter and the origin of the diffusion mechanisms, the environmental effects can be minimized with various approaches, for example, by applying an external magnetic of electric field, employing a two-color excitation scheme, or remote controlling of the nuclear spins [126, 215]. Interestingly, by looking at Eq. (5.23) we can understand that the impact of dephasing and diffusion mechanisms can also be effectively reduced by shortening the lifetime T_1 (so reducing in the impact of T_2^* in deviating from the Fourier limit). Shortening the lifetime of a two-level system can be achieved via cavity quantum electrodynamics (see the description of the Purcell effect in Section 6.4 and Chapter 14), which helped in the observation of high two-photon interference values (for short time separation between the photons) even for non-Fourier-limited emission processes (see Section 5.2.8).

In the specific case of semiconductor quantum dots, dephasing and spectral diffusion often observed in these emitters are due to the coupling of the QD with its mesoscopic environment. These mechanisms can be categorized into three groups, which have different timescales and different impacts on the linewidth (more details can be found in Section 8.3):

- *Phonon interaction*: this results in the appearance of large sidebands around the main transition line (so-called zero-phonon line, ZPL) and in the broadening of the otherwise narrow zero-phonon line. This broadening of the ZPL ($\Delta\nu_{\text{phonon}}$) over the Fourier-limited homogeneous linewidth ($\Delta\nu_{\text{FL}}$) is of the order of $\Delta\nu_{\text{phonon}}/\Delta\nu_{\text{FL}} \ll 1$. Typical timescales of this phonon interaction lie in the picoseconds regime (i. e., smaller than typical QD lifetime, from hundreds of ps to few ns, i. e., homogeneous contribution to Eq. (5.31)).
- *Hyperfine interaction*: fluctuations of the nuclear spins can have an impact on the linewidth of the order of $\Delta\nu_{\text{spin}}/\Delta\nu_{\text{FL}} \le 2$, over typical timescales of the order of microseconds (i. e., inhomogeneous).
- *Charge carrier interaction*: this system-specific behavior has an impact on the linewidth typically of the order of $\Delta\nu_{\text{charge}}/\Delta\nu_{\text{FL}} \le 10$, whereas the timescale of these processes can vary from nano- to milliseconds (i. e., mostly inhomogeneous).

Since these three processes are known to be limiting factors for the achievable photon interference, several approaches have been studied and employed to reduce their effects. More detailed information and a comprehensive list of references on these topics (broadening mechanisms, their measurements, and counteracting methods) can be found in [215] and in Sections 8.3 and 9.4.

The short description previously reported provides processes, linewidths, and respective timescales that are typical for semiconductor quantum dots. It is important to state here that different material systems can be subject to various processes, which can result in the observation of spectral diffusion. However, even in the presence of multiple independent diffusion mechanisms, it is very often the case that the stationary frequency distribution follows a Gaussian behavior (as expected from the central limit theorem). This means that the spectral diffusion can be seen as a classical process rather than a quantum walk. The dynamics can be then described statistically. In the presence of n processes responsible for the spectral shift of the emission frequency, in the limit as $n \to \infty$, an Ornstein–Uhlenbeck (OU) stochastic process can well express the diffusion mechanisms, as described in [212, 213].

Measuring spectral diffusion

Phenomena that are responsible for photon dephasing (phonon interaction, carrier–carrier scattering) and for spectral diffusion (charge or spin noise) typically appear at timescales that are respectively shorter and longer than the radiative lifetime of the investigated transitions. In both cases, these timescales are anyway shorter (or much shorter) than the integration time of microphotoluminescence or Michelson interferometry measurements. Different techniques have been employed to characterize and quantify the presence of spectral diffusion and the mechanisms behind it: spin noise spectroscopy and resonance fluorescence are two exemplary techniques that allow, respectively, for revealing spin and charge noise in charged excitons (see Sec. 7.3) and for deducing the number and location of local charge traps by the line shape of the emission [215]. In the following sections, we will focus on one technique that allows reaching high temporal and spectral resolution in measuring the effects of spectral diffusion while pumping the quantum emitters via pulsed (and even resonant) excitation schemes. Indeed, the quantification of the timescale and amplitude of spectral diffusion mechanisms is appealing from a fundamental physics point of view. In addition, it becomes of central importance when quantum optical experiments are conducted utilizing nonclassical light emitted by the quantum light sources: for this reason, experimental techniques which allow for measuring the spectral diffusion directly on the emitted photons become attractive, in particular, if the emitter can be excited resonantly at the π-pulse to generate on-demand highly indistinguishable photons. The technique described in the following will precisely quantify the effects of spectral diffusion by measuring the deterministically generated photons, providing important insights into the usability of this quantum light. A brief description of another method based on the slow light effect will also be given.

Photon-correlation Fourier spectroscopy

In 2006, Brokmann et al. [20] introduced a novel experimental method for the investigation of spectral fluctuations in the light emission, reaching high temporal and spectral resolution. They named this method *photon-correlation Fourier spectroscopy* (PCFS). This technique combines a Michelson interferometer (see Sec. 5.2.6) with temporal correlations from a Hanbury Brown–Twiss setup (see Sec. 5.2.5): this results in a high spectral resolution, together with a time resolution close to the lifetime of the state under investigation. Interestingly, the probed timescales can span up to several milliseconds. Thanks to this technique, it is possible to record the energy difference between photons ζ with a given temporal separation τ. Then the so-called spectral correlation function $p_\tau(\zeta)$ represents the distribution of the energy differences ζ, and it is connected with the spectrum of the source under investigation. In other words, the PCFS transforms the frequency fluctuations into intensity fluctuations. Eventual spectral diffusion mechanisms can be understood and analyzed by observing the timescales of the recorded intensity fluctuations.

In more detail, PCFS is performed by measuring the intensity at both outputs (labeled 1 and 2) of the Michelson interferometer with two high time resolution single-photon detectors and time correlating these two signals. The second-order correlation function is then measured N times for different positions δ_i of the movable interferometer stage, enabling the observation of spectral fluctuations of the signal. Another key difference from Michelson interferometry is that for every measured position of the movable retroreflector $\{\delta_i\}$ with $i \in \{0, 1, \ldots, N-1\}$, the stage is slowly moved at a velocity v, allowing for using translation stages less accurate and stable as for Michelson measurements. This is shown in Fig. 5.18.

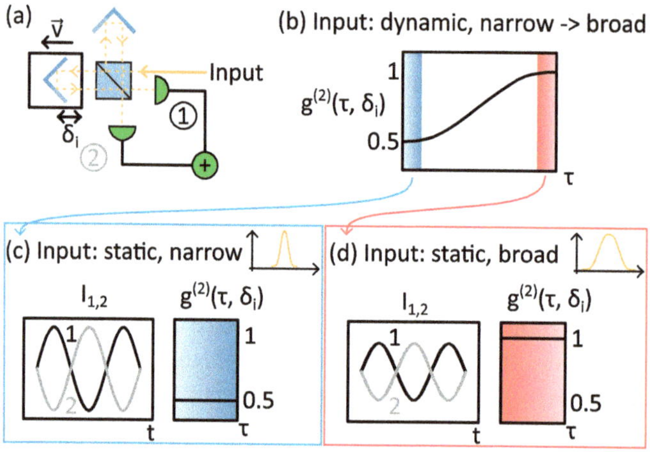

Figure 5.18: (a) Experimental setup, having a Michelson interferometer where one retroreflector set at position δ_i is moved at speed v in the marked direction, as well as the HBT setup correlating the two single-photon detectors at the output of the 50:50 Michelson beamsplitter. The intensity at the output ports, labeled as 1 and 2, is time-correlated. (b) Exemplary expected second-order correlation function for a retroreflector position δ_i over time separation τ. The input signal is considered as dynamics evolving from narrow at short time scales up to broad at large time scales. This temporal dynamics in the spectral evolution is reflected in the behavior of the correlation function. (c) Extreme case of a narrow static spectrum as input: a large interference contrast is observed at the beamsplitter, which reflects in $g^{(2)}(\tau, \delta_i) \approx 0.5$ when correlating the output intensities. (d) Conversely, a static spectrally broad input signal results in lower interference fringes, in contrast to $g^{(2)}(\tau, \delta_i)$ approaching 1.

The spectral information is provided thanks to the use of the Michelson interferometry, whereas the temporal dynamics is studied thanks to the high-resolution detectors in the correlation measurement. To understand this last sentence, let us assume a spectrum with a fast temporal dynamics: at short timescales the spectrum would be narrow, whereas it would be larger at longer timescale (see Fig. 5.18(b) for the expected $g^{(2)}(\tau, \delta_i)$). Consider two cases:

- short timescales (Fig. 5.18(c) shows an exemplary narrow static spectrum): the spectrum that feeds the interferometer is narrow, and for a certain mirror position δ_i,

the oscillation visibility at the beamsplitter is high, because the photons can well interfere at the beamsplitter. Since in PCFS the intensities at the two outputs are cross-correlated, a high visibility translates into low $g^{(2)}(\tau, \delta_i)$ values approaching 0.5 (i. e., more anticorrelated).
- long timescales (Fig. 5.18(d) shows an exemplary broad static spectrum): the spectrum that feeds the interferometer is broad, and for a certain mirror position δ_i, the oscillation visibility at the beamsplitter is lower than for shorter timescales because of lower achieved interference. A lower visibility in the oscillations at the output intensities translates into higher $g^{(2)}(\tau, \delta_i)$ values (approaching 1): conversely, from the previous case, a lower visibility corresponds to less anticorrelated $g^{(2)}(\tau, \delta_i)$ values.

In other words, the second-order cross-correlation $g^{(2)}(\tau, \delta_i)$ provides the signal oscillation visibility at a certain interferometer position δ_i for any arbitrary timescale τ. Additionally to these two extreme cases, the PCFS allows for accessing the dynamics also for all intermediate timescales τ. From the measured second-order correlation functions it is possible to obtain the spectral correlation function $p_\tau(\zeta)$, as explained in the following, which provides information on the investigated spectrum time evolution. Having a closer look at the mathematical formalism, the light intensity at the outputs of the interferometer is given by

$$\begin{gathered} I_1(t, \delta_i) \propto 1 + \cos\left[(2vt + \delta_i)\omega(t)/c\right] \\ \text{and} \\ I_2(t, \delta_i) \propto 1 - \cos\left[(2vt + \delta_i)\omega(t)/c\right], \end{gathered} \tag{5.35}$$

where $I_{1,2}$ are the intensities at the two output ports, $\omega(t)$ is the time dependent angular frequency of the incoming light, v is the velocity at which the movable stage travels for each position δ_i, and c is the speed of light. In case of quantum emitters the angular frequency follows a probability distribution set by the transition lifetime T_1, further broadened by the presence of spectral diffusion mechanisms that can be consider here as a fluctuation $\delta\omega(t)$ around the central angular frequency ω_0. The PCFS setup measures the temporal correlation of the two outputs, providing therefore

$$g^{(2)}(\tau, \delta_i) = \frac{\langle I_1(t, \delta_i) I_2(t + \tau, \delta_i)\rangle_t}{\langle I_1(t, \delta_i)\rangle_t \langle I_2(t + \tau, \delta_i)\rangle_t}; \tag{5.36}$$

this second-order correlation function appears from the measurement conducted with a path length difference in the interferometer δ_i and after having integrated on a time interval from $t = 0$ to $t = t_{\text{end}}$ ($\langle \cdots \rangle_t$ indicates the time average over this interval). The measurement (as in Fig. 5.19) is then repeated for different path length differences, as sketched in Fig. 5.18. The movement of the stage results in the observation of interference fringes at the two output ports (as sketched in Fig. 5.18): similarly to Michelson interferometry, as long as the path difference δ is smaller than the coherence time of the photons, fringes can be well detected. This reflects the spectral information in the observed wavelength distribution of the source, whereas the temporal dynamics of the broadening mechanisms can be extracted thanks to the high time resolution (provided by the single-photon detectors used for the measurement in Eq. (5.36)). To determine the temporal behavior of the spectral signal, we can insert Eq. (5.35) into Eq. (5.36). The time averaging allows dropping from the integral the fast oscillating terms, resulting in

$$g^{(2)}(\tau,\delta_i) = 1 - \frac{1}{2\,t_{\mathrm{end}}}\int_0^{t_{\mathrm{end}}} \cos\left(\frac{2v\tau}{c}[\omega_0 + \delta\omega(t+\tau)] + \frac{2vt}{c}\zeta_\tau(t) + \frac{\delta_i}{c}\zeta_\tau(t)\right)\mathrm{d}t. \tag{5.37}$$

In this equation, we consider that the incoming light has a spectral distribution centered at the frequency ω_0 and fluctuates in time around it by $\delta\omega(t)$. In Eq. (5.37) the random frequency jump $\zeta_\tau(t) = \delta\omega(t+\tau) - \delta\omega(t)$ has been further defined in the time interval $[t, t+\tau]$.[17] It is possible to further approximate Eq. (5.37); being the distance traveled during one measurement smaller than the path length difference ($2vt_{\mathrm{end}} \ll \delta_i$), the second term in the integral can be neglected. Additionally, since the change in optical path $2v\tau$ traveled during the timescale in consideration is small with respect to the coherence time of the emitter, the first term reduces to $2v\tau\omega_0/c$. Therefore we have

$$g^{(2)}(\tau,\delta_i) = 1 - \frac{1}{2t_{\mathrm{end}}}\int_0^{t_{\mathrm{end}}} \cos\left(\frac{2v\tau}{c}\omega_0 + \frac{\delta_i}{c}\zeta_\tau(t)\right)\mathrm{d}t. \tag{5.38}$$

Finally, if the integration time (up to t_{end}) is long enough so that it can be considered much larger than the fluctuation timescales into play, then we can replace the time average with the ensemble average (ergodic theory) over the distribution $p_\tau(\zeta)$ of possible energy shifts ζ:

$$g^{(2)}(\tau,\delta_i) = 1 - \frac{1}{2}\int_{-\infty}^{+\infty} \cos\left(\frac{2v\tau}{c}\omega_0 + \frac{\delta_i}{c}\zeta\right)p_\tau(\zeta)\,\mathrm{d}\zeta. \tag{5.39}$$

In case the fluctuations are invariant upon time inversion, $p_\tau(\zeta) = p_\tau(-\zeta)$, and hence we can rewrite Eq. (5.39) as follows:

$$g^{(2)}(\tau,\delta_i) = 1 - \frac{1}{2}\cos\left(\frac{2v\tau}{c}\omega_0\right)\mathcal{F}(p_\tau(\zeta))_{\delta_i/c}. \tag{5.40}$$

In other words, Eq. (5.40) shows that the correlation function $g^{(2)}(\tau,\delta_i)$ measured between the outputs of the Michelson interferometer consists of an oscillatory term with frequency $2v\omega_0/c$ and the Fourier transform of the distribution of possible energy shifts. When the investigated timescales τ are smaller than the periodic oscillations in

17 In this section, a notation similar to that in [20, 174] has been used, so that it can be easily compared with the original papers if more details on the calculations are needed.

$g^{(2)}(\tau, \delta_i)$, i. e., $\tau \ll \pi c/\omega_0 v$, the cosine can be neglected. This also means that for narrow spectral fluctuations much smaller than ω_0, the condition that $2v\tau$ is much smaller than the coherence length is also fulfilled. Therefore we can rewrite Eq. (5.40) as follows:

$$p_\tau(\zeta) = 2\mathcal{F}^{-1}[1 - g^{(2)}(\tau, \delta_i)]_{\zeta=2\pi c/\delta_i}. \tag{5.41}$$

This important conclusion shows that the distribution of the possible energy shifts can be experimentally obtained from the measurement of the second-order correlation function for various path length differences δ_i. Interestingly, for timescales close to zero, $p_\tau(\zeta)$ is given by the autocorrelation of the homogeneously broadened Fourier-limited spectrum of the investigated emitter. Conversely, for timescales $\tau \to \infty$, the distribution of the energy shifts would contain correlations as from the fully inhomogeneously broadened spectrum, as that recorded from a long acquisition with Fabry–Perot interferometry (see Section 5.2.2). At this point, it is important to remark that the distribution of possible energy shifts $p_\tau(\zeta)$ does not represent directly the spectral distribution of the measured light spectrum. From a formal point of view, $p_\tau(\zeta)$ can be written as a function of the time-resolved spectrum $s_t(\omega)$:

$$p_\tau(\zeta) = \left\langle \int_{-\infty}^{+\infty} s_t(\omega) s_{t+\tau}(\omega + \zeta) \mathrm{d}\omega \right\rangle. \tag{5.42}$$

As we can see from this equation, $p_\tau(\zeta)$ is given by the integral (or, in other words, by the cross-correlation function) between the spectral distribution at time t and its value at $t + \tau$. Thanks to Eq. (5.42), the previous statement becomes now clear, i. e., that as $\tau \to 0$, the spectral distribution $s_{t+\tau}$ will tend to s_t. Therefore the Fourier-limited (homogeneously broadened) spectrum will correlate with itself. Vice versa, as $\tau \to \infty$, the spectrum would suffer all possible frequency fluctuations, and the autocorrelation will be of the fully inhomogeneously broadened spectral distribution. As a final remark, following the arguments in [174], for the case of a semiconductor quantum dot, the time-dependent homogeneous distribution would correspond to the Fourier limit; otherwise said, $s_t(\omega)$ would be a Lorentzian distribution with a linewidth Δv set by the transition lifetime. Then as $\tau \to 0$, $p_{\tau=0}(\zeta)$ would correspond to a Lorentzian distribution with FWHM of $2\Delta v$. As $\tau \to \infty$, the time-averaged spectral distribution would enter in the autocorrelation in $p_{\tau=\infty}(\zeta)$. If we assume a Gaussian broadening having an FWHM set as Δv, then we would still observe a Gaussian distribution in $p_{\tau=\infty}(\zeta)$ but with FWHM of $\sqrt{2}\Delta v$.

Having now discussed the mathematical formalism behind PCFS measurements, it becomes clear how to experimentally conduct these measurements. First, cross-correlations $g^{(2)}(\tau, \delta_i)$ are recorded for different positions of the Michelson stage (i. e., δ_i). From an experimental point of view, the following criteria need to be fulfilled:

(i) integration time and velocity chosen during the acquisition of Eq. (5.36) allow for covering a spatial distance including multiple fringes; in other words, $2vt_{\mathrm{end}} \gg 2\pi c/\omega_0$;

(ii) the interference visibility should be well defined during the measurement of $g^{(2)}(\tau, \delta_i)$, meaning that the distance covered during the measurement should be smaller than the distance between measurements ($2vt_{end} \ll \delta(t_{end}) - \delta(0) = \Delta$);

(iii) the change in optical path within the investigated timescale (which can be referred to as τ) has to be smaller than the coherence length of the photons; otherwise said, $2v\tau \ll cT_2$ with T_2 being the photon coherence time.

Having chosen the velocity, integration time, and optical path according to the aforementioned conditions, the $g^{(2)}(\tau, \delta_i)$ acquired for different δ_i values result in a behavior as in Fig. 5.19(a), where exemplary second-order correlation functions for various path length differences δ_i are plotted over time separation τ. For long τ, $g^{(2)}(\tau) \to 1$, whereas for short timescales, the value should reach 0.5. Deviations from this last expected value

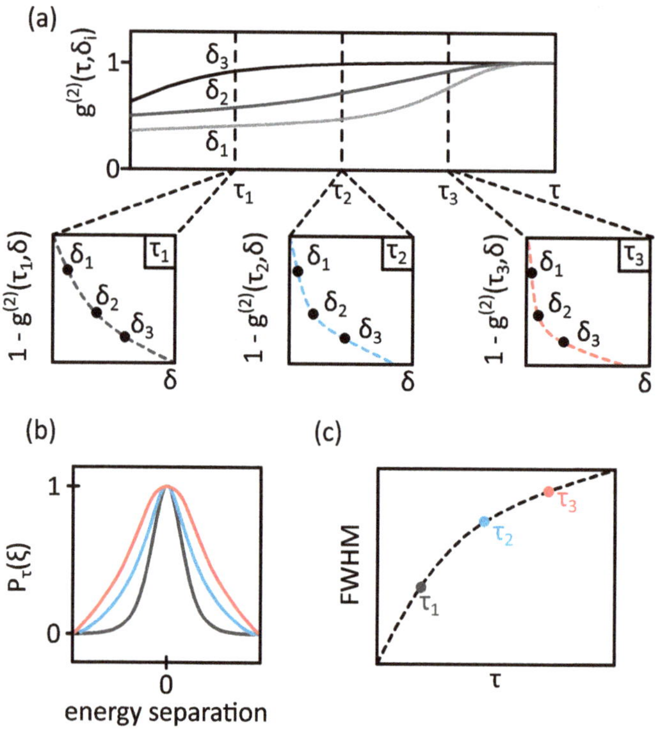

Figure 5.19: (a) Exemplary second-order correlation functions $g^{(2)}(\tau, \delta_i)$ obtained for different interferometer path length differences δ_i as functions of the time separation τ. Below the behavior of $g^{(2)}(\tau_{1,2,3}, \delta)$ versus the path length difference is sketched for three exemplary time separations $\tau_{1,2,3}$. Experimentally, lengths both shorter and longer than zero-path difference (named white fringe) are measured to clearly identify the maxima. (b) Reconstruction of the spectral correlation function via the Fourier transform of $g^{(2)}(\tau_{1,2,3}, \delta)$. (c) Evaluation of the full width at half-maximum as a function of the time separation, obtained from the spectral correlation in (b).

has been observed and attributed to setup imperfection (in particular, a nonideal beam overlap at the beamsplitter), whereas the presence of blinking in the emitter dynamics must be taken into account to reach 1 for long time scales (for more detail, see [174]).

Per each time separation τ_i, it is possible to plot the second-order correlation function over the path length difference, i. e., $g^{(2)}(\tau_i, \delta)$ (see the three graphs on the bottom of Fig. 5.19(a)). The inverse Fourier transform of this correlation function, i. e., $g^{(2)}(\tau_i, \delta)$, provides then $p_{\tau_i}(\zeta)$, from which it is possible to obtain the FWHM (knowing the original lineshape if Lorentzian or Gaussian). In Figure 5.19(b), exemplary curves are sketched for three different τ_i values: this results in the evolution of the full width at half-maximum over the timescale τ, as depicted in Fig. 5.19(c). With this approach, the PCFS provides information on the linewidth and its evolution in the investigated timescales.

Slow-light photon-correlation spectroscopy

As seen in the previous paragraph, photon correlation Fourier spectroscopy is a useful method to determine the evolution of the linewidth on selected timescales. This technique is very powerful, but it still requires the use of a Michelson interferometer, relatively long integration time, and time-consuming data analysis. Interestingly, an alternative method to study the effects of spectral diffusion on quantum emitters was proposed and experimentally demonstrated. The time evolution of the emitter linewidth and respective timescales are experimentally quantified via a technique named "slow-light spectroscopy." First introduced by Vural et al. [213], this method is implemented by replacing the interferometer in the correlation setup with a strongly dispersive medium. During propagation, the photon time-of-flight becomes frequency dependent, therefore enabling the mapping of the frequency domain into time domain. A single autocorrelation measurement suffices for the experimental observation of the emitter's linewidth evolution and respective time scales, from the emitter's lifetime to its stationary limit (see [213] for further details).

5.2.8 Indistinguishability: two-photon interference measurements

Operating with perfectly indistinguishable photons is highly desirable in photonic quantum technology where complex quantum operations are enabled by the capability of two photons to quantum interfere (see also Chapters 4 and 11). As previously discussed, dephasing and spectral diffusion mechanisms can both render the generated photons less indistinguishable, hence impacting their capability to interfere. This is schematically depicted in Fig. 5.20, where both mechanisms are shown.

Interestingly, despite it should be ideally absent, the presence of spectral wandering does not necessarily impact the achievable photon interference under all conditions. Indeed, the onset of spectral wandering mechanisms can be on a longer timescale than the

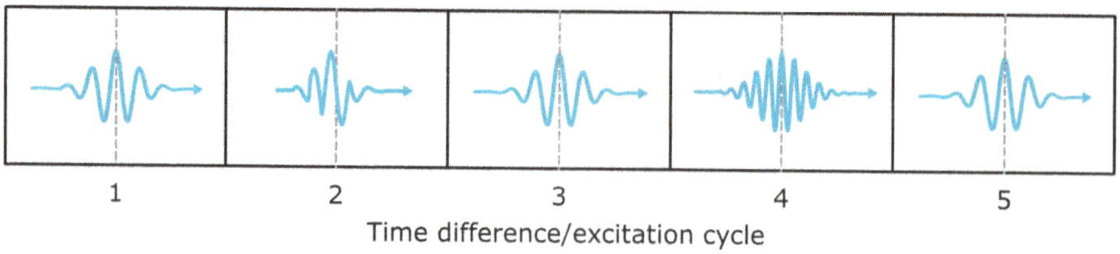

Figure 5.20: Examples of degraded photon indistinguishability due to pure dephasing (depicted as broken line in time slot 2) and to spectral wandering (depicted as a change of photon frequency in time slot 4). Both effects can severely limit the achievable two-photon interference, hence lowering the HOM visibility.

time separation between the generation of two photons brought to interfere: in this case the two emission processes will happen within the same environment, and therefore high indistinguishability is still expected. The impact on the photon interference starts to happen when the time separation between the photon emissions is large enough for the spectral wandering mechanisms to come into play: this results in the generation of distinguishable photons, then reducing their capability to interfere.

In the last years, various independent experiments employing semiconductor QDs as photons sources showed very high degree of photon indistinguishability for a relatively short time separation between the photons (from few nanoseconds to few microseconds). Conversely, the two-photon interference visibility decreased for increasing time separation between the photon emission processes. Therefore it is safe to assume that at short time differences, the frequency change is minimal, whereas it increases for longer time separation, reaching the Gaussian broadened envelope spectrum as in Fig. 5.16. Spectral diffusion plays an even more dramatic role when the photons are generated by distinct sources (see one of the following sections). Even if the emitters are chosen to have similar emission properties (as temporal profile, or equal photon emission frequency and linewidth as measured in high-resolution spectroscopy), the presence of spectral diffusion will severely impact the two-photon interference: when considering two distinct sources, the diffusion processes of the two emitters are fully uncorrelated and independent (for a formal description, see Section 11.3). This explains why high values of two-photon interference visibility are more challenging to be achieved when the two photons stem from different, remote sources.

Additionally, dephasing mechanisms happening at timescales shorter than the radiative lifetime can impact the achievable two-photon interference (TPI) even for arbitrarily short time separation between the photons. In conclusion, for future development of quantum technology implementations involving multiple remote sources, Fourier-limited photons *and* the absence of spectral wandering will be both required, i. e., ideally, neither dephasing nor diffusion mechanisms should be present to ensure maximal visibility in two-photon interference experiments.

The photon indistinguishability can be experimentally probed by interfering two photons, for example, on a beamsplitter, and observing the so-called Hong–Ou–Mandel

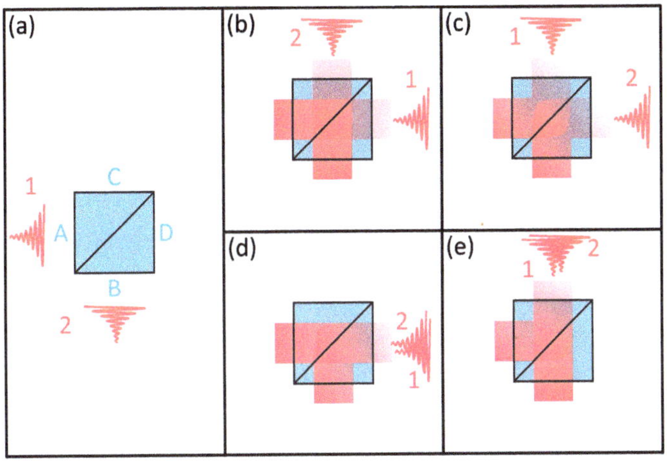

Figure 5.21: (a) Two single photons (labeled as 1 and 2) arriving at the two input ports *A* and *B* of a beamsplitter: the four classical outcomes are further depicted. (b) Photon 1 is transmitted leaving from output port *D*, whereas photon 2 is transmitted leaving from port *C*. (c) Configuration where both photons are reflected. (d) Case with photon 1 being transmitted and photon 2 being reflected: both photons leave the beamsplitter from port *D*. (e) Configuration where photon 1 is reflected and photon 2 is transmitted: photon coalescence at output port *C* is observed.

effect (see Chapters 4 and 11). To understand this quantum interference, let us consider the arrival of two photons at the two entrances of a 50:50 beamsplitter, labeled as A and B (see Fig. 5.21 (a)).

Classically (or also with two perfectly *distinguishable* single photons), four output configurations are possible: both photons are transmitted (Fig. 5.21(b)); both photons are reflected (Fig. 5.21(c)); one is reflected and one transmitted (meaning that photon 1 is transmitted and photon 2 is reflected as in Fig. 5.21(d); and photon 1 is reflected and photon 2 transmitted as in Fig. 5.21(e)). When the photons are perfectly indistinguishable, the equal and opposite amplitudes of the two cases (b) and (c) at the beamsplitter cancel out, leaving only (d) and (e) as possible outcomes of the two-photon interference (see Eqs. (4.13), (4.51), and (4.52)). This means that when two identical photons are impinging on a 50:50 beamsplitter, coalescence of photons is expected, i. e., they will both bunch at the same output (either C or D). A nonperfect two-photon interference would result in the probability of observing photons simultaneously at the two BS output channels. Interestingly, the observation of perfect photon coalescence sets requirements not only on the light source (or multiple sources), which should generate perfectly identical photons; at the same time the experimental setup must allow a perfect temporal and spatial overlap of the arriving photons (see Eqs. (4.44), (4.45), and (4.47)). Therefore effective realizing two-photon interference requires the absence of dephasing and spectral diffusion mechanisms in the emission process, together with a setup enabling the arrival of the two photons at the interference point at the same time, and in addition the two photons having the same spatial mode and polarization.

The experimental quantification of the two-photon interference can be performed by statistically discriminating the events between *distinguishable* and *indistinguishable* photons. To do so, two single-photon detectors are placed at the output channels of the beamsplitter, and the arrival times of photons are measured (more experimental details in the following; see Fig. 5.23). Similarly to HBT experiments, also in this case the interesting information lies in the absence of simultaneous detection events, with one important difference: as explained above, even for distinguishable photons, no simultaneous detection is expected in two cases over four. Therefore it is important to compare the measurement performed with indistinguishable photons (where ideally only photon coalescence is observed, as in Figs. 5.21(d,e)) with respect to the perfectly distinguishable case (where photon coalescence happens statistically in two cases over four).

The difference in coincidence counts at zero time delay, i. e., simultaneous arrival, will therefore measure the two-photon interference visibility V_{HOM} (see Fig. 5.22). Considering all observed coincidences around zero delay for the distinguishable $C_\perp$ and indistinguishable $C_\parallel$ measurements (with the same normalization at the same Poissonian level; see Sec. 5.2.5), the visibility can be defined as

$$V_{\mathrm{HOM}} = \frac{C_\perp - C_\parallel}{C_\perp}. \tag{5.43}$$

When this important parameter needs to be quantified for the investigated light source, the two measurements ($\parallel$ versus $\perp$) are performed: the former requires the simultaneous arrival of the two photons at the beamsplitter with ideally indistinguishable paths, whereas the latter can be realized even in case of perfectly identical photons by altering their properties at the beamsplitter.[18] This is typically done by modifying the respective

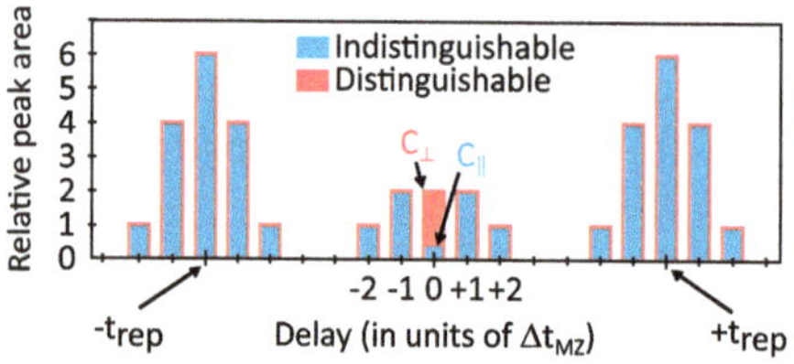

Figure 5.22: Theoretical correlation histogram for a Hong–Ou–Mandel measurement. In red the case for distinguishable photons, in blue for the indistinguishable case. The red histograms have been enlarged for clarity, since all, except the zero-delay one, have the same normalized area than the blue. The areas of Eq. (5.43) are indicated in the figure. The delay is marked in units of time separation between the interfering photons (see Fig. 5.23 and relative experimental discussion). A cluster of five peaks is repeated at each laser repetition.

18 An alternative approach to obtain the two-photon interference visibility is based on the comparison between the coincidence peak at zero delay over the other neighboring coincidence peaks (for nonzero delay), which follow a classical statistics [164].

polarization of two photons, i. e., making one orthogonal to the other, and hence the notation $\parallel$ and $\perp$. This means, once again, that the setup properties themselves have to be well under control.

HOM with unbalanced beamsplitter

Despite all efforts, few setup imperfections are still unavoidable, and hence the observed V_{HOM} could be below 100 % even for perfectly indistinguishable photons (i. e., without impact of dephasing or spectral diffusion). For example, an unwanted deviation of the beamsplitter splitting ratio from 50:50 would result in a degradation of the observed visibility V_{HOM}, and this can be corrected at analysis stage by considering

$$I = V_{\mathrm{HOM}} \frac{R^2 + T^2}{2RT}, \tag{5.44}$$

where R and T are the beamsplitter reflectivity and transmissivity, respectively, such that $T = 1 - R$. Only in the case $R = T = 0.5$ the measured visibility equals the actual photon indistinguishability.[19] In other words, I represents the *actual* photon indistinguishability, only limited by spectral diffusion and dephasing, whereas V_{HOM} provides the *measured* two-photon interference visibility, further including setup imperfections. From a principle point of view, these two quantities provide both valuable information: whereas I provides a measurement of the photon generation "quality", V_{HOM} represents the achievable interference necessary in the implementation of quantum operations.

HOM with multiphoton events

So far, it has been considered the estimation of the photon indistinguishability from the measured HOM visibility including the impact of a nonideal 50:50 beamsplitter. In realistic experiments the estimation of the photon indistinguishability from the visibility V_{HOM} has to take into account the potential contribution of multiphoton events, i. e., the case where $g^{(2)}(0) \neq 0$. Following the arguments and measurements as in [142], a nonperfect single photon source is modeled as two contributions: true single photons with $g^{(2)}(0) = 0$ and a separable noise at the beamsplitter. With this assumption, the HOM visibility at the (ideal) 50:50 beamsplitter becomes

$$V_{\mathrm{HOM}} = I - \left(\frac{1 + I}{1 + M_{\mathrm{sn}}} \right) g^{(2)}(0). \tag{5.45}$$

---

19 In the above equation, it has been considered that the spatial mode overlap at the beamsplitter is unity. In experimental implementation, this can be quantified as the classical interference measured at the beamsplitter, for example, using a narrowband laser and adding the correction factor (assuming a perfect 50:50 BS) $I = (1 - v^2 V_{\mathrm{HOM}})/2$, where v is the classical interference visibility. A general formulation can be found in [164]. It is important to remark that in the cited work, the two-photon emission probability (named g in the paper) is also included (which will become $g = 0$ ($g = 1$) for a single-photon (respectively Poissonian) source).

The aforementioned equation considers a separable noise contributing to the second-order correlation function (with small values, typically, $g^{(2)}(0) < 0.3$ as in the discussed work). With a similar notation as in [142], I stands for the photon indistinguishability, whereas M_{sn} provides the mean packet overlap of the single photon with the noise photons (further with $0 \leq M_{sn} \leq I$). Interestingly, in the case that the noise photons are identical to the actual single photons generated by the source, Eq. (5.45) becomes

$$V_{HOM} = I - g^{(2)}(0), \tag{5.46}$$

since $I = M_{sn}$. Conversely, if the noise photons are completely distinguishable from the single photon under investigation $M_{sn} = 0$, then Eq. (5.45) becomes

$$V_{HOM} = I - (1 + I)g^{(2)}(0). \tag{5.47}$$

These two extreme cases have been experimentally investigated in [142] for the case of distinguishable and indistinguishable noise sources, verifying the validity of the equations above. As discussed, the relation between the experimentally measured visibility V_{HOM} and the photon indistinguishability quantified by I is different depending on the noise source. Therefore it becomes necessary to know what kind of noise provides the multiphoton component. Further experimental demonstrations were performed employing two quantum dot transitions, namely an exciton and a trion (see Chapter 7), under the presence of multiphoton contribution originated from the presence of excitation laser (exciton case) and addition photon emission induced by reexcitation (trion case). Data analysis following Eq. (5.45) showed that in both cases, the noise can be treated as separable and distinguishable, justifying the use of Eq. (5.47) to extrapolate the photon indistinguishability I from the measurement of V_{HOM}.

At this point, it becomes clear how to estimate the photon indistinguishability from the visibility in Hong–Ou–Mandel experiments, where the absence or presence of simultaneous detections at the beamsplitter outputs is the estimation factor. When the photons are partially distinguishable because of dephasing mechanisms or spectral diffusion, or due to setup imperfections, simultaneous clicks on the two detectors can be observed, therefore resulting in an interference visibility $V_{HOM} < 100\,\%$. To experimentally perform two-photon interference at a beamsplitter, it is necessary to inject the two photons at the two BS inputs. To minimize the impact of a nonperfect setup, the experimental configuration must be implemented in a way that the photons arrive at the same time on the beamsplitter, further having the same polarization and spatial mode. In other words, it must be impossible to distinguish between two input paths. The interfering photons can be generated by two distinct sources or by two independent emission processes of the same emitter.

TPI with photons from the same single emitter

In the case that photons are generated by the same emitter, considering on-demand emission processes, the source under investigation needs to be excited twice, and the two emitted photons can be sent one per each BS input channel. In case of optical excitation, this can be experimentally realized employing unbalanced *Mach–Zehnder interferometers* (MZI) (see Fig. 5.23(a)). The first unbalanced interferometer (excitation MZI in the figure) can be used in the optical excitation path, so that two identical pulses with controlled time delay Δt_{MZ} are generated from each excitation source pulse. These can be employed to trigger in the investigated source two emission processes separated by the chosen time delay Δt_{MZ} (for the description of the TPI visibility dependence from excitation time separation, see Section 11.4). The two emitted photons can be then guided (in a free space or via fibers) to the 50:50 beamsplitter. To ensure the simultaneous arrival time for the first and second photons, another unbalanced Mach–Zehnder interferometer can be employed, ensuring that its induced time delay matches that in the excitation path (Δt_{MZ}). From an experimental point of view, the use of a beamsplitter formed by the evanescent coupling of single-mode fibers has the strong advantage of

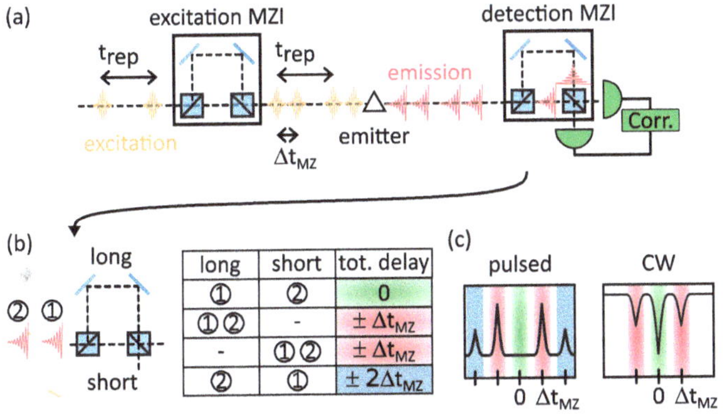

Figure 5.23: (a) Sketch of a possible experimental implementation of two-photon interference using successively generated photons from one source. A first Mach–Zehnder interferometer (excitation MZI) is used to prepare two equal pulses from one laser pulse (for completeness, the repetition rate of the laser is displayed with a time separation marked as t_{rep}) with time separation Δt_{MZ}. These pulses (sketched with a Gaussian profile) are used to excite the emitter, resulting in emission of photons (with exponential profile) with time separation Δt_{MZ} (at each repetition t_{rep}). Polarization control could be included in the MZI to ensure pulses with equal intensity and polarization. The second detection MZI matches the delay of the excitation MZI, ensuring simultaneous arrival of the two photons (the first BS routs the photons in the long or short path, whereas the two photons meet at the second BS). Polarization control (not shown) ensures the arrival of the emitter's photons with the same polarization. (b) The outputs of the second beamsplitter in the detection MZI are equipped with single-photon detectors whose signal is time-correlated. The arrival time of photons on the second beamsplitter depends on the path followed: the total delay is reported in the table for all possible combinations. (c) Exemplary coincidence histograms for indistinguishable photons, under pulsed and CW excitation, are provided: the peaks originate from the respective coincidences at various time delays (color coded).

ensuring an almost ideal overlap of the mode profiles upon interference. Furthermore, a precise control of the light polarization is required to ensure reaching the intended polarization configuration at the BS (either parallel or perpendicular; see Eq. (5.43)).[20] Interestingly, the discussed unbalanced Mach–Zehnder interferometers do not require any phase stabilization procedure, since the chosen time separation Δt_{MZ} is set to be longer than all coherence times involved in the measurements. The distinctive output histogram originating from the observation of the detectors correlation over time difference (Fig. 5.23(c)) can be understood considering all possible time separations between the two photons while traveling in the detection MZI (Fig. 5.23(b)). If the first photon takes the long path while the second follows the short path, then simultaneous arrival is expected. Nevertheless, other path combinations can be followed: in case both photons take the same path, a time separation of $\pm\Delta t_{MZ}$ is maintained (the $\pm$ sign comes from the time inversion between the detectors, i. e., first click in detector A followed by detector B or vice versa). Alternatively, when the first photon takes the short path while the second follows the long path, the delay will increase up to $\pm 2\Delta t_{MZ}$. This explains the observation of a five-peak pattern (see Fig. 5.23(c)), centered around zero time delay and further repeated at each laser repetition (the latter originating from classical statistics of uncorrelated photons; note indeed the difference in the central peak area in Fig. 5.22): for perfectly indistinguishable photons, the peak at zero time delay should be absent (whereas for the classical five-peak pattern, the central one is always present since it does not originate from interference). With these arguments, it becomes clear that the discussed experimental implementation of two-photon interference is inherently probabilistic.

If continuous wave excitation is considered, then the source will not emit photons on-demand, and the two-photon interference pattern will have a completely different behavior (see Fig. 5.23(c) for CW). Since the photons are continuously generated, all time differences will appear in the correlation measurements, reaching the Poissonian level for all time delays except around three time differences: a central dip at zero-delay and two symmetric features at a time separation set by the MZI delay Δt_{MZ}. The levels will be set by their probability of occurrence. The central dip would reach 0 for indistinguishable photons and 0.5 for the distinguishable case. The side peaks are then in both cases at 0.75 for a balanced beamsplitter (hence they can be used for the estimation of the photon indistinguishability). Nevertheless, for arbitrarily high detector resolution,

---

20 The detailed implementation of the fiber control depends on the designed experimental setup. Stress-induced birefringence in the fiber can be used to control the output light polarization. Alternatively, the light can be coupled out from the fiber, sent to a series of quarter- and half-waveplates, and polarizer to select the desired linear polarization and then coupled back in the fiber. Whereas the former has the advantage of limiting residual losses due to bulk optics and fiber out/in coupling, the latter is advantageous because of the precise, reliable, and mechanically stable polarization control achievable. In the free space case the use of only three polarization optics is sufficient for the indented measurement: whereas the polarizer sets linear polarization, the waveplates can compensate for unwanted birefringence by bringing any arbitrary state to a linear one before reaching the polarizer and BS.

the central dip will always reach zero for the indistinguishable case, even for non perfectly identical photons. Nonetheless, CW two-photon interference measurements can still provide a valuable information: the width of the central dip is indeed set by the photon coherence (see [137] and references therein). In other words, a CW source could still be used in time-gated experimental realizations employing two-photon interference within a post-selected time window set by the photon coherence.

As discussed in Chapters 8 and 11, dephasing or spectral diffusion mechanisms can lead to a deviation from the Fourier-limited linewidth, hence diminishing the observed two-photon interference visibility. If such mechanisms cannot be counteracted as explained in Chapter 8, then we could still consider to narrow the photon linewidth by means of a spectral filter (for example, with an etalon as explained in Section 5.2.2). This would allow reaching Fourier-limited linewidth at the expense of the source brightness. Although this approach could find use in specific applications where high indistinguishability from nonideal sources is required, it is hardly scalable. Indeed, the probability of N-photon to interfere scales as $P = (Bs \times I)^N$: if we increase the indistinguishability of two subsequent photons (I) at the expense of the brightness Bs, then the overall success probability of N-photon interference would not improve [51].

Two-photon interference from remote emitters: influence of blinking and photon flux on V_{HOM}

As explained above, two-photon interference represents a cornerstone phenomenon for the investigation of the "performance" of the quantum light source (i. e., the available photon indistinguishability I) and for setting the ground of complex quantum optical implementations (where high values of visibility V_{HOM} are required). At this point, it is worth mentioning that upscaling the experimental complexity, or implementing long-distance quantum operations (like the implementation of a quantum repeater), may require the ability of interfering photons generated by distinct remote nonclassical light sources. Multiphoton (and multisource) experiments showed for several years a challenge in reaching levels of visibility V_{HOM} comparable to the case of two-photon interference with light originated from one source. Although on the one hand, solid-state nanostructures have the great advantage of a deterministic light emission process (key for synchronizing the arrival time of several photons in performing complex experiments), on the other hand, the presence of spectral diffusion mechanisms, which are emitter-dependent (and hence fully uncorrelated to one another for distinct sources), limited the observed V_{HOM}. Although for one-source experiments, near-unity visibilities have been demonstrated (see a comprehensive description in [126]), the values of visibilities $V_{\mathrm{HOM}} \leq 50\,\%$ were observed in case of remote sources experiments, and only recently they reached values above 90 % [235]. Interestingly, in case of remote sources, an MZI is not strictly required since the two photons can be directly sent into the two beamsplitter inputs (with proper synchronization and photon overlap). To evaluate the observed TPI visibility, we can make use of Eq. (5.43) in the form

$$V_{\mathrm{HOM}} = 1 - \frac{C_{\parallel}}{C_{\perp}} = 1 - \frac{g^{(2)}_{\mathrm{HOM}}(0)_{\parallel}}{g^{(2)}_{\mathrm{HOM}}(0)_{\perp}}, \tag{5.48}$$

hence comparing the counts at zero delay for the cases of interfering (∥, indistinguishable) and noninterfering (⊥, distinguishable) photons. However, in this case the analysis of the results can become more complicated by the presence of blinking (see Chapter 8) or any form of intensity mismatch between the photon fluxes emitted by the two sources. Differently from the case of photons generated by one source, in case of interference of photons from remote emitters, a temporal change of the normalization level between these two measurements may lead to an overestimation of the photon indistinguishability: different emission dynamics, source intensities, or the modification of the blinking for any of the sources within the measurements of indistinguishable (parallel) and distinguishable (perpendicular) photons may lead to an inconsistent normalization for the two cases, therefore impacting the correct determination of V_{HOM}. Indeed, whereas the coincidences recorded at zero time delay can only arise from a pair of noninterfering photons, each generated by one source, the coincidences observed for nonzero time delays can have two different causes. This can be understood by looking at Fig. 5.24: the observed coincidences are divided between detection events due two photons generated by the same source (named hereafter autocoincidences) and due to a pair of photons generated by the two sources (one photon each, defined as cross-coincidences). As we can see in Figs. 5.24(a,b), simultaneous detection events (i. e., for zero relative time) can only occur due to cross-coincidences. On the other hand, for nonzero relative time differences, detection events can originate both from auto- and cross-coincidences. As exemplary shown in Fig. 5.24(b) (see measurements in [222]) for distinguishable photons, the presence of temporal correlations in the emission dynamics of even only one source results in the deviation from the Poissonian level at short relative time differences: therefore the detected coincidences at zero relative time are ~0.5 with respect to the coincidences detected at any repetition for a relative time where the Poissonian level is reached (as expected for a pair of distinguishable photons impinging on a 50:50 beamsplitter, where the classical output possibilities expect coincidences only in two cases over four). This ratio no longer holds for short relative times, where the different emission dynamics of one source (for example, in the presence of blinking, i. e., an on-off modulation of the signal) adds additional autocoincidences to the Poissonian level. Additionally, the emission dynamics of the sources may change during the measurements (for example, a displacement of the sample with respect to the excitation beam might alter the emitter's charge environment, further influencing the blinking dynamics). These findings underline even more the importance of determining the correct Poissonian level allowing for the most reliable normalization and evaluation of the experimental results.

So far, we have assumed that both sources employed in the TPI measurements are equally bright. This means that at the Poissonian level, the area of the coincidence peaks

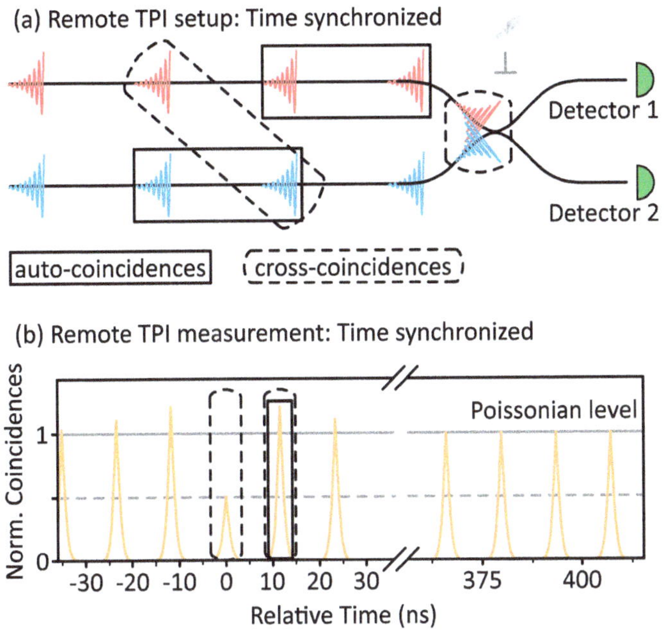

Figure 5.24: (a) Exemplary origin of coincidence cases, i. e., autocoincidences (solid line) and cross-coincidences (dashed line) in a two-photon interference setup for distinguishable (polarization-orthogonal) photons at the beamsplitter. (b) Schematized result of a coincidence measurement of remote TPI for distinguishable photons. Auto- and cross-coincidences are marked: at zero relative time difference, only cross-coincidences are expected, whereas for the other peaks, both contributions are present (with blinking on one source). For actual data, we refer to [222].

is due in equal parts to auto- and cross-coincidences (see [222] for more detail). The situation changes when the sources have different brightness. If we consider the case that the observed count rates for the two sources are respectively c_1 and $c_2 \leq c_1$, then the auto- and cross-coincidences peak area take the form

$$A_{\text{auto}} = (c_1^2 + c_2^2)RT, \tag{5.49}$$

$$A_{\text{cross}} = c_1 c_2 (R^2 + T^2) \tag{5.50}$$

for the general case of a beamsplitter with reflectivity and transmissivity R and T.[21] In case of a 50:50 splitting ratio, while monitoring the source countrate, it becomes possible to extract the value $g^{(2)}_{\text{HOM}}(0)_{\perp,\text{extra}}$ from a single measurement even in the presence of

21 For the case where $c_2 \leq c_1$, employing a 50:50 beamsplitter, $A_{\text{auto}} \geq A_{\text{cross}}$: in the extreme case where $c_2 = 0$, the experimental result would be equal to the autocorrelation of only source 1. In case of a non-50:50 beamsplitter, cross-coincidences will be dominating over autocoincidences.

blinking or countrate fluctuation (for the description generalized to uneven splitting ratios, see Supplemental Material of [222]). This takes the form

$$g^{(2)}_{\mathrm{HOM}}(0)_{\perp,\mathrm{extra}} = \frac{2c_2/c_1}{(c_2/c_1+1)^2}. \tag{5.51}$$

This extrapolated value, together with the measured parallel case $g^{(2)}_{\mathrm{HOM}}(0)_{\parallel}$, allows us to obtain the measured visibility as in Eq. (5.48). Nevertheless, this determination might not be always straightforward, since temporal dynamics can happen on large relative time differences (which are experimentally challenging to observe, requiring high computational power, especially for time-tagging based measurements) or even at multiple timescales (this means that parallel and perpendicular measurements might be inconsistently normalized).

To avoid complications in the determination of the correct Poissonian level, a measurement method has been proposed and experimentally demonstrated for the case of remote-source experiments [222]. Instead of sending the two photon streams directly to a beamsplitter, a Mach–Zehnder interferometer (employed for one-source measurements) is fed by the two photons originating from the two remote sources under investigation. In this way, it is possible to temporally separate between coincidences originating from autocoincidences and cross-coincidences: the observed five-peak pattern centered around zero-delays will contain only cross-coincidences, whereas the other (classical) peaks will still include auto- and cross-coincidences. Nevertheless, the central peaks are sufficient for the correct determination of the two-photon interference visibility, even in case of blinking or intensity mismatch between the two sources. This alternative interferometer configuration provides coincidence peaks around zero delay, which are independent from temporal and intensity fluctuations, making the determination of the two-photon interference visibility more reliable for the appealing case of remote sources experiments.

Role of wavepacket dispersion on TPI

Increasing efforts are spent to bring quantum technologies out of the lab, and when discussing photonic approaches, the possibility of employing single-mode optical fibers becomes highly attractive to cover long distances. For this scope, it is necessary to operate with nonclassical light sources operating in the telecommunication bands, in particular, within the telecom O-band (centered around 1310 nm) and C-band (centered around 1550 nm). These two wavelength regions become appealing since they represent, respectively, the absolute minimum in photon wavepacket dispersion and in light absorption. On the one hand, loss of light in quantum implementations represents a dramatic problem since the quantum information cannot be amplified as for classical signals. On the other hand, a change of the photon wavepacket due to dispersion while propagating in the fibers may result in a decrease of the achievable two-photon interference because of the distinguishability arising from the dispersion of one wavepacket over the other:

in realistic implementations the fiber length between multiple nodes is likely to be different, resulting in a propagation length larger for some photons with respect to others (see [221]). Circumventing losses is not an easy task and may require the use of quantum repeaters to increase the distance of transmission of quantum information. Nevertheless, wavepacket distortion due to dispersion can in principle be compensated knowing the length (and so the effect) of the fiber link. This led to considering O-band wavelengths as suitable for relatively short fiber networks based on quantum light (as the ones in cities), whereas for longer distances, the dispersion could be compensated, and the lowest losses provided by the propagation within the telecom C-band could allow larger separation between nodes.

5.2.9 Entangled photons properties of real emitters

Several quantum optical implementations, from communication to computation, benefit and require the use of entangled photons. Entanglement has been first theorized almost 100 years ago in 1935 [41], and it is now the heart of implementations in the second quantum revolution. In a nutshell, two particles are entangled when their wavefunction cannot be represented as product of the wavefunctions of the constituting two particles. In other words, the wavefunction is *not separable*. The first example of entangled particles is here provided by the polarization state of two single photons. A classical description can be successfully employed if the polarization property of the two photons can be described independently for each of the particles. On the contrary, when the description needs to include the states of both particles, entanglement comes into play. Here measurements conducted on one of the photons will instantaneously affect the state of the second particle, disregarding how far apart they are: the two particles are described by one wavefunction making them one entire entity. Causality will in any case not be violated, since no information is transmitted between the two involved entangled photons. Having the possibility to manipulate this entangled entity by acting only on one particle becomes very attractive for a broad range of experiments and implementations, from quantum communication to computation, imaging, and sensing, only to name a few. This explains why such a kind of quantum states are the subject of intense research, even if they are difficult to produce and manipulate. Indeed, these types of states can be easily impacted by decoherence induced by interaction with the environment, which can render the entangled particles two classical objects. Therefore, on the one hand, it would be ideal to implement entangled states that are fully decoupled from the surrounding environment. On the other hand, this unwanted interaction becomes necessary when manipulation of the wavefunction needs to be applied to *use* the properties of entangled particles.

As it will also be discussed later on, the Schrödinger equation describes how the quantum state of a physical system develops. Let us consider the Hamiltonian $\hat{H}$ having the two quantum states $|\psi_1\rangle$ and $|\psi_2\rangle$ as eigenstates of the physical system. We can then write

$$\hat{H}\,|\psi_1\rangle = E_1\,|\psi_1\rangle \quad \text{and} \quad \hat{H}\,|\psi_2\rangle = E_2\,|\psi_2\rangle\,, \tag{5.52}$$

knowing that all coherent superpositions of these two eigenstates, $|\Psi\rangle = \alpha\,|\psi_1\rangle + \beta\,|\psi_2\rangle$, will also be themselves eigenstates of the quantum system.

Now let us consider two particles (labeled A and B), each having a state vector $|\psi_A\rangle$ and $|\psi_B\rangle$ (elements of the respective Hilbert spaces $\mathbb{H}_{A,B}$). The compound system is said to be *separable* if the state can be written as a product $|\Psi\rangle = |\psi_A\rangle \otimes |\psi_B\rangle$.[22] Then considering the superposition between having particles A and B in states $|\psi_A\psi_B\rangle$ and $|\phi_A\phi_B\rangle$, we get

$$|\Psi\rangle = \frac{1}{c}(|\psi_A\psi_B\rangle + |\phi_A\phi_B\rangle), \tag{5.53}$$

where the term c constitutes a normalization factor from which $\langle\Psi|\Psi\rangle = 1$. If the two-particle state (5.53) cannot be written as a product of states (of the two respective Hilbert spaces), i. e., it is not separable, it is said to be *entangled* [190]. This type of state, until the interaction with a measurement device happens, shows quantum correlations, which have no classical analog.

If the system is found in a *pure* or *mixed* state,[23] the *density operator* provides a common description of the interacting system [108]. This operator has the form

$$\hat{\rho} = \sum_{\kappa} P_\kappa |\psi_\kappa\rangle\langle\psi_\kappa|, \tag{5.54}$$

where P_κ (which is ≥ 0) represents the probability of the system to be in the state $|\psi_\kappa\rangle$ (which is not necessarily orthogonal to the other states $|\psi\rangle$). This Hermitian operator has further unitary trace, meaning that $\sum_\kappa P_\kappa = 1$, as we would expect interpreting P_κ as the probability of the system to be found in state $|\psi_\kappa\rangle$.

Thanks to this operator, we have access to the maximum amount of information on the quantum state under investigation. In both cases of a system in a pure or mixed state, it can be demonstrated that the expectation value of every operator can be calculated knowing the density matrix as the trace of the product between the matrices representing the operators as

$$\langle\hat{A}\rangle = \mathrm{Tr}(\hat{\rho}\hat{A}) = \sum_{\psi_\kappa} P_{\psi_\kappa}\langle\psi_\kappa|\hat{A}|\psi_\kappa\rangle. \tag{5.55}$$

22 A more compact notation uses the abbreviation $|\psi_A\rangle \otimes |\psi_B\rangle = |\psi_A\rangle\,|\psi_B\rangle = |\psi_A\psi_B\rangle$.

23 When the system is found in a pure state, this means that the quantum state of the system can be expressed as a linear superposition of the basis vector. Nevertheless, quantum systems are sometimes not simply describable as pure states, being statistical ensembles of pure states or due to their interaction with other systems, and they have to be represented as mixed states [108].

If we consider the operator $\hat{A}$ as a physical observable, from the equation above we can see that the ensemble average of its expectation values on the pure states $|\psi_\kappa\rangle$ provides the expectation value $\langle\hat{A}\rangle$ of the operator $\hat{A}$. For further information on this demonstration and evaluation of the system in a pure state, we can refer to [108], Chapter 1.

Let us now discuss the density matrix in the particular example of a two-photon polarization state, which is very interesting for various case studies. Indeed, in Section 7.3.2, it is described how the biexciton-exciton-ground state cascade in a semiconductor quantum dot can be employed for generating polarization entangled photon pairs. This density matrix will then be Hermitian with only nonnegative eigenvalues. The form of this matrix is very different in case of uncorrelated, classical, or quantum correlations. Considering here the photon polarization and taking into account two photons that can have either horizontal or vertical linear polarization, when no polarization is preferred, i. e., uncorrelated, the density matrix would have the form

$$\hat{\rho} = \frac{1}{4}(|HH\rangle\langle HH| + |HV\rangle\langle HV| + |VH\rangle\langle VH| + |VV\rangle\langle VV|). \tag{5.56}$$

For clarity, the matrix representation of the density operator $\hat{\rho}$ is given as $\hat{\rho}_{nm} = \langle n|\hat{\rho}|m\rangle$ with $|n\rangle$ being the set of orthonormal basis vectors. In the present case the simplest basis set becomes

$$|\phi_1\rangle \equiv |HH\rangle, \quad |\phi_2\rangle \equiv |HV\rangle, \quad |\phi_3\rangle \equiv |VH\rangle, \quad |\phi_4\rangle \equiv |VV\rangle. \tag{5.57}$$

Therefore the density matrix would take the form

$$\hat{\rho} = \begin{array}{c} \\ |HH\rangle \\ |HV\rangle \\ |VH\rangle \\ |VV\rangle \end{array} \begin{array}{c} \begin{array}{cccc} \langle HH| & \langle HV| & \langle VH| & \langle VV| \end{array} \\ \begin{pmatrix} a_{11} & a_{12} & a_{13} & a_{14} \\ a_{12}^* & a_{22} & a_{23} & a_{24} \\ a_{13}^* & a_{23}^* & a_{33} & a_{34} \\ a_{14}^* & a_{24}^* & a_{34}^* & a_{44} \end{pmatrix} \end{array} \tag{5.58}$$

with real diagonal elements. In the case of an *uncorrelated state* as described in Eq. (5.56), matrix (5.58) takes the form of Fig. 5.25(a) with only diagonal elements different from zero, each with amplitude 1/4 ($a_{11} = a_{22} = a_{33} = a_{44} = 1/4$). The behavior of the density matrix changes drastically in case of *classically correlated polarization states*. In this case, two photons are considered having the same polarization, i. e., both horizontal or vertical, with the overall two-photon polarization state being in a statistical mixture of $|\Psi\rangle_1 = |HH\rangle$ and $|\Psi\rangle_2 = |VV\rangle$, so that the density operator becomes

$$\hat{\rho} = \frac{1}{2}(|\Psi\rangle_1\langle\Psi|_1 + |\Psi\rangle_2\langle\Psi|_2) = \frac{1}{2}(|HH\rangle\langle HH| + |VV\rangle\langle VV|). \tag{5.59}$$

This results in the density matrix shown in Fig. 5.25(b), where only two diagonal elements ($a_{11} = a_{44} = 1/2$) are present with an amplitude of 1/2. Even being different from

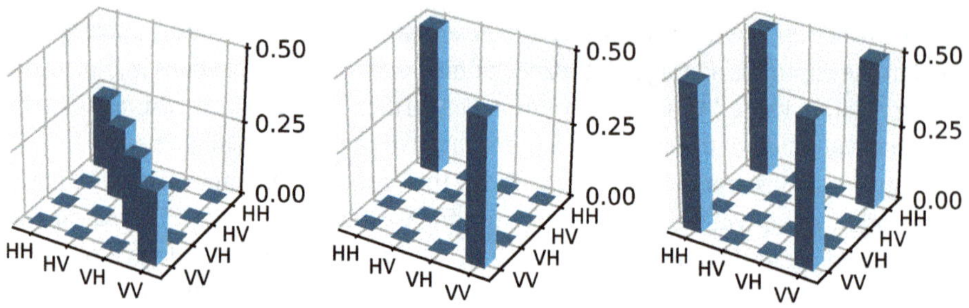

Figure 5.25: Density matrix for (a) uncorrelated, (b) classically-correlated, and (c) quantum-correlated states (in polarization).

the case of uncorrelated two photon states, this case still does not represent the expectation of polarization-entangled photon pairs. In the case of an entangled state, $|\Phi^+\rangle$ (named Bell state) of the form

$$|\Phi^+\rangle = \frac{1}{\sqrt{2}}(|HH\rangle + |VV\rangle), \tag{5.60}$$

the density matrix becomes

$$\hat{\rho} = \frac{1}{2}(|HH\rangle + |VV\rangle)(\langle HH| + \langle VV|). \tag{5.61}$$

Having available the Bell state in Eq. (5.60), we can still obtain any other Bell state (described with R and L polarization), simply employing birefringent elements (see the following discussion). For this reason, polarization-entangled photon pairs are of interest from an experimental point of view, since the photon polarization can be relatively easily experimentally manipulated. By comparing (b) and (c) in Fig. 5.25 we can observe that correlations coming from entanglement are more than those observed for a classical polarization correlation: although the diagonal elements $|HH\rangle\langle HH|$ and $|VV\rangle\langle VV|$ (i. e., the so-called populations) are present in both cases, only for entangled two-photon polarization states, the outer diagonal elements $|HH\rangle\langle VV|$ and $|VV\rangle\langle HH|$ (also called coherences) are present, and they are fully inherent to the quantum nature of entanglement ($a_{11} = a_{44} = a_{14} = a_{14}^* = 1/2$). These latter elements provide a description of the phase relation between different bases and differentiate coherent superposition over statistical mixture of states.

Let us now consider the case of a semiconductor quantum dot emitting photon pairs via the biexciton-exciton-ground cascade. We will show in Section 7.3.2 that when it is not possible to distinguish between the paths $|XX\rangle \rightarrow |{+1}\rangle \rightarrow |0\rangle$ and $|XX\rangle \rightarrow |{-1}\rangle \rightarrow |0\rangle$, the state is described as[24]

24 Subscripts X and XX refer to the exciton and biexciton photons being in a certain polarization state: H (V) as linear horizontal (vertical) and L (R) as linear-polarized (circular-polarized).

$$|\Psi^{+}\rangle = \frac{1}{\sqrt{2}}(|L_{XX}R_{X}\rangle + |R_{XX}L_{X}\rangle) = \frac{1}{\sqrt{2}}(|H_{XX}H_{X}\rangle + |V_{XX}V_{X}\rangle), \tag{5.62}$$

and therefore we would expect a density matrix as in Fig. 5.25(c). This is not necessarily true in the presence of a fine-structure splitting larger than the homogeneously broadened exciton line, which would allow for distinguishing between the two cascaded paths reported above. In case of a nonvanishing FSS, an additional phase term appears in Eq. (5.62) (see Eq. (7.26)). As described in Section 7.3.2, entangled photon pairs are still generated by the emitter, but the state is now dependent on the emission time delay τ between the two photons. As discussed in the following, in quantum state tomography, time-integrated measurements will be performed integrating over all possible times τ, resulting in the degradation of the entanglement in case of nonzero FSS. Alternatively, using fast detectors could allow for resolving the beating originating from the FSS (see Sec. 7.3.2) and therefore quantifying the entanglement without degradation. Nevertheless, a vanishing fine-structure splitting would allow the quantum dot to generate polarization-entangled photon pairs beyond classical correlation.

It is worth mentioning that several sample-related processes (e. g., spin flips between bright exciton states, phase shifts induced by a certain fine structure splitting) and setup imperfections can impact the entanglement. These can lead to a deviation of the experimentally reconstructed density matrix from the ideal case as depicted in Fig. 5.25(c). For example, the inner diagonal elements may not be zero because of spin scattering processes, which result in a flip of the exciton spin during the evolution of the exciton–photon state or uncorrelated background emission. Additionally, unwanted birefringence in the elements constituting the experimental setup may even produce outer off-diagonal elements in the imaginary part of the reconstructed density matrix. A realistic density matrix is exemplary shown in Fig. 5.26 (see [65] for more detail).

The quantification of the entanglement or, in other words, the amount of entangled photon pairs present in the measured density matrix can be performed employing the commonly named *entanglement measures*. They map the density matrix to a monotonic

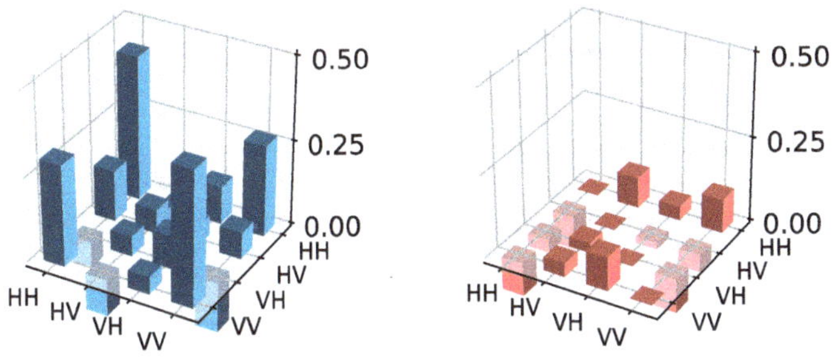

Figure 5.26: Exemplary real (left) and imaginary (right) parts of the density matrix as observed in actual experiments (see [65]), showing nonzero inner and off-diagonal elements. It is worth mentioning that in some cases, the axes of the density matrix are differently ordered, so the position of populations and coherences is not necessary in the same place as here displayed, but in other corners.

function within [0, 1]. With these measures, uncorrelated or classically correlated photons would result in 0, whereas 1 would represent maximally entangled states (given by a specified pure state). The first of these measures is the *concurrence* C [13, 27, 64, 79] defined as

$$C(\hat{\rho}) = \max\{\sqrt{r_1} - \sqrt{r_2} - \sqrt{r_3} - \sqrt{r_4}, 0\}, \tag{5.63}$$

where r_i are the eigenvalues of the operator

$$\hat{R} = \hat{\rho}(\hat{\sigma}_y \otimes \hat{\sigma}_y)\hat{\rho}^*(\hat{\sigma}_y \otimes \hat{\sigma}_y) \quad \text{with } \hat{\sigma}_y = \begin{pmatrix} 0 & -i \\ i & 0 \end{pmatrix}. \tag{5.64}$$

Since the concurrence takes values between 0 and 1, the eigenvalues of the above operator are real positive numbers ordered from the largest to smallest. Alternatively, we can use the *tangle* T [64, 225] defined as

$$T(\hat{\rho}) = C^2(\hat{\rho}). \tag{5.65}$$

Another entanglement measure often used is the *negativity* N [64, 83, 148, 208] given by

$$N = \frac{\|\hat{\rho}^{T_A}\| - 1}{2}, \tag{5.66}$$

having defined the trace norm $\|\hat{\rho}^{T_A}\|$ of $\hat{\rho}^{T_A}$, which is the partial transpose of $\hat{\rho}$ over a subsystem A.

In recent years, to provide a value easily comparable amongst different groups, the characterization of the entanglement has been given by reconstructing the density matrix and providing then the concurrence, tangle, or negativity as the final measured result. It is worth mentioning that the density matrix can also be characterized employing the *fidelity*, which quantifies how the measured state resembles the expected one (with values of 0.5 for classical correlation and up to 1 for fully entangled). Let us assume that the emitted signal can be described by the pure state $|\Psi_s\rangle$ and, after transmission, the state would be another pure one $|\Psi_r\rangle$. The fidelity is then defined as

$$f(|\Psi_s\rangle\langle\Psi_s|, |\Psi_r\rangle\langle\Psi_r|) = |\langle\Psi_s|\Psi_r\rangle|^2. \tag{5.67}$$

In the particular case that the expected state is the aforementioned Bell state $|\Phi^+\rangle$ and the measured state is described by the density matrix $\hat{\rho}$, the fidelity can be defined as

$$f(|\Phi^+\rangle\langle\Phi^+|, \hat{\rho}) = \langle\Phi^+|\hat{\rho}|\Phi^+\rangle. \tag{5.68}$$

In the specific case of a quantum dot with a nonvanishing fine-structure splitting, the last equation would provide an effective fidelity, where the Bell state would include a phase term proportional to the FSS.

The experimental reconstruction of the density matrix and the estimation of the fidelity are performed employing cross-correlations as measurement technique (as also mathematically demonstrated in [3]). In the setup shown in Fig. 5.27(a), each of the two photons is independently analyzed in polarization: in this experimental configuration, two sets of quarter-waveplates, half-waveplates, and polarizers set the projection measurements in any selected linear or circular polarization basis required to reconstruct the density matrix. After this selection, the photons are sent to two detectors, and their signal is time correlated. In this way, polarization-dependent cross-correlation functions

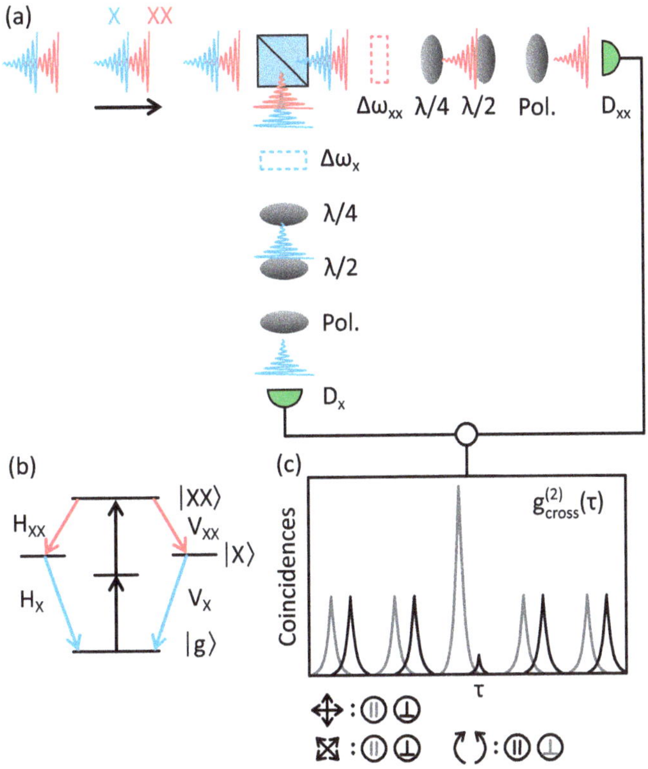

Figure 5.27: (a) Exemplary cross-correlation setup employed for quantum state tomography and fidelity evaluation. The two photons, originating from the biexciton-exciton-ground cascade (b), are sent one to each analysis arm, so that their polarization can be independently manipulated. In the sketch, after the 50:50 beamsplitter, two filters ($\Delta\omega_X$ and $\Delta\omega_{XX}$) ensure that only exciton photons or biexciton photons reach the respective detector. Alternatively, the beamsplitter can be replaced by wavelength selective elements (as volume Bragg gratings or transmission gratings, i. e., polarization-insensitive elements) to reflect and transmit the respective photons. A quarter-waveplate, half-waveplate, and polarizer project the photons into the intended basis. (b) Exemplary biexciton-exciton-ground state cascade, where the biexciton level is reached via two-photon excitation (see Section 10.2.3.4). (c) Exemplary cross-correlation coincidences for linear, diagonal, and circular bases. Parallel and perpendicular measurements are artificially shifted in time for clarity. For more information, see [64] and [132].

$g_{ij}^{(2)}(\tau)$ can be directly measured. At this stage, the emission of polarization-entangled photon pairs from a semiconductor quantum dot is exemplarily discussed (via the biexciton-exciton-ground cascade as in Fig. 5.27(b)).

Measuring experimentally the fidelity can be performed by quantifying the degree of polarization between the two photons in three bases: linear, diagonal, and circular. This requires six cross-correlation measurements between the photons as schematically shown in Fig. 5.27(c). The six cross-correlation measurements are shown in Fig. 5.28. The linear basis events would be present in both classically correlated or entangled cases, but the correlations observed for a basis change, as for the circular ones, would prove the presence of entanglement. The fidelity can then be extracted from measurements as in Fig. 5.28 using [85]

$$C_v = \frac{g_{i,j}^{(2)}(0) - g_{i,j'}^{(2)}(0)}{g_{i,j}^{(2)}(0) + g_{i,j'}^{(2)}(0)} \tag{5.69}$$

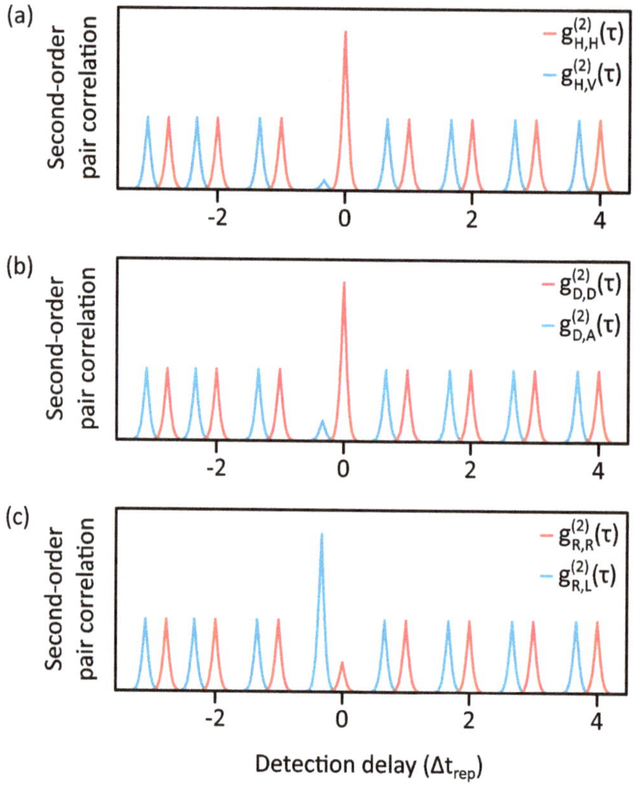

Figure 5.28: Sketch of the second-order cross correlations in the six different bases (linear, diagonal, and circular). Coincidences at nonzero time values provide the Poissonian level (detection events of uncorrelated photons coming from different pairs; in the case of a QD, different cascades). Parallel and perpendicular measurements are artificially shifted in time for clarity. For more detail, we refer to [132] and [64].

with subscript v indicating the basis (linear, diagonal, or circular), whereas i, j and i, j' indicate parallel $\{(H,H),(D,D),(R,R)\}$ and crossed $\{(H,V),(D,A),(R,L)\}$ projections, respectively. Integrating the counts around zero time delay for the different measurements, we can obtain C_{lin}, C_{dia}, and C_{cir}, deriving the fidelity to the predicted Bell state $|\Phi^+\rangle$ as

$$f_{\Phi^+} = \frac{1 + C_{\text{lin}} + C_{\text{dia}} - C_{\text{cir}}}{4}. \tag{5.70}$$

If $f_{\Phi^+} > 0.5$, then the classical limit is exceeded. Alternatively, using the same six measurements, we can estimate the *Bell parameter*. Following [26, 125], the relevant three parameters become

$$S_B^{lc} = \sqrt{2}(C_{\text{lin}} - C_{\text{cir}}), \tag{5.71}$$
$$S_B^{\text{dc}} = \sqrt{2}(C_{\text{dia}} - C_{\text{cir}}), \tag{5.72}$$
$$S_B^{ld} = \sqrt{2}(C_{\text{lin}} + C_{\text{dia}}). \tag{5.73}$$

The case $S_B \leq 2$ would imply violation of the Bell inequality[25] and therefore the presence of quantum correlations. The estimation of the fidelity only requires six correlations over the 16 necessary for reconstructing the density matrix (see the following discussion), but this comes at the expense that it provides only an estimate of the state. Additionally, it is important to remark that unwanted birefringence in the setup may reduce the fidelity, whereas this has no effects on measures like concurrence. For this reason, the density matrix is typically reconstructed to provide a complete knowledge on the state and to derive the described measures (concurrence, negativity, tangle, etc.).

In the most general case the reconstruction of the density matrix can be performed via the quantum state tomography. In the case of a two-photon polarization state, the density matrix depends on 16 parameters. The measurement bases here are as follows:
- horizontal $|H\rangle$ and vertical $|V\rangle$,
- diagonal $|D\rangle = (|H\rangle + |V\rangle)/\sqrt{2}$ and antidiagonal $|A\rangle = (|H\rangle - |V\rangle)/\sqrt{2}$,
- right-circular $|R\rangle = (|H\rangle - i|V\rangle)/\sqrt{2}$ and left-circular $|L\rangle = (|H\rangle + i|V\rangle)/\sqrt{2}$,

following the definition on the Poincaré sphere (see Fig. 5.29). The 16 measurements sets can be performed projecting the first and second photons, respectively, onto [87, 193]

$$\begin{aligned} &\{|H\rangle, |V\rangle, |R\rangle, |D\rangle\}, \\ &\{|H\rangle, |V\rangle, |L\rangle, |R\rangle, |D\rangle\}, \end{aligned} \tag{5.74}$$

25 The Bell inequality [11] in the Clauser–Horne–Shimony–Holt (CHSH) form is written as $|S_B| \leq 2$ with $S_B = C_{\alpha,\beta} - C_{\alpha,\beta'} + C_{\alpha',\beta} + C_{\alpha',\beta'}$, where the different correlations are expressed in the basis configuration $\{\alpha, \alpha', \beta, \beta'\}$ [125, 132].

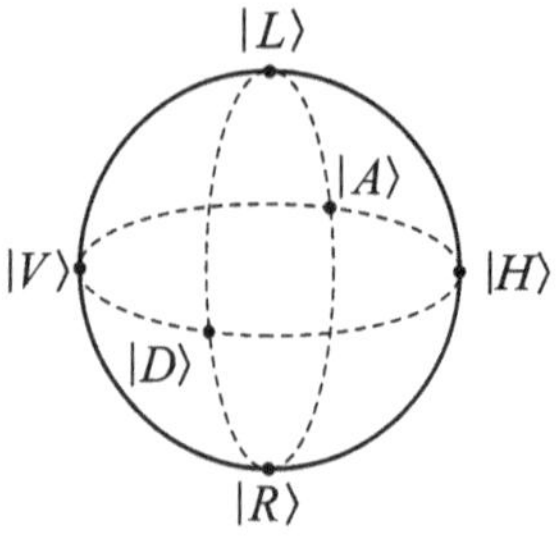

Figure 5.29: Poincaré sphere for polarization qubits.

for example, having

$$\{|HH\rangle, |HV\rangle, |VV\rangle, |VH\rangle, |RH\rangle, |RV\rangle, |DV\rangle, |DH\rangle, \tag{5.75}$$

$$|DR\rangle, |DD\rangle, |RD\rangle, |HD\rangle, |VD\rangle, |VL\rangle, |HL\rangle, |RL\rangle\}, \tag{5.76}$$

as summarized in Table 5.1 (other sets of measurements can also be found, depending on the experimental situation, as discussed in [87]).

Table 5.1: Exemplary complete set of the tomographic state analysis.

Meas. number	1	2	3	4	5	6	7	8	9	10	11	12	13	14	15	16
Photon 1	$\|H\rangle$	$\|H\rangle$	$\|V\rangle$	$\|V\rangle$	$\|R\rangle$	$\|R\rangle$	$\|D\rangle$	$\|D\rangle$	$\|D\rangle$	$\|D\rangle$	$\|R\rangle$	$\|H\rangle$	$\|V\rangle$	$\|V\rangle$	$\|H\rangle$	$\|R\rangle$
Photon 2	$\|H\rangle$	$\|V\rangle$	$\|V\rangle$	$\|H\rangle$	$\|H\rangle$	$\|V\rangle$	$\|V\rangle$	$\|H\rangle$	$\|R\rangle$	$\|D\rangle$	$\|D\rangle$	$\|D\rangle$	$\|D\rangle$	$\|L\rangle$	$\|L\rangle$	$\|L\rangle$

Extracting the coincidences from the $g_{ij}^{(2)}$ with (i,j) as the measurement basis discussed above and employing an algorithm minimizing the likelihood function allows for reconstructing the density matrix [3, 87].

Polarization-entangled photons represent a very powerful tool in several aspects of fundamental quantum physics and in the implementation of quantum technologies. Few exemplary topics covering both fundamental and applied research aspects are quantum communication, where quantum repeaters based on teleportation and entanglement swapping are currently investigated, and quantum information, where entanglement can find place in quantum cryptography protocols as the BBM92 [12]. A strong advantage is that it is easy to manipulate the polarization of light. Several experiments have successfully attempted to share entangled photon pairs over long distances, even via satellites [203, 231], demonstrating the interest of using these quantum states of light for studying the foundation of physics and for technological implementations. Particularly intriguing for long distance communication is the use of optical fibers. Silica fibers represent, so far, the backbone of the global communication infrastructures, being a known and reliable technology. Therefore it seems natural to assume that photonic quantum implementations, like quantum communication and information over long distances, will require the use of these silica fibers. Apart from the choice of a suitable photon wavelength, normally within the telecom O- (centered around 1310 nm, absolute minimum of photon wavepacket dispersion) or C-band (centered around 1550 nm, absolute minimum of absorption), it is necessary to consider the impact of polarization mode dispersion due to random birefringence in the fiber.[26] Indeed, even single-mode fibers sustain the propagation of two distinct polarization modes. The appearance of birefringence within the fiber length (due to local stress, bends, and twist even on microscale, slightly elliptical cross-section) results in polarization mode dispersion: in the time domain the light would experience a different delay as a function of the polarization (in the frequency domain, a change of the output frequency for a fixed input polarization). Even in nonpolarization maintaining single-mode fibers principal states of polarization exist, so that the output polarization does not change if the light is sent in these states (at least at the first order) [57]. Nevertheless, when fibers are deployed out of the lab, covering long distances, local environmental changes will influence the birefringence impacting the polarization state. In the case of polarization-entangled photons, this can have a detrimental impact on the entanglement, even bringing it below the classical limits [4]. To avoid entanglement degradation, measurements performed on long fiber networks employing polarization entangled photons included stabilization mechanisms compensating for eventual unwanted changes in the birefringence [228].

Alternatively, when coming to distributing entanglement over long distances via optical fibers, the use of time-bin instead of polarization entanglement has been proposed [126]. This other approach is not affected by any random change of birefringence, but it comes at the expense of more complicated setups employing phase-stable interferometers. Time-bin entanglement relates on the quantum correlation in the creation time of the photons. The time-bins are, for example, two quasi-orthogonal temporal wavepackets, as shown by the two pulses in Fig. 5.30. If an interferometer is used to generate a coherent superposition of an early pulse $|E\rangle$ and a late pulse $|L\rangle$ as

$$|\Psi\rangle_{\text{pulse}} = \frac{1}{\sqrt{2}}\left(|E\rangle + e^{i\phi_P}|L\rangle\right) \tag{5.77}$$

with variable phase ϕ_P, then this can be used to excite the system for the generation of time-bin entangled photon pairs (see Fig. 5.30(a)). For this scope, it is important that the quantum system emits a single photon pair in each time-bin with 50 % probability.

26 It is important to mention that light in telecommunication bands is not only suitable for fiber-based experiments, but also for satellite-based implementations: exemplarily, operating at 1550 nm allowed for performing experiments in broad daylight (benefiting from high atmospheric transmission, low background light, and Rayleigh scattering, which scales as $1/\lambda^4$) [112].

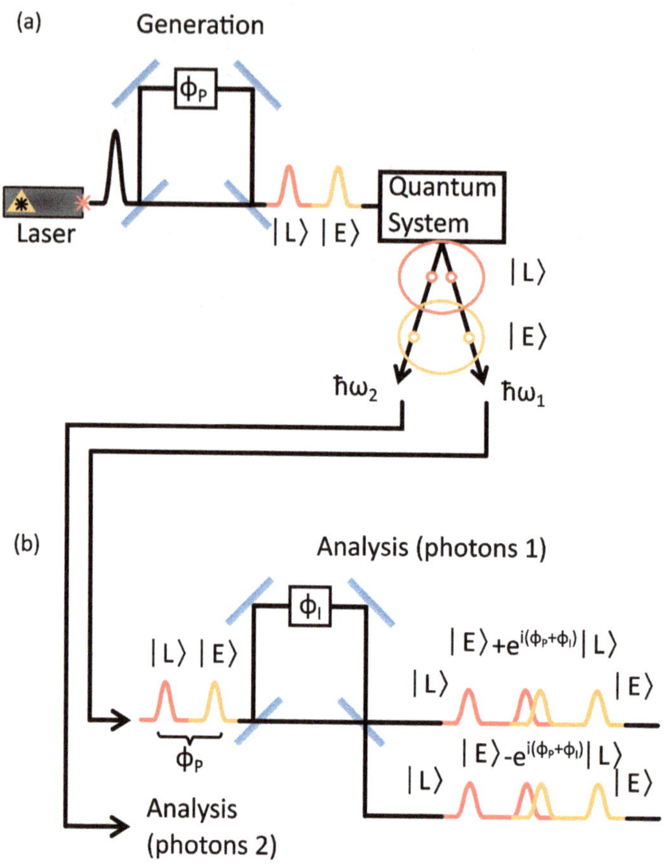

Figure 5.30: (a) Sketch of the generation of time-bin entanglement exciting the quantum system with a coherent superposition of two pulses (early and late). The delay set by the interferometer between early and late time bins has to be longer than the coherence time of the photons into play and the temporal width of the laser pulse. The phase between the two excitation pulses, ϕ_P, generated with the first interferometer, is transferred to the quantum system. The quantum system generates photon pairs (two photons, labeled 1 and 2, with energies $\hbar\omega_1$ and $\hbar\omega_2$), having equal probability of generating the pair in either the first or the second excitation pulse. Setup from [132]. (b) Single qubit interferometer with respective phase ϕ_I and equal time delay as the generation interferometer. A second analysis setup is used in the following for photon 2 as well. At the output of each interferometer, three sets of temporally separated peaks can be observed: the first (third) one is due to the early (late) photon taking the short (long) path. The central peak is due to interference between the early photon following the long path and the late photon following the short path (horizontally shifted for clarity).

If the phase is well defined and transferred to the quantum system, then the time-bin-entangled state can be written as

$$|\Phi\rangle_{\text{time-bin}} = \frac{1}{\sqrt{2}}(|E\rangle_1|E\rangle_2 + e^{i\phi_P}|L\rangle_1|L\rangle_2), \tag{5.78}$$

having defined two photons with subscripts 1 and 2 with distinct energies $\hbar\omega_1$ and $\hbar\omega_2$. As we can see by the previous equations, the phase ϕ_P defined in the interferometer generating the two excitation pulses has to be stable (or at least it has to be stable with respect to the phase in the interferometer used for the analysis ϕ_I). The projection measurements required for the reconstruction of the density matrix can be performed using a 1-bit delayed interferometer (see Fig. 5.30(b)) [193]. With strong similarities with respect to the polarization entangled case, for time-bin entanglement, the basis set can be provided by the vectors

$$\begin{aligned} &|E\rangle, \\ &|L\rangle, \\ &|Z^{\pm}\rangle = \frac{1}{\sqrt{2}}(|E\rangle \pm |L\rangle), \\ &|Y^{\pm}\rangle = \frac{1}{\sqrt{2}}(|E\rangle \pm i|L\rangle), \end{aligned} \tag{5.79}$$

where their position on the Poincaré sphere is similar to the polarization case (compare Fig. 5.29 with Fig. 5.31).

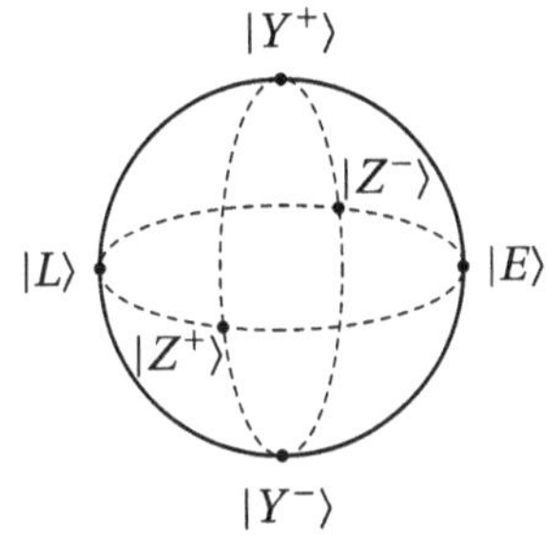

Figure 5.31: Poincaré sphere for time-bin qubits.

As we can see, there is a strong similarity between the basis set for polarization entanglement and time-bin entanglement with $|E\rangle \leftrightarrow |H\rangle$, $|L\rangle \leftrightarrow |V\rangle$, $|Z^+\rangle \leftrightarrow |D\rangle$, $|Z^-\rangle \leftrightarrow |A\rangle$, $|Y^+\rangle \leftrightarrow |L\rangle$, and $|Y^-\rangle \leftrightarrow |R\rangle$.[27] Therefore, similarly to Eq. (5.74), in the reconstruction of the quantum state tomography for time-bin-entangled photons, the projection measurements are $\{|E\rangle, |L\rangle, |Y^+\rangle, |Y^-\rangle, |Z^+\rangle\}$. Injecting the time-bin qubit via the 1-bit delayed interferometer of Fig. 5.30(b) would result in three distinct time slots at each beamsplitter output, which can be distinguished by measuring the arrival time with a single-photon detector. The first and third time slots, marked in the figure as $|E\rangle$ and $|L\rangle$, are referred to as *time basis*, whereas the second time slot is called the *energy*

27 Be careful not to confuse the left circular L in polarization entanglement and the late time bin L in the current description.

basis. These correspond to nonorthogonal measurements bases: the time basis provides the projection on $|E\rangle$ or $|L\rangle$, whereas the energy basis provides projections in the superposition states $|Z^{\pm}\rangle$ and $|Y^{\pm}\rangle$. Indeed, whereas the detection in the first (third) time slot corresponds to the generation of a photon in the early (late) pulse, the second time slot is provided by the superposition of the case with a photon generated via the early pulse, which took the long interferometer arm, and a photon generated by the late pulse, which took the short interferometer arm. This provides the projection along the state

$$|E\rangle + e^{i\phi_I}|L\rangle \tag{5.80}$$

or $|E\rangle - e^{i\phi_I}|L\rangle$ in the second beamsplitter output, accounting for the additional phase acquired by the reflection. Controlling the phase ϕ_I of the interferometer allows for spanning the states in the meridian of the Poincaré sphere with the particular cases as $\phi_I + \phi_P = 0$, resulting in the projection along $|Z^{\pm}\rangle$ basis (in the two BS outputs), and $\phi_I+\phi_P = \pi/2$, projecting onto $|Y^{\pm}\rangle$ basis (here, together with the interferometer phase ϕ_I, we also consider the phase from the first interferometer ϕ_P used to generate the superposition of the early and late excitation pulses; since the first generation interferometer can be taken as reference time difference, the phase ϕ_P can be taken as zero [119]). Now that the single qubit projection is clear, it can be employed for the reconstruction of the full density matrix in a similar fashion as in the polarization quantum state tomography. So the coherent superposition of early and late pulses of Eq. (5.77) can be used to generate the time-bin-entangled state of Eq. (5.78) utilizing single qubit projections on photons 1 and 2 for the quantum state tomography (see Fig. 5.30). This would need the control in the two 1-bit delayed interferometers of the two phases ϕ_{I_1} and ϕ_{I_2} to project over 16 different bases (see Fig. 5.30, where one analysis interferometer is sketched). The reconstruction of the full density matrix can then be performed via two single qubit projections, i. e., one per each photon, and the peculiarity of the time-bin analysis explained before can be taken into account. Indeed, it has been explained that time and energy bases can be distinguished by the arrival time of the photons on the detector. Therefore, for each configuration of the two interferometers, i. e., for each choice of ϕ_{I_1} and ϕ_{I_2}, four cases will always contain the classical correlations in the time basis (for example, $|E\rangle, |E\rangle$). The other four will provide the correlation between time and energy basis and the last one the correlation between energy bases, for a total of nine different basis combinations per each configuration of the two interferometers. This means that the 16 required measurements are [193]

$$\{ |EE\rangle, |EL\rangle, |LE\rangle, |LL\rangle, |LZ^+\rangle, |EZ^+\rangle, |Z^+Z^+\rangle, |Y^+Z^+\rangle, \tag{5.81}$$

$$|Y^+E\rangle, |Y^+L\rangle, |Y^+Y^+\rangle, |EY^+\rangle, |LY^+\rangle, |Z^+Y^+\rangle, |Z^+E\rangle, |Z^+L\rangle\}, \tag{5.82}$$

summarized in Table 5.2.

They can be obtained using only the following four combinations of the two phases ϕ_{I_1} and ϕ_{I_2} of the respective interferometers [132]:

Table 5.2: Exemplary complete set of the tomographic state analysis.

Meas. number	**1**	**2**	**3**	**4**	**5**	**6**	**7**	**8**	**9**	**10**	**11**	**12**	**13**	**14**	**15**	**16**
Phot. 1	$\lvert E\rangle$	$\lvert E\rangle$	$\lvert L\rangle$	$\lvert L\rangle$	$\lvert L\rangle$	$\lvert E\rangle$	$\lvert Z^+\rangle$	$\lvert Y^+\rangle$	$\lvert Y^+\rangle$	$\lvert Y^+\rangle$	$\lvert Y^+\rangle$	$\lvert E\rangle$	$\lvert L\rangle$	$\lvert Z^+\rangle$	$\lvert Z^+\rangle$	$\lvert Z^+\rangle$
Phot. 2	$\lvert E\rangle$	$\lvert L\rangle$	$\lvert E\rangle$	$\lvert L\rangle$	$\lvert Z^+\rangle$	$\lvert Z^+\rangle$	$\lvert Z^+\rangle$	$\lvert Z^+\rangle$	$\lvert E\rangle$	$\lvert L\rangle$	$\lvert Y^+\rangle$	$\lvert Y^+\rangle$	$\lvert Y^+\rangle$	$\lvert Y^+\rangle$	$\lvert E\rangle$	$\lvert L\rangle$

$$\{0,0\},\{0,\pi/2\},\{\pi/2,0\},\{\pi/2,\pi/2\}. \tag{5.83}$$

In practical experiments, the triple coincidences between the detectors and the (synchronizing) excitation laser allows for determining the 16 parameters. Similarly to the polarization entanglement case, the quantum state tomography can be performed by reconstructing the density matrix with the same algorithm [87, 193]. It is now clear that the respective phases in the analysis interferometers have to be stable and reliably controllable during the measurements so that the quantum state tomography can be performed. This can be experimentally obtained by actively stabilizing each interferometer, sensibly increasing the experimental complexity. Alternatively, an interesting approach is based on the implementation of the three required interferometers (one excitation and two analysis) into a single setup (see Fig. 5.32). Using trihedral retroreflectors, the excitation laser and two photons follow parallel paths within the same interferometer: in this way, any unwanted phase change will affect all three paths, simultaneously keeping unaltered the relative phase. This experimental configuration, despite very interesting for fundamental physics studies, has an intrinsic limitation: all interferometers are placed in the same location and therefore are not usable for long-distance experiments

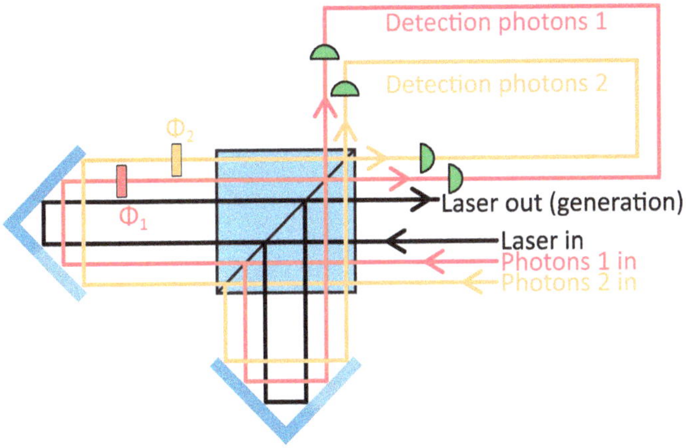

Figure 5.32: Sketch of a setup employing three paths (one excitation and two 1-bit analysis) in the same interferometer thanks to trihedral retroreflectors. The phase for the two photons can be independently adjusted. Unwanted overall phase change would affect all three paths simultaneously avoiding the necessity of highly stabilized interferometers [132].

employing optical fibers (for which interferometers at distant locations are needed). More information on this experimental setup and the postselection process employed for the data analysis can be found in [88, 132]. These experiments have been conducted employing resonant two-photon excitation to populate the biexciton transition in a semiconductor quantum dot. This allows for coherent transfer of the excitation phase to the created biexciton state (see Section 7.3.2). This is crucial for the generation of the time-bin entanglement, but it becomes impossible to ensure the required presence of the photons with 50 % probability to be in the early or late time-bin. Indeed, resonant (and two-photon excitation) is known to be a reliable way to coherently control the population of the excited state (see Chapter 6). Exemplary, in the present case, exciting at the π-pulse would generate a photon pair (from the biexciton-exciton-ground cascade) for each pulse. So far, most of the experiments circumvented this limitation lowering the excitation power, forgoing a fully deterministic photon generation process, nevertheless providing interesting investigations of fundamental properties of time-bin entanglement generated via a semiconductor quantum dot [88]. Alternatively, it has been proposed to employ a long-lived metastable state for the deterministic generation of time-bin-entangled photon pairs using a semiconductor QD [146, 184]. Once this long-lived state $|M\rangle$ has been initialized (i. e., populated), a first excitation pulse tuned on the transition $|M\rangle - |XX\rangle$ is sent (yellow pulse in Fig. 5.33(a)). Keeping the pulse area at $\pi/2$ ensures a 50 % probability of populating the biexciton state. If this happens, then the QD will generate a photon pair, relaxing to the ground state (the early photon pair, Fig. 5.33(b)). The arrival of the second pulse (red in Fig. 5.33(a)) will then have no effect,

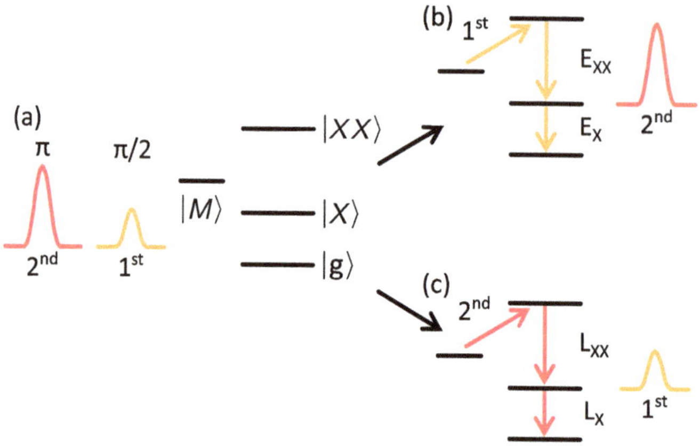

Figure 5.33: (a) Two excitation pulses resonant with the transition $|M\rangle - |XX\rangle$ (labeled as 1st and 2nd, with pulse areas $\pi/2$ and π, respectively) are sent to the quantum dot. The first pulse can either excite the QD to the biexciton state or be simply transmitted. If it excites the QD to the biexciton state, then the second pulse is not interacting with the dot and so transmitted, as sketched in (b). Alternatively, if the first pulse does not excite the QD, then the second excites the biexciton with 100 % probability as in (c). This scheme ensures the required presence of photons in the early or late time-bins with 50 % probability.

since it is detuned from any QD transition. Conversely, if the first pulse would not produce the excitation of the $|XX\rangle$ state, then the second pulse (with area π) will excite the biexciton state from the metastable state with 100 % probability (Fig. 5.33(c)). This would produce the generation of the late photon pair. This approach would allow us to produce *deterministically* time-bin entangled photon pairs (with 50 % probability to be found in the early or late time-bin), therefore being very attractive for practical implementations as well. Nevertheless, finding a suitable metastable state is not trivial: the idea of employing dark excitons for this purpose is very appealing due to their long lifetime (longer than the XX). However, an inherent challenge is their effective initialization, since they lack optical activity. Despite that, several efforts are currently conducted by multiple research groups to circumvent these challenges.

5.3 Summary

- Experimental characterization of the physical properties of semiconductor nanostructures can be obtained with various spectroscopic and time resolved techniques, as *photoluminescence* (and *microphotoluminescence*), *photoluminescence excitation*, high-resolution spectroscopy (as in *Fabry–Perot interferometry*), and *time-resolved spectroscopy*.
- Realistic emitters of quantum light can be categorized in terms of their *brightness*, their capability of emitting *single photons*, and their respective *coherence* and *indistinguishability*.
- Photon *indistinguishability* can be limited by the presence of *dephasing mechanisms*, faster than the emitter's transition lifetime, or by *spectral diffusion* (typically slower than the lifetime).
- Photon *coherence* can be measured via *Michelson interferometry*, alternatively observing the impact of dephasing on the interference of two photons (the *Hong–Ou–Mandel* interference).
- *Spectral diffusion mechanisms*, their impact on the emitter's emission linewidth, and their respective timescales can be experimentally measured via *Photon-correlation Fourier spectroscopy* or *slow-light spectroscopy*. The presence of spectral diffusion is also observed in the degradation of the two-photon interference visibility in Hong–Ou–Mandel experiments.
- *Entangled photons* show a degree of correlation beyond classically correlated photons. Polarization-entangled or time-bin-entangled photons find applications in various implementations of quantum physics and technology. The quantification of the degree of entanglement can be obtained employing *entanglement measures* as *concurrence*, *tangle*, or *negativity*. The *density matrix* can be reconstructed via *quantum state tomography*.
- The degree of entanglement is sometimes also quantified via the *fidelity*, i. e., how much the measured state resembles the expected one. This requires a lower number

of cross-correlation measurements with respect to other measurements of the entanglement (as the concurrence), but it is also prone to reduction due to unwanted setup birefringence.

- Alternative to polarization entanglement, *time-bin* represents a valuable tool for long-distance entanglement distribution, since it is not affected by unwanted birefringence in long fibers deployed outside the lab environment (therefore less controllable). Nonetheless, it requires multiple phase-stable interferometers, which complicate the experimental implementation.

6 Cavity quantum electrodynamics

6.1 Introductory remarks

This section is devoted to the discussion on some of the basic notions of cavity quantum electrodynamics. We will begin with the interaction between a model two-level emitter and an external field. The first case study will be the interaction of this two-level system (TLS) with a classical field. The time evolution of the two levels will be discussed, with particular emphasis on the population inversion. The important notion of the Rabi frequency will be introduced. Then the discussion will be followed by the description of the TLS interacting with a quantized field, i. e., the Jaynes–Cummings model. The population inversion will be shown for the case of a quantized field in a Fock, coherent, or chaotic state. The first glimpse of cavity quantum electrodynamics will be given by the observation of periodic and reversible two-level emitter spontaneous decay for an excited TLS placed into an empty cavity, a behavior which cannot be observed for a TLS driven by a classical field. Finally, the two-level system will be placed into a cavity in the presence of dissipation, from the emitter and the cavity itself, and the dynamics of the two-level system will be described. These topics will constitute the basis for the description of enhanced device concepts discussed in Part III.

6.2 A two-level system and a monochromatic field

Before discussing cavity quantum electrodynamics, let us first begin with the description of the interaction of a two-level system with a classical, monochromatic field. This will be useful for defining the notation employed in the following description. The arguments follow in part the discussion in [108], where more details on the calculations can be found. The two-level system is here described via two states, ground $|g\rangle$ and excited $|e\rangle$, with respective energies (as in Fig. 6.1):

$$E_g = \hbar\omega_g \quad \text{and} \quad E_e = \hbar\omega_e > E_g. \tag{6.1}$$

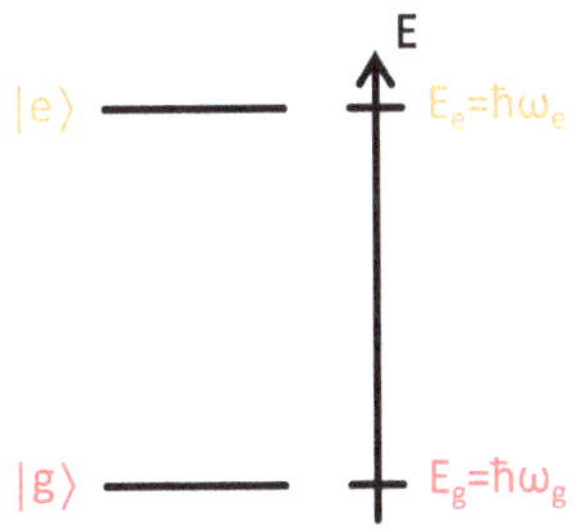

Figure 6.1: Sketch of the states in the two-level system and their respective energies.

https://doi.org/10.1515/9783110703412-006

This means that the frequency of the transition $|g\rangle \leftrightarrow |e\rangle$ is given by $\omega_{eg} = \omega_e - \omega_g$. The system Hamiltonian is composed by two terms:

$$\hat{H} = \hat{H}_{\text{TLS}} + \hat{V}_I(t), \tag{6.2}$$

the TLS and two-level emitter-field interaction terms, respectively. They can be written as

$$\hat{H}_{\text{TLS}} = \hbar\omega_g \,|g\rangle\,\langle g| + \hbar\omega_e \,|e\rangle\,\langle e|\,, \tag{6.3}$$

$$\hat{V}_I(t) = -\hat{d}\cdot\vec{E}(t) = -\hat{d}_{\vec{e}}\,E(t). \tag{6.4}$$

In the last equality, $\hat{d}_{\vec{e}} = \hat{d}\cdot\vec{e}$, the projection of the dipole operator $\hat{d}$ in the direction $\vec{e}$ of the electric field ($\vec{E}(t) = \vec{e}\,E(t)$ at the position of the point dipole). This represents the direction of the electric field polarization with time-dependent amplitude as

$$E(t) = \mathcal{E}e^{-i\omega t} + \mathcal{E}^* e^{i\omega t} = 2|\mathcal{E}|\cos(\omega t - \varphi), \tag{6.5}$$

where $\mathcal{E}$ represents the magnitude, ω is the frequency, and φ is the phase of the field.[1] Furthermore, it is important to remark that the electric field frequency ω is chosen close to the transition frequency of the two-level system, as it will become clear further. As approximated in [108], the field polarization is here chosen linear, but it can be generalized to circular one as well. A more complete description involving the density matrix can be found in the book of Lambropoulos and Petrosyan [108], but a description based on the wavefunction will suffice here. At this point the TLS is described quantum mechanically, whereas the classical monochromatic field is treated as a time-dependent term. The two-level system wavefunction can be written as

$$|\Psi(t)\rangle = c_g(t)\,|g\rangle + c_e(t)\,|e\rangle\,. \tag{6.6}$$

The TLS eigenstates $|g\rangle$ and $|e\rangle$ have in the previous equation the time-dependent (complex) amplitudes c_g and c_e. The time evolution of the system is described by the Schrödinger equation

$$i\hbar\frac{\partial}{\partial t}\,|\Psi(t)\rangle = \hat{H}\,|\Psi(t)\rangle\,. \tag{6.7}$$

At this point the initial conditions $\Psi(0)$ can be set so as the emitter is in the ground state, therefore having $c_g(0) = 1$ and $c_e(0) = 0$. Inserting the wavefunction defined in Eq. (6.6) into the Schrödinger equation (6.7) and using the Hamiltonian terms (6.3) and (6.4), we get

---

1 This definition of the electric field is commonly used in quantum theory. In other works the electric field is defined as $E(t) = \mathcal{E}_0\cos(\omega t) = 1/2\mathcal{E}_0(e^{i\omega t} + e^{-i\omega t})$, e. g. [48]. As a consequence, a factor of 1/2 enters in the interaction term of the Hamiltonian and the subsequent differential equations. However, this does not modify the description of the physical processes.

$$\frac{\partial}{\partial t}c_g = -i\omega_g c_g + ic_e\frac{d_{ge}}{\hbar}(\mathcal{E}\,e^{-i\omega t} + \mathcal{E}^* e^{i\omega t}), \tag{6.8}$$

$$\frac{\partial}{\partial t}c_e = -i\omega_e c_e + ic_g\frac{d_{eg}}{\hbar}(\mathcal{E}\,e^{-i\omega t} + \mathcal{E}^* e^{i\omega t}). \tag{6.9}$$

For this result, we first have to consider the inner product with $|g\rangle$ and then $|e\rangle$, their orthonormality, and the matrix elements of the dipole operator $\hat{d}$, $d_{ij} = \langle i|\,\hat{d}_{\hat{e}}\,|j\rangle$ so that $d_{gg} = d_{ee} = 0$ and $d_{ge} = d^*_{eg}$ (often with real terms). In the interaction picture the coefficients $c_{g,e}$ can be written in terms of slow-varying amplitudes $\tilde{c}_{g,e}$:

$$c_{g,e}(t) = \tilde{c}_{g,e}(t)e^{-i\omega_{g,e}t}. \tag{6.10}$$

This allows writing

$$\frac{\partial}{\partial t}\tilde{c}_g = i\tilde{c}_e\frac{d_{ge}}{\hbar}\left(\mathcal{E}\,e^{-i(\omega+\omega_{eg})t} + \mathcal{E}^* e^{i(\omega-\omega_{eg})t}\right) \simeq i\tilde{c}_e\frac{d_{ge}}{\hbar}\mathcal{E}^* e^{i\Delta t}, \tag{6.11}$$

$$\frac{\partial}{\partial t}\tilde{c}_e = i\tilde{c}_g\frac{d_{eg}}{\hbar}\left(\mathcal{E}\,e^{-i(\omega-\omega_{eg})t} + \mathcal{E}^* e^{i(\omega+\omega_{eg})t}\right) \simeq i\tilde{c}_g\frac{d_{eg}}{\hbar}\mathcal{E}e^{-i\Delta t}. \tag{6.12}$$

The introduced term $\Delta = \omega - \omega_{eg}$ represents the detuning of the electric field frequency ω with respect to one of the TLS transitions ω_{eg}. The last equivalence is allowed considering the most interesting near-resonant terms, $\Delta \ll \omega_{eg} \sim \omega$, whereas the terms oscillating with sum frequencies $\pm(\omega + \omega_{eg})$ are neglected (the rotating wave approximation). Solving the differential equations above allows for finding the slow-varying amplitudes as

$$\tilde{c}_g(t) = e^{i\frac{\Delta}{2}t}\left[\cos(\Omega_{\text{eff}}\,t) - i\frac{\Delta}{2\Omega_{\text{eff}}}\sin(\Omega_{\text{eff}}\,t)\right], \tag{6.13}$$

$$\tilde{c}_e(t) = ie^{-i\frac{\Delta}{2}t - i\varphi}\frac{\Omega}{\Omega_{\text{eff}}}\sin(\Omega_{\text{eff}}\,t), \tag{6.14}$$

where $\Omega_{\text{eff}} = \sqrt{\Omega^2 + (\Delta/2)^2}$ is called the *effective Rabi frequency*, which reduces, for $\Delta = 0$, to the *Rabi frequency*:

$$\Omega = \frac{d_{ge}|\mathcal{E}|}{\hbar}. \tag{6.15}$$

Interestingly, for $t = 0$, we get $\tilde{c}_g(0) = 1$ and $\tilde{c}_e(0) = 0$, which is consistent with the initial conditions of having the emitter in the ground state $|g\rangle$ and the normalization $|\tilde{c}_g(t)|^2 + |\tilde{c}_e(t)|^2 = 1$ for all $t \geq 0$. Particularly, the case of $\Delta = 0$ is interesting, for which Eqs. (6.13) and (6.14) take the form

$$\tilde{c}_g(t) = \cos(\Omega t), \quad \tilde{c}_e(t) = ie^{-i\varphi}\sin(\Omega t). \tag{6.16}$$

Thanks to these equations, we can understand the meaning of the Rabi frequency Ω, which provides the oscillation frequency between the two states of a two-level system

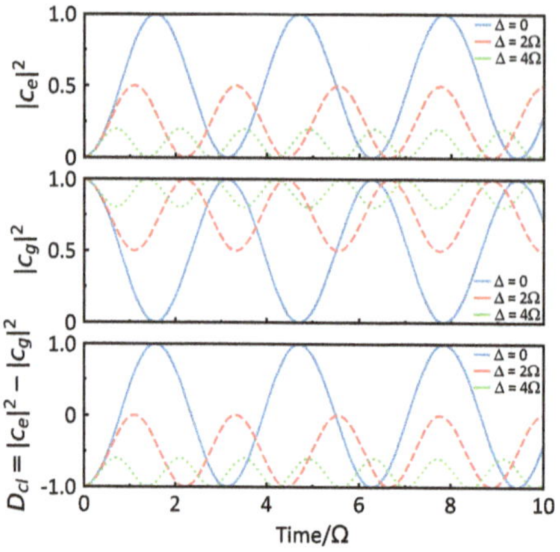

Figure 6.2: Time dependence of excited and ground state populations $|\tilde{c}_e(t)|^2$ and $|\tilde{c}_g(t)|^2$ and population inversion for different values of detuning $\Delta = 0$ (solid blue), $\Delta = 2\Omega$ (dashed red), and $\Delta = 4\Omega$ (dotted green). The figure depicts the exemplary case of a TLS in the ground state at $t = 0$ [108].

under an external driving field. Furthermore, it becomes clear that the driven oscillation of the two-level system is more efficient when the two frequencies are matching, $\omega = \omega_{eg}$, or otherwise said, $\Delta = 0$ (see Fig. 6.2). While the phase difference is $\frac{\pi}{2} - \varphi$ for $\Delta = 0$, for $\Delta \neq 0$, an additional phase term Δt between the oscillations of $\tilde{c}_j(t)$ amplitudes is present.

Conversely, if the initial state of the two-level system is set to be in the upper, excited state $|e\rangle$, then for $\Delta = 0$, the complex slow-varying amplitudes (6.16) take the form

$$\tilde{c}_g(t) = ie^{i\varphi}\sin(\Omega t), \quad \tilde{c}_e(t) = \cos(\Omega t). \tag{6.17}$$

An important quantity is the *population inversion* defined as

$$D_{\mathrm{cl}}(\varepsilon, t) = \left|c_e(t)\right|^2 - \left|c_g(t)\right|^2. \tag{6.18}$$

Making use of Eqs. (6.13) and (6.14), the population inversion becomes

$$D_{\mathrm{cl}}(\mathcal{E}, t) = \frac{\Omega^2 - (\frac{\Delta}{2})^2}{\Omega^2 + (\frac{\Delta}{2})^2}\sin^2(\Omega_{\mathrm{eff}}\, t) - \cos^2(\Omega_{\mathrm{eff}}\, t) \tag{6.19}$$

$$= -\left(\frac{\Omega}{\Omega_{\mathrm{eff}}}\right)^2 \cos(2\Omega_{\mathrm{eff}}\, t) - \left(\frac{\Delta}{2\Omega_{\mathrm{eff}}}\right)^2. \tag{6.20}$$

For perfect resonance ($\Delta = 0$), this becomes

$$D_{\mathrm{cl}}(\mathcal{E}, t) = -\cos(2\Omega t). \tag{6.21}$$

The subscript “cl” stands for classical, as for the treatment of the electric field so far. The field magnitude and the dipole matrix elements d_{ge} (the coupling constant), as they enter in the Rabi frequency $\Omega = d_{ge}|\mathcal{E}|/\hbar$, set how effective are the oscillations between the two levels. Note that in case of experiments where the electric field is constituted by a laser pulse (with finite length), the Rabi frequency will be time-dependent. For this reason, the so-called *pulse area* is often used. It is defined as

$$\theta(z) = \frac{2d_{ge}}{\hbar} \int_{-\infty}^{\infty} \mathcal{E}(z,t)\, dt. \tag{6.22}$$

In the particular case of a rectangular pulse of duration τ, the pulse area would take the form $\theta = 2\Omega\tau$. Otherwise said, the pulse area can be seen as the time-integrated Rabi frequency, meaning that a population oscillation is observed for increasing pulse area (see Fig. 6.3). At the so-called π-pulse, the population is inverted, i. e., it is fully found in the excited state. Population inversion in pulsed operation is very attractive for experimental realizations: this operation condition ensures the triggered excitation of the two-level system, which can eventually be followed by the emission of a photon.[2]

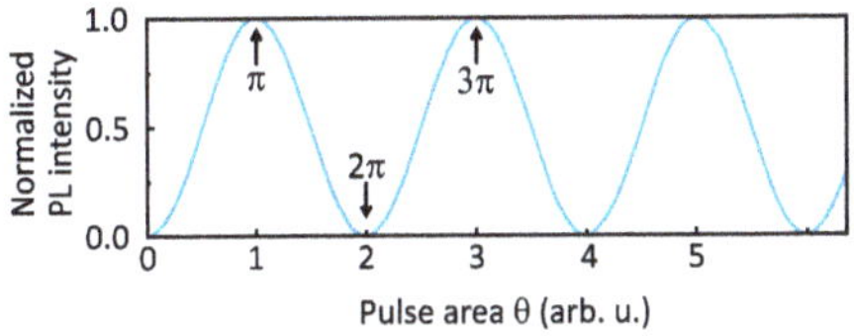

Figure 6.3: Exemplary Rabi oscillations as a function of the pulse area, showing the coherent population control.

6.3 The Jaynes–Cummings model

Differently from the previous section, let us now consider the case of a quantized field interacting with the two-level system. The exemplary case is the field in a cavity mode with frequency ω near to the ground-excited transition frequency ω_{eg}. All other cavity modes are far detuned in frequency and can therefore be neglected. This is called the *Jaynes–Cummings model*, i. e., the coupling between a two-level system and a single-mode electromagnetic field. Here the most important concepts are described; for more detail on the calculations, we again refer to [108]. The total Hamiltonian of the system can be written as the sum of three terms:

$$\hat{H} = \hat{H}_{\mathrm{TLS}} + \hat{H}_F + \hat{V}_I. \tag{6.23}$$

2 When considering the field as described in footnote 1, the oscillation frequency in the amplitudes varies as $\Omega t/2$, hence the pulse area can be defined as $\theta(z) = \frac{d_{ge}}{\hbar}\int_{-\infty}^{\infty} \mathcal{E}(z,t)\, dt$, e. g. [48]. The definition of the π-pulse still indicates the condition of fully inversed population.

The first one can be rewritten from Eq. (6.3), choosing the zero-point energy exactly in between the states $|e\rangle$ and $|g\rangle$: this means that their respective energies will be $E_e = \frac{1}{2}\hbar\omega_{eg}$ and $E_g = -\frac{1}{2}\hbar\omega_{eg}$. The two-level system Hamiltonian becomes

$$\hat{H}_{\text{TLS}} = \frac{1}{2}\hbar\omega_{eg}\,|e\rangle\,\langle e| - \frac{1}{2}\hbar\omega_{eg}\,|g\rangle\,\langle g| = \frac{1}{2}\hbar\omega_{eg}\hat{\sigma}_z, \tag{6.24}$$

where the two-level system operator is defined as $\hat{\sigma}_z = |e\rangle\,\langle e| - |g\rangle\,\langle g|$ (i. e., the Pauli spin operator, since the Hilbert space can be reduced to two dimensions, being a two-level system under investigation). The second term of Eq. (6.23) represents the cavity Hamiltonian defined as

$$\hat{H}_F = \hbar\omega\hat{a}^\dagger\hat{a}, \tag{6.25}$$

having as commonly defined $\hat{a}^\dagger$ and $\hat{a}$ as creation and annihilation operators of the cavity field mode. The third term of Eq. (6.23) can be written in the dipole approximation as

$$\hat{V}_I = -\hat{d}\hat{E}(z_0, t), \tag{6.26}$$

and it represents the interaction between the two-level system and the field. The electric field in the position z_0, where the emitter is located, is defined as (fundamental mode of a cavity with dependence $\sin(kz_0)$):

$$\hat{E}(z_0, t) = \varepsilon_\omega(\hat{a}(t) + \hat{a}^\dagger(t))\sin(kz_0), \tag{6.27}$$

where the field per photon inside the cavity volume V has been defined as $\varepsilon_\omega = \sqrt{\frac{\hbar\omega}{\varepsilon_0 V}}$ (see also Section 2.4). The dipole moment operator can be written in terms of the atomic operators $\hat{\sigma}_+ = |e\rangle\,\langle g|$ and $\hat{\sigma}_- = |g\rangle\,\langle e|$:

$$\hat{d} = |g\rangle\,\langle g|\,d\,|e\rangle\,\langle e| + |e\rangle\,\langle e|\,d\,|g\rangle\,\langle g| = d_{ge}\hat{\sigma}_- + d_{eg}\hat{\sigma}_+. \tag{6.28}$$

The (real) atom-cavity field coupling constant is given by

$$g = -\left(\frac{d_{ge}\varepsilon_\omega}{\hbar}\right)\sin(kz_0), \tag{6.29}$$

which results in

$$\hat{V}_I = \hbar g(\hat{\sigma}_- + \hat{\sigma}_+)(\hat{a} + \hat{a}^\dagger). \tag{6.30}$$

Once again considering the rotating wave approximation, the equation above can be written as

$$\hat{V}_I = \hbar g(\hat{\sigma}_+\hat{a} + \hat{a}^\dagger\hat{\sigma}_-). \tag{6.31}$$

Interestingly, two terms are present: $\hat{\sigma}_+\hat{a}$, which corresponds to the two-level system transition from ground to excited together with the absorption of a photon and its consequent destruction in the electromagnetic mode, and $\hat{a}^\dagger\hat{\sigma}_-$, represents the transition from excited to ground state of the TLS, together with the emission of a photon.[3] With the explicit form of the three terms, the total system Hamiltonian can be written as

$$\hat{H} = \frac{1}{2}\hbar\omega_{eg}\hat{\sigma}_z + \hbar\omega\hat{a}^\dagger\hat{a} + \hbar g(\hat{\sigma}_+\hat{a} + \hat{a}^\dagger\hat{\sigma}_-). \tag{6.32}$$

Differently from the classical description of the field (see Section 6.2), here both the TLS and the cavity field are quantized. This means that the state of the full system should include both TLS and field. Together with the ground and excited states of the TLS, the cavity field is described by the basis of number states $|n\rangle$, $n = 0, 1, 2, 3, \ldots$. If we consider the case of the TLS in the excited state $|e\rangle$ and a field described with $|n\rangle$, then the total state $|e, n\rangle$ is coupled by the interaction term $\hat{V}_I$ of the Hamiltonian to the state $|g, n+1\rangle$. Therefore we can write

$$\langle g, n+1|\,\hat{H}\,|e, n\rangle = \langle e, n|\,\hat{H}\,|g, n+1\rangle = \hbar g\sqrt{n+1}. \tag{6.33}$$

These two states have the respective energies

$$E_{e,n} = \langle e, n|\,\hat{H}\,|e, n\rangle = \hbar\left(\omega n + \frac{1}{2}\omega_{eg}\right), \tag{6.34}$$

$$\begin{aligned} E_{g,n+1} &= \langle g, n+1|\,\hat{H}\,|g, n+1\rangle = \hbar\left[\omega(n+1) - \frac{1}{2}\omega_{eg}\right] \\ &= E_{e,n} + \hbar\Delta. \end{aligned} \tag{6.35}$$

As before, the term corresponding to the detuning between the field and a two-level atom is given by $\Delta = \omega - \omega_{eg}$. The Hamiltonian (6.32) can be divided into a sum of Hamiltonians:

$$\hat{H}_n = E_{e,n}\begin{bmatrix} 1 & 0 \\ 0 & 1 \end{bmatrix} + \hbar\begin{bmatrix} 0 & g\sqrt{n+1} \\ g\sqrt{n+1} & \Delta \end{bmatrix}, \tag{6.36}$$

where the total Hamiltonian is given by $\hat{H} = \sum_n \hat{H}_n$. Each Hamiltonian $\hat{H}_n$ acts on the Hilbert subspace $\mathbb{H}_n = \{|e, n\rangle, |g, n+1\rangle\}$, which is mutually decoupled from the others for $n = 0, 1, 2, 3, \ldots$. The diagonalization of the matrix above gives the eigenvalues

$$\lambda_\pm^{(n)} = \frac{E_{e,n}}{\hbar} + \frac{\Delta}{2} \pm \Omega_{\text{eff}}, \tag{6.37}$$

where $\Omega_{\text{eff}} = \sqrt{g^2(n+1) + (\Delta/2)^2}$. Therefore the eigenstates can also be written as

3 In. [108] a more complete calculation is given. There it is explicitly remarked that two additional terms oscillating at sum frequencies are neglected (instead of near frequency as the ones described) and further do not conserve the energy. They represent the case of atom and field excited or deexcited simultaneously.

$$|\pm_n\rangle = \frac{1}{\sqrt{N_\pm}}\left[\left(\Omega_{\text{eff}} \mp \frac{\Delta}{2}\right)|e,n\rangle \pm g\sqrt{n+1}\,|g,n+1\rangle\right]. \tag{6.38}$$

The last equation has been normalized via the constants $N_\pm = (\Omega_{\text{eff}} \mp \Delta/2)^2 + g^2(n+1)$. It is trivial to find out that for the resonant case $\Delta = 0$, we have

$$\lambda_\pm^{(n)} = \frac{E_{e,n}}{\hbar} \pm g\sqrt{n+1}, \tag{6.39}$$

taking into account the form of Ω_{eff} and $N_\pm$ for $\Delta = 0$, and

$$|\pm_n\rangle = \frac{1}{\sqrt{2}}(|e,n\rangle \pm |g,n+1\rangle). \tag{6.40}$$

The states $|\pm_n\rangle$ are referred to as the *dressed states* of the Hamiltonian $\hat{H}_n$. In these two states, we have the symmetric or antisymmetric superposition of the overall system bare states $|e,n\rangle$ and $|g,n+1\rangle$. The splitting between the two dressed states amounts of $2g\sqrt{n+1}$.[4]

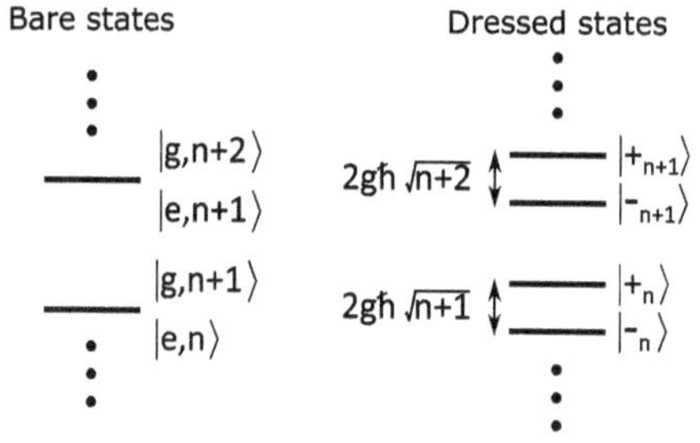

Figure 6.4: Comparison of system bare states (left) and dressed states (right). In both cases, we consider $\Delta = 0$.

As an interesting example, consider the case of the two-level system being in the ground state $|g\rangle$, with the cavity field in th Fock state $|n\rangle$, $n = 1, 2, \ldots$. Following the previous notation, the overall system can be found in the state $|g,n\rangle$. The total system Hamiltonian (6.32) couples the overall system to the state $|e,n-1\rangle$, where the two-level system is then found in the excited state and the cavity field has one photon less:

$$\langle e,n-1|\,\hat{V}_{\text{I}}\,|g,n\rangle = \langle g,n|\,\hat{V}_{\text{I}}\,|e,n-1\rangle = \hbar g\sqrt{n}. \tag{6.41}$$

The state vector for the TLS-cavity field system has the form

$$|\tilde{\Psi}(t)\rangle = \tilde{c}_{g,n}(t)\,|g,n\rangle + \tilde{c}_{e,n-1}(t)\,|e,n-1\rangle \tag{6.42}$$

4 In the text the eigenvalues are normalized by $\hbar$, following the notation used in [108]. In Fig. 6.4, since energies are discussed, the $\hbar$ is explicitly written.

in the interaction picture, and further for the resonant case ($\Delta = 0$), we can find (for more detail on the calculations, see [108]) that the probability amplitudes have the forms

$$\tilde{c}_{g,n}(t) = \cos(g\sqrt{n}t), \quad \tilde{c}_{e,n-1}(t) = -i\sin(g\sqrt{n}t). \tag{6.43}$$

This shows a strong similarity with the case of a two-level system interacting with a classical field (for $\Delta = 0$). In particular, the overall system shows an oscillation between $|g, n\rangle$ and $|e, n-1\rangle$ global states, with the Rabi frequency defined as

$$\Omega = g\sqrt{n} = -\frac{d_{ge}}{\hbar}\varepsilon_\omega\sqrt{n}\sin(kz_0). \tag{6.44}$$

Interestingly, for times when both amplitudes are $\tilde{c}_{g,n}(t) \neq 0$ and $\tilde{c}_{e,n-1}(t) \neq 0$, the state vector (6.42) is nonseparable (between atom and field components), and therefore the system is in an entangled state. The atomic population inversion takes the form

$$D_{\text{Fock}}(n,t) = -\cos(2g\sqrt{n}t), \tag{6.45}$$

the same as in Eq. (6.21). The strong similarity with the classical field case ends if we consider the two-level system to be in the initial state $|e\rangle$. Now the field is still in the Fock state $|n\rangle$, but now the case of zero photon in the cavity field is also possible, i. e., $n = 0, 1, 2, \ldots$. As before,

$$\langle g, n+1|\,\hat{V}_{\mathrm{I}}\,|e, n\rangle = \hbar g\sqrt{n+1} \tag{6.46}$$

provides the coupling between the two states $|e, n\rangle$ and $|g, n+1\rangle$. The state vector has now the form

$$|\tilde{\Psi}(t)\rangle = \tilde{c}_{e,n}(t)\,|e, n\rangle + \tilde{c}_{g,n+1}(t)\,|g, n+1\rangle, \tag{6.47}$$

resulting in the probability amplitudes of

$$\tilde{c}_{e,n}(t) = \cos(g\sqrt{n+1}t) \quad \text{and} \quad \tilde{c}_{g,n+1}(t) = -i\sin\big(g\sqrt{n+1}t\big). \tag{6.48}$$

This implies that the two-level population inversion becomes

$$D_{\text{Fock}}(n,t) = \cos(2g\sqrt{n+1}t). \tag{6.49}$$

This result is very different from the classical case: being now the Rabi frequency $g\sqrt{n+1}$ (instead of $g\sqrt{n}$ as before), an interesting behavior is observed for the case of an empty cavity. Even for $n = 0$, the state decays from the excited $|e\rangle$ to the ground state $|g\rangle$, and then it returns periodically to the excited state. The observed *periodic* and *reversible* spontaneous decay represents the first step in cavity quantum electrodynamics (cQED) since this behavior only appears when the field is treated quantum mechanically. If instead of considering the cavity field being in a Fock state but rather in a coherent state $|\alpha\rangle$, the overall state can then be described as

$$|g,\alpha\rangle = e^{-\frac{1}{2}|\alpha|^2} \sum_{n=0}^{\infty} \frac{\alpha^n}{\sqrt{n!}} |g,n\rangle \tag{6.50}$$

if the two-level system is at the ground state for $t = 0$ or

$$|e,\alpha\rangle = e^{-\frac{1}{2}|\alpha|^2} \sum_{n=0}^{\infty} \frac{\alpha^n}{\sqrt{n!}} |e,n\rangle \tag{6.51}$$

if the initial state is the excited one. Without going into too much detail, the population inversion is very different from the case described with the cavity field in a Fock state, even if the coherent state is described as a sum of number states $|n\rangle$ with different amplitudes $\alpha^n/\sqrt{n!}$ (see also Chapter 3). Every component of $|\alpha\rangle$, i. e., $|n\rangle$, would drive the two-level system with its own Rabi frequency resulting in the observation of oscillations in the atomic population inversion, their disappearing and reappearing (see Chapter 3 of [108]).

In the last case, if the cavity boundaries would have a finite temperature T, then the cavity field can be considered chaotic. In this scenario the atomic population inversion would have no recognizable pattern, very differently from the two previous cases [108].

6.4 Two-level system in a cavity in the presence of dissipation – the Purcell effect

Thanks to the Jaynes–Cummings model discussed in the previous section, it is now clear that an excited two-level system in an empty cavity is subject to a periodic, and reversible, spontaneous decay. In the ideal case where no dissipation is present, the excitation is periodically exchanged between the TLS and the cavity field, resulting in the so-called vacuum Rabi oscillations. Assuming the initial state of the system to be $|e,0\rangle$, i. e., TLS in the excited state and vacuum field, the system has been so far considered to evolve to the state $|g,1\rangle$, having these two states described via the Hamiltonian (6.32). On the other hand, in realistic systems, two decay channels have to be incorporated, one for the TLS and the other for the cavity system. In presence of these decay channels, the state $|e,0\rangle$ can also evolve to $|g,0\rangle$, as for the state $|g,1\rangle$, which can also decay to $|g,0\rangle$.

The emitter decay: considering a two-level system in the free space, the excited TLS $|e\rangle$ can decay to $|g\rangle$ with the *irreversible* emission of a photon within the full solid angle 4π. This spontaneous decay is here denoted as Γ_{bulk} (Fig. 6.5(left)). The situation can be very different when the TLS is placed into a cavity. Depending on the resonator geometry, the two-level system can interact with one (or a subset) of the free-space modes: ultimately, the emission can happen only in the actual cavity mode (Γ_c) while having the spontaneous decay into the other modes $\Gamma \to 0$. Still, when this is not happening, the

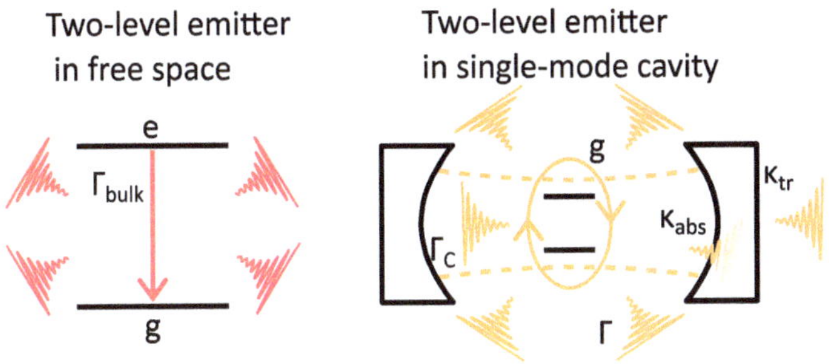

Figure 6.5: Sketch of the two-level system when in a free space (left) and when placed in a cavity (right). All parameters are reported, in particular, the TLS decay rate in the cavity mode Γ_c, the spontaneous decay in the other modes Γ, the TLS-field coupling rate g, the cavity decay rates κ_{tr} and κ_{abs} (for transmission and absorption, respectively, with the total cavity decay rate $\kappa = \kappa_{abs} + \kappa_{tr}$). Typically in optical microcavities, κ_{tr} is larger than κ_{abs}, since it is often useful that the photons leave the cavity in specific directions, while efforts are made to minimize absorption losses.

presence of $\Gamma \neq 0$ couples the state $|e,0\rangle$ to $|g,0\rangle$, the photon being not emitted in the cavity field (Fig. 6.5(right)). For simplicity, in the following, we will consider a single-mode cavity, where any other mode is spectrally far-detuned from the two-level transition.

The cavity decay: this dissipation channel is typically denoted as κ. Two terms contribute to this decay channel, the absorption κ_{abs} (for example, absorption in mirror material) and the transmission κ_{tr} (e. g., transmission through the mirrors with reflectivity $R < 1$) losses, having $\kappa = \kappa_{abs} + \kappa_{tr}$. This term enters into a very well-known cavity parameter, the so-called quality factor (see Sec. 13.3), which can be defined in various forms. Here, for example, in analogy to the cavity dissipation, it can be defined as $Q = \omega_c/\kappa$ (where ω_c is the frequency of the cavity mode). Due to κ, the system can decay from $|g,1\rangle$ to $|g,0\rangle$. In more detail, we can consider the two-level system to be placed into a single-mode cavity, as depicted in Fig. 6.5(right).

Following the convention adopted in [108], the Hamiltonian (6.32) can be rewritten in the rotating frame of the cavity mode as

$$\hat{H} = \hbar\Delta\hat{\sigma}_-\hat{\sigma}_+ + \hbar g(\hat{\sigma}_+\hat{a} + \hat{a}^\dagger\hat{\sigma}_-) \tag{6.52}$$

$$= -\hbar\Delta\hat{\sigma}_+\hat{\sigma}_- + \hbar g(\hat{\sigma}_+\hat{a} + \hat{a}^\dagger\hat{\sigma}_-) \tag{6.53}$$

with $\Delta = \omega_c - \omega_{eg}$ (i. e., detuning between the cavity mode frequency ω_c and TLS frequency ω_{eg}) and again $\hat{\sigma}_+ = |e\rangle\langle g|$ and $\hat{\sigma}_- = |g\rangle\langle e|$. In the presence of atomic and cavity decays, their impact on the system can be considered adding two terms to the previous equation, resulting in the effective (non-Hermitian) Hamiltonian

$$\hat{H}_{\text{eff}} = \hat{H} - i\frac{\hbar}{2}\Gamma\hat{\sigma}_+\hat{\sigma}_- - i\frac{\hbar}{2}\kappa\hat{a}^\dagger\hat{a}. \tag{6.54}$$

Applying the Schrödinger equation (which includes $\hat{H}_{\text{eff}}$) to the state vector

$$|\Psi(t)\rangle = c_{e,0}(t)\,|e,0\rangle + c_{g,1}(t)\,|g,1\rangle\,, \tag{6.55}$$

we get

$$\frac{\partial}{\partial t}c_{e,0} = -\frac{1}{2}\Gamma c_{e,0} - igc_{g,1}, \tag{6.56}$$

$$\frac{\partial}{\partial t}c_{g,1} = -\left(i\Delta + \frac{1}{2}\kappa\right)c_{g,1} - igc_{e,0}. \tag{6.57}$$

As said at the beginning, the initial condition of an excited TLS in an empty cavity is given by $c_{e,0}(0) = 1$ and $c_{g,1}(0) = 0$. Without describing the calculation based on the Laplace transformation (which can be found in [108]), in the case study of perfect spectral resonance between two-level system and cavity mode, i. e., $\Delta = 0$, it is possible to distinguish between two regimes, depending on the values of Γ and κ with respect to the coupling constant g.

The *strong coupling* regime is found when $g > (\kappa - \Gamma)/4$; considering that typically $\kappa > \Gamma$, the condition can be rewritten as $g > \kappa/4$. Then the complex amplitudes take the forms

$$c_{e,0}(t) = e^{-\frac{1}{4}(\kappa+\Gamma)t}\left[\cos(gt) + \frac{\kappa - \Gamma}{4g}\sin(gt)\right] \tag{6.58}$$

$$\approx e^{-\frac{1}{4}(\kappa+\Gamma)t}\cos(gt), \tag{6.59}$$

$$c_{g,1}(t) = -ie^{-\frac{1}{4}(\kappa+\Gamma)t}\sin(gt). \tag{6.60}$$

These lead to an interesting behavior of the probabilities of the TLS to be found in the excited state ($|\langle e|\Psi(t)\rangle|^2 = |c_{e,0}(t)|^2$) or the cavity to contain one photon ($|\langle 1|\Psi(t)\rangle|^2 = |c_{g,1}(t)|^2$). As we can see in Fig. 6.6(a), an oscillatory behavior is still observed with Rabi oscillations between the states $|e,0\rangle$ and $|g,1\rangle$ but with an additional damping. This is because both states can decay to $|g,0\rangle$ due to Γ and κ. This damping rate can be inferred by the square of the norm of the state vector as

$$\langle\Psi(t)|\Psi(t)\rangle = |c_{e,0}(t)|^2 + |c_{g,1}(t)|^2 \tag{6.61}$$

$$\approx e^{-\frac{1}{2}(\kappa+\Gamma)t}, \tag{6.62}$$

hence being $\frac{1}{2}(\kappa + \Gamma)$.

When $g < \kappa/4$, the system is said to be in the *weak coupling* regime. The system dynamics is very different from the case of strong coupling as shown in Fig. 6.6(b), where no oscillations are observed. To reach an analytical expression, we can consider $g \ll k$ and $\Gamma < \kappa$, so that the complex amplitudes become [108]

$$c_{e,0}(t) = e^{-\frac{1}{2}(\Gamma+\Gamma_c)t}, \tag{6.63}$$

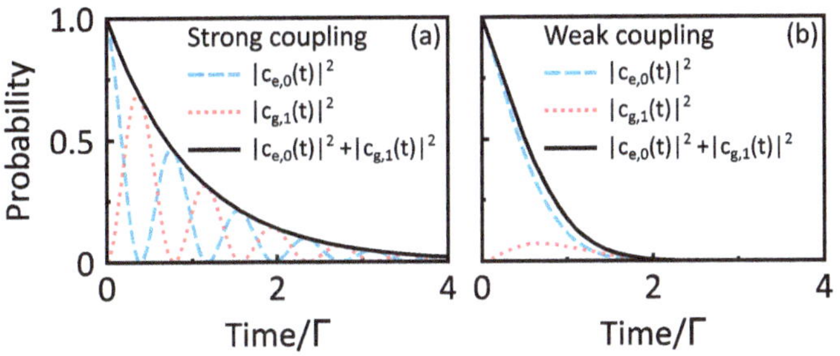

Figure 6.6: Exemplary dynamics of a two-level system placed into a cavity in the presence of dissipation. The dashed lines represent the probability of finding the TLS in the excited state having no photon in the cavity, whereas the dotted lines represent the case of the two-level system state being in the ground state and the cavity containing one photon. The solid lines represent the square of the norm of the state vector. (a) *Strong coupling* case, where the Rabi oscillations are damped ($g = 4\Gamma$, $\kappa = \Gamma$, as in [108]). (b) *Weak coupling* case, where no oscillations are observed ($g = \Gamma$, $\kappa = 4\Gamma$, as in [108]).

$$c_{g,1}(t) = i\frac{2g(\kappa-\Gamma)}{(\kappa-\Gamma)^2-4g^2}\left[e^{-\frac{1}{2}\kappa t} - e^{-\frac{1}{2}(\Gamma+\Gamma_c)t}\right] \tag{6.64}$$

$$\approx i\frac{2g}{\kappa-\Gamma}\left[e^{-\frac{1}{2}\kappa t} - e^{-\frac{1}{2}(\Gamma+\Gamma_c)t}\right], \tag{6.65}$$

where we introduced the term

$$\Gamma_c = \frac{4g^2}{\kappa-\Gamma}. \tag{6.66}$$

This represents the contribution due to the presence of the cavity to the TLS decay rate. Under the reasonable assumption that $\kappa \gg \Gamma$, we can write the cavity contribution over the free space decay rate as (see Section 5.2.4)

$$F_P = \frac{\Gamma_c}{\Gamma_{\text{bulk}}} = Q\left(\frac{6\pi}{Vk^3}\right) = \frac{3Q(\frac{\lambda}{n})^3}{4\pi^2 V}. \tag{6.67}$$

As predicted by the early works of Purcell, the presence of a cavity can drastically accelerate the emission into the cavity mode over the spontaneous emission into the free-space modes, the so-called *Purcell effect*. This ratio is defined as the *Purcell factor* and depends on the ratio Q/V of the cavity quality factor to the mode volume. This is one reason why when characterizing photonic cavities, Q/V is often provided (see also Chapter 13 for further details). In other words, for high Q factors and/or small mode volume, a strong reduction of the observed decay time of the two-level atom is expected, with respect to the case of a TLS in a free space. In a more general case, Eq. (6.67) should be written as [95]

$$F_P = \frac{3Q(\frac{\lambda}{n})^3}{4\pi^2 V}\cdot\frac{(\frac{\kappa}{2})^2}{(\omega_{eg}-\omega_c)^2+(\frac{\kappa}{2})^2}\cdot\frac{|\vec{E}(\vec{r})|^2}{|\vec{E}_{\text{MAX}}|^2}\cdot\left(\frac{\vec{d}\cdot\vec{E}(\vec{r})}{dE}\right)^2. \tag{6.68}$$

Indeed, Eq. (6.67) assumes that i) the two-level system is perfectly placed in the spatial maximum of the cavity field, ii) the detuning between cavity field and TLS transition is zero, i. e., there is a perfect spectral matching between two-level system and cavity mode, iii) the TLS is in the spatial maximum of the cavity electric field, and iv) the dipole orientation is along the field polarization. These assumptions are not considered in the general form (6.68). The quantity F_P as in Eq. (6.67), which can be thought as the Purcell factor only dependent on the cavity parameters, i. e., with perfect spatial matching, spectral matching, and dipolar orientation of the TLS with respect to the cavity field, now also includes three additional terms: the first one accounts for the effect of spectral detuning on the Purcell factor. When this term is <1, the spontaneous decay is inhibited, i. e., the decay time gets longer. This inhibition can be limited, in realistic implementations, by the presence of other modes where the emitter can couple [95]. The second additional term accounts for the spatial position of the atom with respect to the cavity field $\vec{E}(\vec{r})$ (with maximum $\vec{E}_{\text{MAX}}$). The third additional term considers the dipole orientation with respect to the direction of the field.

As discussed in Section 5.2.4, embedding a two-level system into a cavity and operating in the *weak coupling* regime currently represents a widely used approach to strongly enhance the light extraction, therefore realizing bright sources of quantum light: additionally, the shortening of the radiative lifetime allows operating the source at higher excitation rates, increasing the amount of photons usable in dedicated experiments [136].

6.5 Summary

- Under the drive of an external coherent field, the TLS population oscillates between its two levels, excited and ground, with the so-called *Rabi frequency*. Its generalized expression considers the case of the cavity field outside of the condition of exact resonance with the two-level transition frequency, i. e., $\Delta \neq 0$.
- The oscillations of the two-level populations are most efficient when the driving field is at exact resonance with the atomic transition, i. e., $\omega = \omega_{eg}$, for which $\Delta = 0$.
- The concept of population inversion has been introduced, providing the description of the two-level system dynamics under the driving of classical or quantized field. Under pulsed excitation, the population inversion becomes a key in experimental implementations, since it ensures a triggered excitation of the TLS, eventually followed by the spontaneous emission of a photon.
- The interaction between a two-level system with a quantized field, i. e., the Jaynes–Cummings model has been introduced. The periodic oscillatory behavior between excited and ground states even for a quantized field with zero photons shows a first glimpse of cavity quantum electrodynamics, since these results cannot be obtained considering a classical driving field.
- The case study of a TLS placed into a damped cavity is analyzed. Dissipation mechanisms are considered: the first is given by the cavity decay κ itself, whereas the

second is represented by the spontaneous TLS decay in free space modes other than the cavity mode (Γ versus Γ_c).

- Two regimes are distinguished: for a TLS-cavity field coupling $g > (\kappa - \Gamma)/4$, the system is said to be in the *strong coupling*. Damped Rabi oscillations are observed in the probability of finding the TLS in the excited state (or having the cavity field containing one photon). The second regime is called the *weak coupling*, and it is reached when $g < \kappa/4$, for which no Rabi oscillations are observed.
- Analytical expressions obtained for $g \ll \kappa$ and $\Gamma < \kappa$ allow us to quantify the Purcell effect, i. e., allow for estimating the shortening of the spontaneous decay of the two-level system when placed into a high-quality factor cavity, with respect to the case of a TLS placed in a free space. The Purcell factor is shown to depend on the ratio of the cavity quality factor and the mode volume, i. e., $\propto Q/V$. A general formula for the Purcell factor is also provided, including the general case where nonideal placement of the two-level system is considered (i. e., nonperfect overlap with the cavity mode, nonexact resonance or orientation of the dipole).

Part II: **Physics of semiconductor quantum dots**

7 Basic physical properties of quantum dots

7.1 Introduction

Semiconductor quantum dots (QDs) are nanometer-sized three-dimensional structures that confine electrons and holes in all three directions. Their size is in the order of the De Broglie wavelength of electrons and holes in these structures, giving rise to a quantized and discrete energy spectrum. Many different types of QDs have already been developed, such as chemically synthesized QDs, naturally formed QDs by interface fluctuations, electrostatically and lithographically defined QDs, and epitaxially grown self-assembled QDs. The latter possess the advantage that their size, shape, composition, and location can be tailored to a large extend covering the spectral range from the blue region (~400 nm) up to the telecom C-band (around 1550 nm). In addition, they also possess ultra-high quantum efficiencies close to one, and they can be naturally embedded during growth into more complex structures, e. g., between Bragg reflectors and into p–i–n structures, making them attractive as highly efficient quantum light emitters in photonic devices.

In this chapter, we present the general electronic and optical properties of direct bandgap semiconductor quantum dots (e. g., InAs/GaAs), which are epitaxially grown by molecular beam epitaxy (MBE) or metal organic vapor phase epitaxy (MOVPE). The notation InAs/GaAs signifies an InAs QD surrounded by GaAs material. We focus here on the most important aspects necessary to understand the physics of quantum light generation in quantum dots. More sophisticated discussions on the physics of semiconductors and quantum dots can be found in respective textbooks, e. g., in [61] and [124–126].

7.2 Electronic properties and carrier confinement

The motion of electrons in a crystal matrix is typically discussed in the Born–Oppenheimer approximation. In this framework, the motion of the nuclei and one of the electrons can be separated. In addition, the motion of the nuclei is at first neglected. Furthermore, applying the Hartree–Fock approximation, a single electron can be considered to perceive an average field generated by the nuclei and all other electrons. Within this approximation, the Schrödinger equation for the electron reads [36]

$$\hat{H}\Psi(\vec{r}) = \left[-\frac{\hbar^2}{2m}\Delta + V(\vec{r})\right] = E\Psi(\vec{r}), \tag{7.1}$$

where $V(\vec{r})$ is the effective interaction potential (mean field theory). It shows the same symmetry properties as the crystal, i. e., $V(\vec{r}) = V(\vec{r} + \vec{t})$, where $\vec{t} = t_1\vec{a}_1 + t_2\vec{a}_2 + t_3\vec{a}_3$ represents a basic translation between lattice sites, and $\vec{a}_i$ are the basis vectors of the crystal.

https://doi.org/10.1515/9783110703412-007

Within the Bloch theorem, due to the periodic structure of the crystal, the electronic states are described by the so-called Bloch functions

$$\Psi_{n,\vec{k}}(\vec{r}) = u_{n,\vec{k}}(\vec{r})e^{i\vec{k}\vec{r}}, \tag{7.2}$$

where $u_{n,\vec{k}}(\vec{r}) = u_{n,\vec{k}}(\vec{r} + \vec{t})$ refers to the atomic part, it is periodic with the lattice and is indexed with a quantum number n (band index) and the electron wave vector $\vec{k}$. Using periodic boundary conditions, the k_i values are restricted to

$$k_i = \frac{2\pi}{a_i}\frac{h_i}{N_i} \quad (h_i \in \mathbb{Z},\ i = x, y, z)$$

with $-\frac{N_i}{2} < h_i < \frac{N_i}{2}$ (N_i is the number of unit cells in the i-direction). We therefore get a quasi-continuum of allowed states in a three-dimensional crystal due to the large number of unit cells $N = N_x N_y N_z$. The eigenenergies $E_n(\vec{k})$ of the eigenstates $\Psi_{n,\vec{k}}(\vec{r})$ define the energy bands. In a semiconductor the highest unoccupied energy band is called the conduction band, which is separated by the bandgap from the highest occupied band, the so-called valence band. The bandgap is typically in the range from one to several electronvolts.

A missing electron in the valence band is called a hole state. It can be created, for example, by the excitation of a single electron from the valence band into the conduction band. The hole h is a quasi-particle with positive charge $+e$, which represents the remaining attractive ensemble of $N-1$ valence band electrons. For a detailed discussion of their physical properties, we can refer to standard semiconductor textbooks. In the context of this book, mainly the charge, angular momentum, and spin structure of the hole are of importance and will be discussed later.

A further very useful simplification is achieved by introducing the notion of the effective mass approximation. Here the influence of the crystal on the motion of carriers (electron and holes) is put into the effective mass of the carrier, and therefore it allows for a treatment of single-carrier motion as quasi-free particles. The effective mass is defined as

$$\left(\frac{1}{m^*}\right)_{ij} = \frac{1}{\hbar^2}\frac{\partial^2 E(\vec{k})}{\partial k_i \partial k_j}, \tag{7.3}$$

which is, in the most general case, a tensor element. The reciprocal effective mass is determined by the curvature of the energy surface. Close to the center of the first Brillouin zone ($-\frac{\pi}{a} \leq k \leq \frac{\pi}{a}$) at $k \approx 0$, the dispersion of the conduction band of typical direct group III–V semiconductors is isotropic and can be approximated by a parabola. Consequently, the effective electron mass is a constant, and the energy of the electron converts into the simple form

$$E_{n,\vec{k}} = \frac{\hbar^2 \vec{k}^2}{2m^*} = E_n + \frac{\hbar^2}{2m^*}(k_x^2 + k_y^2 + k_z^2),; \quad m^* = \begin{cases} m_e^* & \text{for } e^- \quad \text{(electrons)}, \\ m_h^* & \text{for } h^+ \quad \text{(holes)}, \end{cases} \tag{7.4}$$

where E_n is the minimum of the band n ($k = 0$). This is formally the dispersion of a free particle of mass m^*.

In a bulk semiconductor the aforementioned dispersion relation results in a density of states $D^{3D}(E) \sim \sqrt{E}$ (see Fig. 7.1(a)). Significant changes to the carrier eigenstates and the density of states come into play when the carrier motion is confined on a scale smaller than its De Broglie wavelength

$$\lambda_{e,h}^{\text{DeBroglie}} = \frac{2\pi\hbar}{\sqrt{3m_{e,h}^* k_B T}}, \tag{7.5}$$

where k_B is the Boltzmann constant, and T is the temperature of the system. In group III–V semiconductors, typical values of the De Broglie wavelength of electrons and holes are in a regime of a few nanometers to a few tens of nanometers.

We can distinguish three types of so-called low-dimensional semiconductor heterostructures considering the three different quantum confinement directions (x, y, z) [36].

- In a quantum well structure (2D system), electrons and holes are confined in one direction (z) and have free motion in the other directions (x and y). The eigenenergies are given by $E = E_{z,n} + \frac{\hbar^2}{2m^*}(k_x^2 + k_y^2)$, where $E_{z,n}$ is the confinement energy related to the confinement potential in the z-direction. The corresponding density of states D^{2D} is shown in Fig. 7.1(b). It has a stair-like behavior, where the stair positions in energy are given by $E_{z,n}$.
- In a quantum wire structure (1D system), electrons and holes are confined in two directions (y and z) and have free motion in the x-direction. The eigenenergies are given by $E = E_{z,n} + E_{y,m} + \frac{\hbar^2}{2m^*}(k_x^2)$, where $E_{n,m} = E_{z,n} + E_{y,m}$ is the confinement energy given by the two corresponding potentials in the z- and y-directions. For each energy $E_{n,m}$, the density of states is given by $D(E) \sim 1/\sqrt{E - E_{n,m}}$ (see Fig. 7.1(c)).
- In a quantum dot (0D system), electrons and holes are confined in all three directions, and, as a consequence, we get fully localized states with a density of states expressed by the Dirac functions $D^{0D}(E) \sim \sum_{n,m,l} \delta(E - E_{(n,m,l)})$ at each confinement energy $E_{n,m,l} = E_{z,n} + E_{y,m} + E_{x,l}$ (see Fig. 7.1(d)).

In the expression above, n, m, and l represent the appropriate set of quantum numbers, which classify the respective eigenstates of the explicit dot confinement potential. Epitaxially grown semiconductor quantum dots exhibit a large variety of different shapes, sizes, and symmetries leading to different effective carrier confinement potentials. Numerous numerical approaches of different degrees of sophistication have been developed in the past to calculate the respective electron and hole states in real QDs. Within

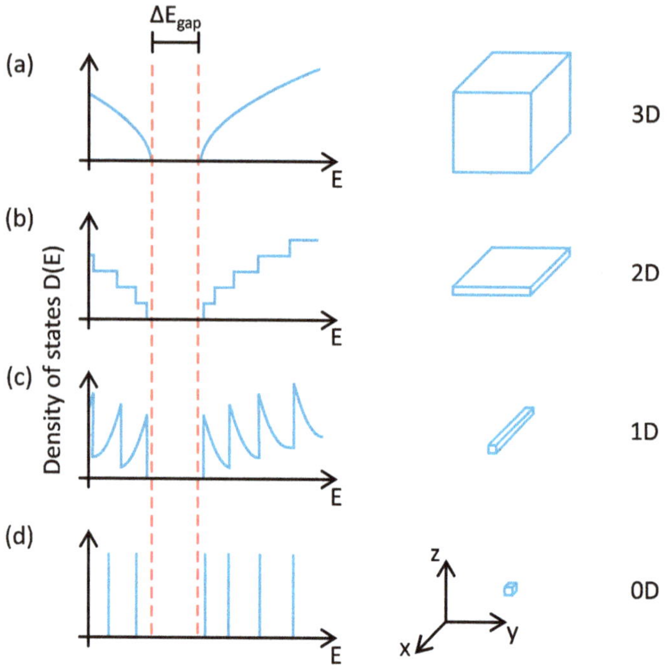

Figure 7.1: Schematic comparison of the density of states in semiconductors for various dimensions.

this textbook, the electron and holes states of epitaxially grown self-assembled semiconductor quantum dots; e. g., InAs/GaAs or CdSe/ZnSe QDs are discussed. These QDs exhibit very often a typical lens shape in the *xy*-plane with a certain height in the *z*-direction, considerably smaller than the diameter of the lens. For example, typical InAs/GaAs QDs emitting around 900 nm possess lateral dimensions of 20–50 nm and heights of 2–5 nm. This shape results in a strong vertical carrier confinement (i. e., large energy quantization), which is typically modeled with a sharp stepwise potential in the *z*-direction (growth direction) and a weak in-plane parabolic confinement parallel to the growth direction. In the *z*-direction, typically, only the lowest state is confined due to the large energy quantization and the finite potential height given by the band offset of the QD and barrier material. Due to the weaker lateral confinement, the first higher energy levels are determined by the in-plane confinement.

7.2.1 Single-particle states in a harmonic potential

In the following, we will discuss the electronic structure of single-particle states for such a harmonic 2D-confinement potential ($V(\rho) = \frac{1}{2}m^*\omega_0^2\rho^2$) with rotational symmetry in the *xy*-plane and show analytic solutions. Such a model system has been calculated by Fock and Darwin under consideration of an optional magnetic field $\vec{B}$ applied along the $\vec{z}$-direction, which conserves the axial symmetry. The respective Hamiltonian of the time-independent Schrödinger equation is given by

$$\hat{H}_{\text{Fock–Darwin}} = \frac{1}{2m^*}\{\vec{p} - e\vec{A}(\vec{r})\}^2 + \frac{1}{2}m^*\omega_0^2(x^2 + y^2), \tag{7.6}$$

where the magnetic field $\vec{B}(\vec{r}) = \vec{\nabla} \times \vec{A}(\vec{r})$ is introduced by the corresponding vector potential $\vec{A}(\vec{r}) = \frac{1}{2}|\vec{B}|(y, -x, 0)$. The eigenenergies of the single carrier states are

$$E_{(n,l)} = (2n + |l| + 1) \cdot \hbar\sqrt{\omega_0^2 + \frac{\omega_c^2}{4}} + \frac{1}{2}\hbar\omega_c\, l, \tag{7.7}$$

where $\omega_c = \frac{eB}{m^*}$ is the cyclotron frequency. Each eigenstate is characterized by a set of quantum numbers, i. e., a radial quantum number $n = 0, 1, 2 \ldots$ and an azimuthal quantum number (angular momentum) $l = 0, \pm1, \pm2, \ldots$, respectively. In the absence of a magnetic field the energies are given by

$$E_{(n,l)} = (2n + |l| + 1) \cdot \hbar\omega_0, \tag{7.8}$$

which represent a series of equidistant states with energy distance $\Delta E = \hbar\omega_0$. A similar convention for the nomenclature of atomic shells has been introduced for QDs, where $s = 2n + |l| = 0, 1, 2, \ldots$ is referred to the shell index. The further characterization of the states is also done in terms of the angular momentum number $l = 0, \pm1, \pm2, \ldots$, with states of the same $|l|$ being degenerate in energy. Considering the Pauli exclusion principle for the electron and hole spin degrees of freedom, the degeneracy of a QD shell is given by $d = 2(s + 1) = 2, 4, 6, \ldots$ for $s-, p-, d-, \ldots$ shells, respectively (see Fig. 7.2). As in atomic physics, the $|l|$ degeneracy can be lifted by the application of a magnetic field (see Eq. (7.6)).

In a real QD the finite potential height due to the fixed band offset between dot and barrier material leads to a limited number of confined states. Depending on the actual potential confinement, single InAs/GaAs-based QDs typically show two or three confined states for electrons and holes.

For nonclassical light sources such as single- and entangled-photon sources, only interband electronic transitions resulting in the radiative recombination of an electron–hole pair are considered. We can therefore deduce the electronic states of the respective QD from the electronic properties near the Γ point of the respective semiconductor materials. In the following, these properties are discussed for the most studied semiconductor QD system, InAs/GaAs QDs. GaAs and InAs are direct bandgap semiconductors that crystallize in the zincblende structure and have bandgap energies of $E_g = 1.43$ eV and 0.35 eV at room temperature, respectively.

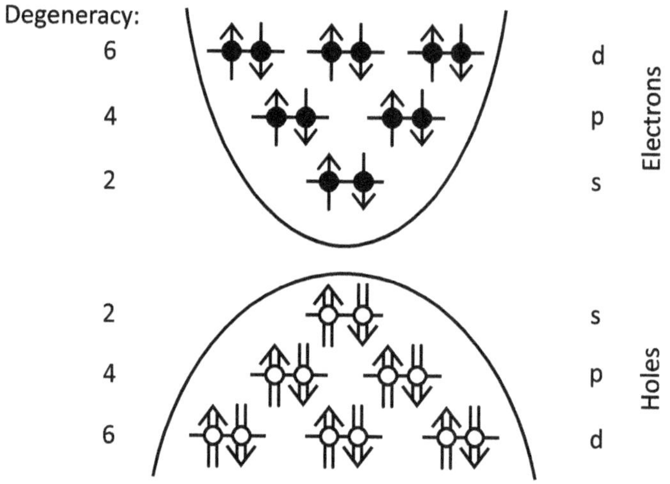

Figure 7.2: Shell structure for electron and holes in a harmonic confinement potential with their respective nomenclature and degeneracy.

7.2.2 Total angular momentum of electronic states in bulk material

For a detailed description of the electronic states in the full Brillouin zone in semiconductors, the so-called Luttinger Hamiltonian has to be solved. This goes well above the scope of this book, and the interested readers can refer to the corresponding semiconductor textbooks. Here we restrict the discussion to states close to the center of the Brillouin zone. Exemplarily, a schematic of the band structure of bulk GaAs is shown around the Γ point for propagation in the z-direction in Fig. 7.3(a). The bandstructure of InAs is similar to that of GaAs, yet with different values of the band splittings and curvatures. For small k_z-values, the dispersion relation of the electron and three-hole bands can be approximated by parabolas. Consequently, the carrier motion can be described by an effective mass m^* (see Eq. (7.3)) for the respective band. In the following, mainly symmetry arguments will be discussed, which are finally relevant to describe the different allowed radiative transitions. The electronic states of these semiconductors at the Γ point present the same symmetry as the atomic orbitals s, p_x, p_y, and p_z.

The lattice-periodic part $u_{n,\vec{k}}(\vec{r})$ (see Eq. (7.2)) of the conduction-band Bloch function exhibits s-type symmetry. As a consequence, the conduction band states possess a two-fold spin-degeneracy with the total spin $S = 1/2$ and its projection on the z-direction of $S_z = \pm 1/2$ (in short notation, $(S, S_z) = (\frac{1}{2}, \pm\frac{1}{2})$).

The valence band states hybridize from p-type orbitals and thus carry an angular momentum of $L = 1$ at $\vec{k} = 0$. Taking also into account the two possible spin states, this results in the formation of six possible hole states, namely the heavy-hole band $(J, J_z) = (\frac{3}{2}, \pm\frac{3}{2})$, the light-hole band $(J, J_z) = (\frac{3}{2}, \pm\frac{1}{2})$, and the split-off band $(J, J_z) = (\frac{1}{2}, \pm\frac{1}{2})$ with total angular momentum J, and J_z are the z-projections of the total angular momentum.

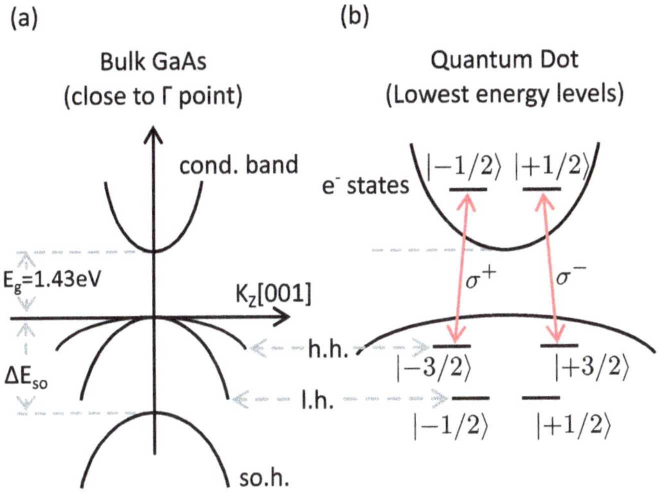

Figure 7.3: (a) Schematic band structure of GaAs for electron propagation in the *z*-direction ([001]) close to the Γ point. (b) QD single particle levels and optical transitions between electron and heavy-hole levels.

Due to a strong spin-orbit interaction, the states $J = \frac{3}{2}$ are split from the states $J = \frac{1}{2}$ by the split-orbit energy Δ_{SO} (see Fig. 7.3(a)). The highest two valence bands are therefore four-fold degenerate at $\vec{k} = 0$.

7.2.3 Total angular momentum of electronic states in a quantum dot

As discussed above, the carrier motion in a QD is confined in all three directions, leading to discrete energy states for the carriers. The lattice-periodic part $u_{n,\vec{k}}(\vec{r})$ of the QD wavefunction is assumed to be weakly perturbed by the nanostructure and therefore retains the symmetry from the corresponding bulk material. Because a typical QD is composed of 10^4–10^5 atoms, the confinement potential slowly varies at the atomic scale, and the carrier wavefunction can therefore be written in the envelope function approximation as

$$\Psi_{n,\vec{k}}(\vec{r}) = u_{n,\vec{k}}(\vec{r})\phi_n(\vec{r}), \tag{7.9}$$

where $\phi_n(\vec{r})$ is a slowly varying envelope function, which fulfills the Schrödinger equation within a confinement potential of cylindrical symmetry $V(\vec{r}) = V(\rho, z)$, the in-plane part of the potential is the previously discussed symmetric and harmonic 2D-confinement potential ($V(\rho) = \frac{1}{2}m^*\omega_0^2\rho^2$), and the *z*-component (growth direction) is given by the sharp stepwise potential given by the band offset between QD and barrier material.

The conduction band ground states consists of two degenerate states $|s, s_z\rangle = |\frac{1}{2}, \pm\frac{1}{2}\rangle$ (see Fig. 7.3(b)). In an InAs/GaAs QD the spin-orbit coupling and the strain induced lattice mismatch between the InAs wetting layer, and the GaAs substrate lifts the degeneracy of the valence bands with the heavy-hole band $|j, j_z\rangle = |\frac{3}{2}, \pm\frac{3}{2}\rangle$ as the highest energy valance band [61]. The light-hole states $|j, j_z\rangle = |\frac{3}{2}, \pm\frac{1}{2}\rangle$ are split-off in energy by at least several meV and can therefore usually be neglected in optical experiments, especially for radiative transitions, typically occurring from the lowest energy states of a QD. The $|j, j_z\rangle = |\frac{1}{2}, \pm\frac{1}{2}\rangle$ states are further split-off as discussed for the bulk case and play no role for the optical properties. In strain-free QDs (as, for example, GaAs QDs in AlGaAs), the highest energy valence band is still mostly heavy-hole like due to the small aspect ratio of QDs.

7.2.4 Interband optical transition and selection rules in single quantum dots

The confined electronic states of a QD strongly interact with light, which is one reason for the excellent optical properties of single-photon sources based on single QDs. A radiative transition of an electron–hole pair decays exponentially with rate Γ_{rad} determined by Fermi's golden rule. In the following, the discussion on the electronic part of the optical matrix element is restricted to focus on the optical selection rules in QDs. To evaluate the selection rules, we have to consider the matrix element $\langle\Psi_f|\hat{H}'|\Psi_i\rangle$, where $\hat{H}'$ is the light–matter interaction Hamiltonian triggering the transition from the initial state $|\Psi_i\rangle$ to the final state $|\Psi_f\rangle$. In the envelope function approximation (see Eq. (7.9)), the initial and final states can be written as $\Psi_{i,\vec{k}}(\vec{r}) = u_{i,\vec{k}}(\vec{r})\phi_i(\vec{r})$ and $\Psi_{f,\vec{k}}(\vec{r}) = u_{f,\vec{k}}(\vec{r})\phi_f(\vec{r})$, respectively. In the electric dipole approximation, the light–matter interaction Hamiltonian is given as $\hat{H}' = \vec{e} \cdot \hat{\vec{p}}$, where $\vec{e}$ is the polarization of the electric field, and $\hat{\vec{p}}$ is the momentum operator. Thus for interband optical transitions in the QD, the matrix element in the electric dipole approximation can be expressed as

$$\langle\Psi_{f,\vec{k}}(\vec{r})|\vec{e} \cdot \hat{\vec{p}}|\Psi_{i,\vec{k}}(\vec{r})\rangle = \vec{e} \cdot \langle u_{f,\vec{k}}|\hat{\vec{p}}|u_{i,\vec{k}}\rangle \int_V \phi_f^*(\vec{r})\phi_i(\vec{r})d^3\vec{r}, \tag{7.10}$$

where V represents basically the QD volume. The first term of this interband matrix element gives rise to characteristic selection rules, which are determined by the total angular momentum configurations of electrons and holes in their respective bands states. The heavy-hole states $|j_z\rangle = |\pm\frac{3}{2}\rangle$ can be coupled to the electron states $|j_z\rangle = |s_z\rangle = |\pm\frac{1}{2}\rangle$ via a photon of $\sigma^\pm$ circular polarization (see Fig. 7.3(b)).

It is important to note that the above-mentioned strict selection rules are lifted to a certain extent, since the above-discussed heavy-hole states possess to a certain part also light–hole character and vice versa due to heavy-light-hole mixing. This mixing can be explained by a more complete discussion of the valence band states in QDs, which goes well beyond this textbook. The mixture is in the order of 10^{-3}. As a consequence, sometimes forbidden optical transition can be observed in QD spectra.

The second term is responsible for the selection rules among the confined states of the QD, for which only electrons and holes having the same envelope functions symmetry are radiatively coupled. This means that the allowed optical interband transitions correspond to the radiative recombination of electrons and holes that occupy levels with the same shell index s ($\Delta s = 0$), e. g., s-shell to s-shell or p-shell to p-shell transitions thus are allowed. Typical radiative recombination rates for QDs are discussed in the next section after discussing the implication of Coulomb correlations on this property.

Furthermore, it is important to note that the used dipole approximation works well for quantum emitters that are much smaller than the wavelength of light. This is typically true for self-assembled grown InAs/GaAs QDs, where the experimental observations are remarkably well explained by the presented theory. However, a breakdown of the dipole theory was presented where QDs are placed near a metal interface. Here also magnetic-dipole and electric-quadrupole contributions have been observed (see Tighineanu et al., Chapter 5 in [126]).

7.2.5 Coulomb interaction between charge carriers

For a comprehensive description of the excited states of a QD, Coulomb interactions between charge carriers have to be taken into account. Considering different excitation conditions, optical or electrical excitation, and material parameters like doping, various different population and charging configurations (number of electron and holes) can occur in a QD, and the Coulomb interaction between the charge carriers will influence the total energy of the resulting multiparticle state. These multiparticle states within a QD are then referred to as excitons, e. g., neutral exciton (one bound electron–hole pair (X)), charged excitons (e. g., trion: one electron–hole pair plus an extra carrier, i. e., electron (X^-) or hole (X^+)), biexcitons (two electron-hole pairs (XX)), or other multiexciton states, whereas this terminology is not strictly correct in the sense of excitons in bulk material as will be discussed below.

The discussion will start with a neutral exciton in bulk material. Such a system constitutes a hydrogen-like system in a solid-state environment, i. e., a 3D exciton with the associated Hamiltonian

$$\hat{H}_{e,h} = \left\{-\frac{\hbar^2}{2m_e^*}\nabla_e^2 - \frac{\hbar^2}{2m_h^*}\nabla_h^2\right\} - \frac{e^2}{4\pi\varepsilon_0\varepsilon_r|\vec{r}_e - \vec{r}_h|} \tag{7.11}$$

with the dielectric constant ε_r accounting for screening effects within the crystal host matrix. The eigenstates of the two-body system (3D exciton) can be derived from a separation of coordinates into center-of-mass translation with the total mass of the 3D exciton $M^* = m_e^* + m_h^*$ and an internal motion with reduced effective mass $\mu^* = m_e^* m_h^* / (m_e^* + m_h^*)$ as

$$E_{X,n}(\vec{K}_X) = E_g - \frac{\mu^*}{m_0 \varepsilon_r^2} \cdot \frac{m_0 e^4}{2(4\pi\varepsilon_0 \hbar^2)^2} \cdot \frac{1}{n^2} + \frac{\hbar^2 K_X^2}{2M^*}, \tag{7.12}$$

$$E_{X,n}(\vec{K}_X) = E_g - \frac{\mu^*}{m_0 \varepsilon_r^2} \cdot \frac{R_y}{n^2} + \frac{\hbar^2 K_X^2}{2M^*}, \tag{7.13}$$

with m_0 the free electron mass, n the principal quantum number, and $\vec{K}_X = \vec{k}_e + \vec{k}_{hh}$ the center-of-mass exciton wave vector. The third term describes the center-of-mass motion, and the second term describes the exciton binding energy with the atomic Rydberg energy R_y (13.6 eV). The exciton binding energy in the semiconductor is scaled by $\mu^*/(m_0 \varepsilon_r^2)$, and it is typically in the order of $\sim 10^{-3}$, smaller than the atomic Rydberg energy. The Bohr radius of the 3D exciton is given by

$$a_B^* = \frac{4\pi\varepsilon_0 \varepsilon_r \hbar^2 n^2}{\mu^* e^2} = n^2 \frac{m_0}{\mu^*} \varepsilon_r a_0 \tag{7.14}$$

with the hydrogen Bohr radius a_0. For example, the exciton binding energy and the Bohr radius of the 3D exciton ($n = 1$) in GaAs are 4.2 meV and 13 nm, respectively.

For excitons in QDs, the calculation of Coulomb interactions is more complicated due to the mesoscopic confinement potential of the QD, which enforces the binding of the participating carriers. On the one hand, the Coulomb energy scales inversely with the QD size $1/R_{\mathrm{dot}}$, and therefore the Coulomb and exchange interactions are enhanced compared to bulk. On the other hand, the quantum confinement energy scales as $1/R_{\mathrm{dot}}^2$. Thus it is convenient to discuss excitons in QDs for two different regimes of spatial confinement, the weak- and strong-confinement regimes.

In the weak-confinement regime ($a_B^* < R_{\mathrm{dot}}$), the exciton Bohr radius is assumed to be significantly smaller than the QD size R_{dot}, thus leading to a quantization of the exciton center-of-mass motion. This regime can only be reached with unusually large diameter QDs or monolayer-fluctuation quantum dots. Achieving this regime is interesting for some applications because such QDs exhibit a giant dipole moment, giving rise to a very strong light–matter coupling strength. With elongated InGaAs QDs, the so-called strong coupling regime of cavity quantum electrodynamics has been achieved (see Chapter 6), and ultrafast radiative recombination times have been demonstrated due to the enhanced dipole moment in these QDs. Furthermore, in GaAs QDs obtained by droplet etching of nanoholes in AlGaAs followed by GaAs filling, short exciton lifetimes of ~250 ps have been reported.

However, most of the QDs exhibit sizes smaller than the corresponding exciton Bohr radius. This means that the strong three-dimensional confinement of the quantum dot (e. g., for In(Ga)As/GaAs QDs), induces a strong carrier localization, which is then responsible for the electron–hole pair formation. In this case of strong-confinement regime ($a_B^* > R_{\mathrm{dot}}$), the energy correction to the binding energy of the electron–hole pair is thus dominated by the quantum confinement energy: this is of the order of several hundreds of meVs, whereas the correction energy arising from the Coulomb interaction only amounts to several meVs. Nevertheless, the Coulomb interactions between the carriers are significant due to the small confinement length, and as a result, different charge configurations of QD excitons exhibit distinct different transition energies (see Fig. 7.4).

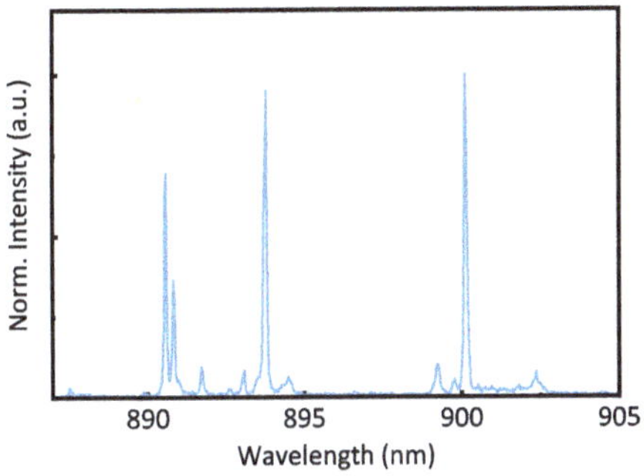

Figure 7.4: Microphotoluminescence spectrum (see Chapter 5) of a single quantum dot after nonresonant optical excitation (see Chapter 10). The spectrum shows multiple exciton lines, i. e., X, XX, X^-, and X^+.

Most of the investigated QD systems are in the strong-confinement regime. Then the electron and hole can be regarded in a first approximation as noninteracting particles. The oscillator strength is given by $f = 2m\omega d^2/(e^2\hbar)$ with d the QD dipole moment, m the electron mass, and ω the angular frequency of the transition (see Reitzenstein and Forchel, Chapter 8 in [125]). The typical oscillator strength of an exciton, e. g., in an InAs/GaAs QD is $f = 10$, which is at least one order of magnitude larger than for an atom. The spontaneous transition rate in a homogeneous medium of refractive index n is then given by

$$\gamma_{\text{rad}} = \frac{1}{\tau_{\text{rad}}} = \frac{|d|^2\omega^3 n}{3\pi\varepsilon_0\hbar c^3}, \tag{7.15}$$

which results in a typical exciton lifetime for an InAs-QD in the order of 1 ns. A compilation of radiative lifetime data for different QD systems can be found in Table 14.3 in Part III. Because of the short radiative lifetimes, QDs are excellent ultra-high repetition rate quantum photonic emitters. Since the radiative recombination rate can be further enhanced by utilizing the Purcell effect in microcavities, repetition rates in the 1–10 GHz regime are easy reachable.

7.3 Excitonic states in quantum dots and quantum light generation

In the following, we will discuss the most important excitonic configurations in a QD, i. e., the neutral exciton (X), the negatively (positively) charged trion X^- (X^+) state, and the neutral biexciton (XX). Furthermore, we will discuss their role for the generation of nonclassical light states, i. e., single-photon states, entangled photon-pair states, and photonic cluster states. We will see that mainly the angular momentum and spin structure of these states determine their applicability with respect to the generation of the above-mentioned light states and not their absolute energy.

7.3.1 Neutral exciton and single-photon generation

The simplest exciton state in a QD is the neutral exciton, where one electron–hole pair is excited or captured within the QD (see Fig. 7.5).

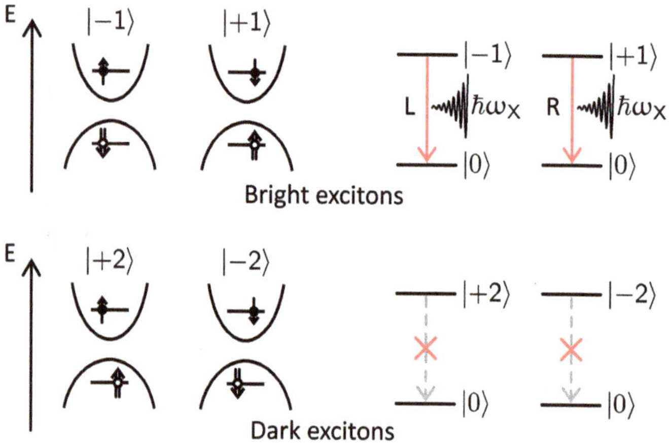

Figure 7.5: (Left) Schematic single-particle picture of the excitonic ground state in a quantum dot. All states are labeled according to their angular momentum projection, where small arrows indicate the involved possible spin states of the electrons ($s = \pm 1/2$) and the holes ($s = \pm 3/2$). In total, four different spin configurations, i. e., two bright and two dark excitons, are possible (see text). The corresponding eigenenergies require corrections resulting from the Coulomb interaction of the involved carriers. (Right) Excitonic picture, indicating a radiative transition of a bright exciton with circular polarization (R = right circular, L = left circular), which results in the emission of a single photon.

As discussed in Section 7.2.3, from the consideration of the single-particle states, the electron can occupy one of the two possible spin configurations $|s, s_z\rangle = |\frac{1}{2}, \pm\frac{1}{2}\rangle$. The highest energy state for the hole is the heavy-hole state. Thus the hole can occupy one of the two possible spin configurations $|j, j_z\rangle = |\frac{3}{2}, \pm\frac{3}{2}\rangle$. Since the exciton consists of an electron–hole pair, there are four different excitonic configurations with total angular momentum projections $M_J = s_z + j_z = \pm 1, \pm 2$, respectively. The selection rule for optical dipole transition is given by

$$\Delta M_J = \pm 1, \tag{7.16}$$

so that the $|\pm 2\rangle$ states are dipole forbidden and hence are called dark excitons. The states $|\pm 1\rangle$ can decay optically with an angular momentum transfer of $\hbar$ and are therefore called bright excitons. The emission of the state $|+1\rangle$ ($|-1\rangle$) results in the emission of a circular right R- (left L-)polarized photon.

In the picture of single-particle levels, the four possible excitonic states $M_J = s_z + j_z = \pm 1, \pm 2$ are energetically degenerate. Due to exchange interaction between the electron and the hole, which couples their different spin configurations, this degeneracy is lifted. For further discussion, it is convenient to divide the exchange interaction [124] (chapter by M. Bayer) into short-range and long-range parts. The short-range part is sensitive to the symmetry of the underlying lattice, whereas the long-range part is sensitive to the geometry of the envelope part of the excitonic wave function. The effect of the short-range part is twofold. First, it introduces a splitting between the bright and dark excitons due to the asymmetry of the underlying crystal. Second, it lifts the degeneracy of the two dark states $|\pm 2\rangle$ by mixing them into new eigenstates that are superpositions of the original states (see Fig. 7.6):

$$|X^{\pm}_{\text{dark}}\rangle = \frac{1}{\sqrt{2}}(|+2\rangle \pm |-2\rangle). \tag{7.17}$$

The long-range exchange interaction also contributes to the splitting between the dark and the bright excitons, and it introduces a splitting of the bright excitonic states $|\pm 1\rangle$ in

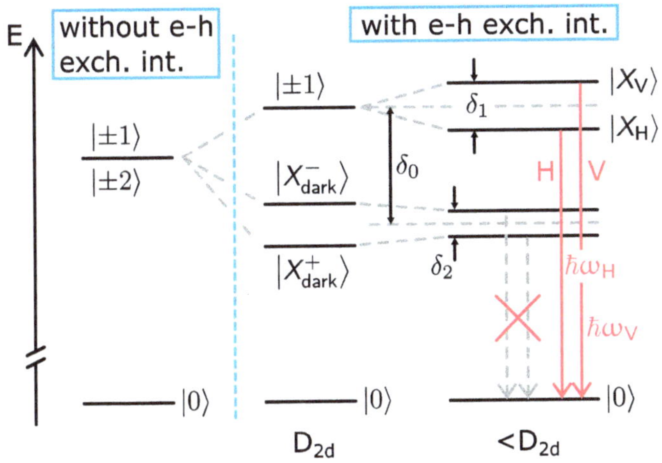

Figure 7.6: Energy level scheme of the excitonic ground state in a quantum dot with and without the inclusion of exchange interaction between the electron and hole spins. The initial fourfold degeneracy for noninteracting spins is lifted when electron–hole exchange is considered. This leads to optically allowed "bright" ($|\pm 1\rangle$) and forbidden "dark" exciton states ($|X^{\pm}_{\text{dark}}\rangle$) (see text). The bright states are further split into a doublet ($|X_{H,V}\rangle$) under the conditions of lowered in-plane symmetry ($<D_{2d}$). The arrows indicate the allowed optical transitions with their respective polarizations (H, V).

case the in-plane rotational symmetry D_{2d} of the QD is broken. As a result, new eigenstates are formed as superpositions of the original states:

$$|X_{H,V}\rangle = \frac{1}{\sqrt{2}}(|{+1}\rangle \pm |{-1}\rangle). \tag{7.18}$$

The new eigenstates give rise to two transition dipoles oriented along orthogonal directions (H, V) in the QD plane, which are typically found along the principal axes of the crystal matrix. Accordingly, the radiative transition of the state $|X_H\rangle$ ($|X_V\rangle$) results in the emission of a linear H- (V-)polarized photon. The H- and V-photons differ in energy according to the so-called fine-structure energy δ_1. Typical splitting energies for δ_0, δ_1, and δ_2 (see Fig. 7.6) for InGaAs/GaAs QDs have been reported in the order of 100–200 µeV, 0–50 µeV, and several µeV, respectively. However, also distinctly larger values of a few meV for δ_0 have been reported in the literature. It is important to note that nearly all QDs exhibit a certain fine-structure splitting. Nevertheless, several methods to modify the fine structure, like applying strain and magnetic and electric fields have been presented to change the excitonic fine-structure splitting, especially to tune it to zero, and even its sign has been reversed. In the case of $\delta_1 = 0$, circular polarized photons (R, L) are emitted from the states $|{+1}\rangle$ and $|{-1}\rangle$, which are then the eigenstates of the system. The different fine structure manipulation methods will be discussed in more detail in Chapter 9.

The excitation of a single electron–hole pair in a QD into a bright exciton state therefore gives rise to the emission of exactly a single-photon state. The one-photon wavefunction of such an excitonic photon is described by the photon-wavepacket creation operator and reads (see Eq. (4.21))

$$|\Psi_X\rangle = \int_{t=0}^{\infty} dt\,\xi(t)\hat{a}_X^\dagger(t)|0\rangle, \tag{7.19}$$

where $\hat{a}_X^\dagger(t)$ is the time domain creation operator that creates the exciton photon in its photonic mode, and $\xi(t)$ is the one-photon wavepacket. It is given by

$$\xi(t) = \sqrt{\gamma_X}e^{-i\omega_X t}e^{-\gamma_X t/2}, \tag{7.20}$$

where ω_X is the angular frequency of the exciton transition, and γ_X is the decay rate of the exciton state.

The dark exciton states might influence the radiative quantum efficiency of a QD. The electron or the hole of an exciton in the bright state can undergo a spin-flip via carrier-phonon interactions together with the spin-orbit coupling or the short-range exchange interaction, which leaves the QD in a dark configuration, i. e., with a dark exciton state $|X_{\text{dark}}^{\pm}\rangle$. This means that the exciton emission is quenched, as a spin-flip has to occur prior to radiative recombination. This might take distinctly longer than the repetition rate of a pulsed excitation laser if $\delta_0 > k_B T$. However, spin flips between bright and dark exciton states typically occur on a tens-of-nanosecond or even longer time scale, i. e., at least one order of magnitude slower than the radiative recombination time of bright exciton states, which is of the order of 1 ns.

7.3.2 Neutral biexciton and entangled photon-pair generation

The neutral biexciton (XX) consists of two electrons and two holes with antiparallel spins, which occupy the first quantized state (s-shells) of the conduction and the valence band in the QD, respectively (see Fig. 7.7). Due to the Coulomb interaction between the four charged carriers involved, the XX exhibits a so-called biexciton binding energy, which in most cases is positive:

$$E_{XX}^{B} = 2E_X - E_{XX}. \tag{7.21}$$

Typical values are 2–3 meV for In(Ga)As/GaAs QDs and 4–6 meV in InP/GaInP QDs. There are also reports of negative biexciton binding energies. For example, −5 meV for InGaN/GaN QDs and up to −30 meV for GaN/AlGaN QDs ([125], Chapter 3). A negative binding energy is only possible due to the strong 3D carrier confinement in the QD, whereas in higher-dimensional systems (3D, 2D, and 1D), only positive values are possible. Biexcitons form a spin-singlet, and therefore this state does not have any degeneracy, and, furthermore, the direct recombination into the crystal ground state is dipole forbidden. However, it is possible that the two excitons forming the biexcitonic state decay consecu-

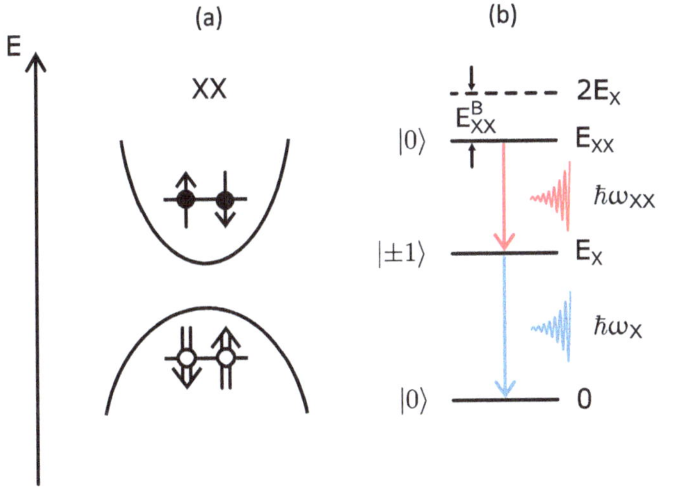

Figure 7.7: (a) Single particle picture of the biexciton. (b) Excitonic picture with a simplified scheme of the radiative biexciton-exciton cascade neglecting the excitonic fine structure splitting. E_{XX}^{B} denotes the biexciton binding energy.

tively. Due to the biexciton binding energy, the recombination energy of the first exciton is lowered by E_{XX}^{B} with respect to the second electron–hole pair (see Fig. 7.7(b)).

Since the decay of the biexciton (decay of the first exciton) projects the QD into an excitonic configuration, resulting in the successive decay of an exciton, we speak of the biexciton–exciton cascade. The first emitted photon a biexcitonic photon, and the second is called an excitonic photon. This cascade is of particular interest, as it allows for the production of polarization-entangled photon pairs and time-bin-entangled photon pairs as we will discuss in Section 5.2.9.

For a more detailed discussion of the radiative biexciton-exciton cascade, the exciton fine structure has to be considered since the exciton represents the intermediate state of the cascade. Thus the polarizations of the biexcitonic and excitonic photons are determined by the excitonic states, and they are perfectly correlated. In practice, different scattering effects may induce a transition between the excitonic states, which would reduce the perfect polarization correlation. Figure 7.8 shows the schematic sketch of the biexciton–exciton cascade for two different cases, i. e., zero and a distinct nonzero excitonic fine-structure splitting.

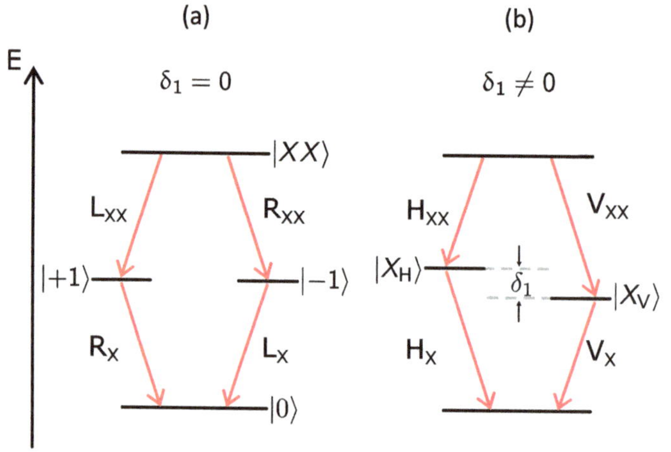

Figure 7.8: Energy scheme of the biexciton–exciton system with the respective radiative transitions and their corresponding polarization. (a) For zero excitonic fine-structure splitting ($\delta_1 = 0$), there exist two possible decay channels with opposite circular polarizations (left hand (L) and right hand (R)). (b) For nonzero excitonic fine-structure splitting ($\delta_1 \neq 0$), two equally linear polarized (horizontal (H) and vertical (V)) biexciton–exciton cascades exist. For a detailed description, see text.

In the case of a *vanishing fine-structure splitting* ($\delta_1 = 0$), the exciton states $|\pm 1\rangle$ are eigenstates of the system, and the polarization of the emitted photons are determined by the respective selection rules discussed above. The biexcitonic state $|XX\rangle$ is a degenerate spin-singlet state ($J = 0$). If the biexciton decays into the $|+1\rangle$ state, then an L_{XX}-photon is emitted followed by an R_X-photon, and if it decays into the $|-1\rangle$ state, then an R_{XX}-

photon is emitted followed by an L_X-photon. Since there is no way to distinguish which of the two paths the system followed, other than measuring the emitted polarizations, the generated two-photon state is described by the superposition of the two paths, i. e., a polarization-entangled state (see also Eq. (5.62)):

$$|\Psi^+\rangle = \frac{1}{\sqrt{2}}(|L_{XX}R_X\rangle + |R_{XX}L_X\rangle), \tag{7.22}$$

which is called a *Bell* state. The state is written in the circular basis. Using the expressions

$$|R\rangle = \frac{1}{\sqrt{2}}(|H\rangle + |e^{\frac{-i\pi}{2}}V\rangle), \quad |L\rangle = \frac{1}{\sqrt{2}}(|H\rangle + |e^{\frac{i\pi}{2}}V\rangle) \tag{7.23}$$

and

$$|D\rangle = \frac{1}{\sqrt{2}}(|H\rangle + |V\rangle), \quad |A\rangle = \frac{1}{\sqrt{2}}(|H\rangle - |V\rangle), \tag{7.24}$$

the state can also be written in the rectilinear polarization basis, i. e., horizontally and vertically polarized states (H, V), and in the diagonal polarization basis, i. e., diagonally and antidiagonally polarized states (D, A):

$$\begin{aligned} |\Psi^+\rangle &= \frac{1}{\sqrt{2}}(|L_{XX}R_X\rangle + |R_{XX}L_X\rangle) \\ &= \frac{1}{\sqrt{2}}(|H_{XX}H_X\rangle + |V_{XX}V_X\rangle) \\ &= \frac{1}{\sqrt{2}}(|D_{XX}D_X\rangle + |A_{XX}A_X\rangle). \end{aligned} \tag{7.25}$$

Obviously, the state appears entangled in each of the three polarization bases. Furthermore, we can learn from Eq. (7.25) that the biexcionic and excitonic photons will have uniform polarization correlations if measured in rectilinear or diagonal bases and opposite polarization correlations if measured in circular polarization.

In the case of a *nonvanishing fine-structure splitting* ($\delta_1 \neq 0$), the intermediate excitonic states $|X_H\rangle$ and $|X_V\rangle$ are superposition states of the $|\pm 1\rangle$ states (Eq. (7.18)) and are split by the fine-structure splitting (see Fig. 7.8). For the description of the biexciton–exciton system, we now have also to consider the phase evolution of the system during the decay. The recombination of the first electron–hole pair in the biexciton state creates either a horizontally or vertically polarized biexcitonic photon (H_{XX} or V_{XX}) and leaves the QD in the exciton states $|X_H\rangle$ or $|X_V\rangle$, respectively. Consequently, the photon/exciton system is projected into the entangled state $(|H_{XX}X_H\rangle + |V_{XX}X_V\rangle)/\sqrt{2}$. As the two exciton states are energetically split by δ_1, they will experience a phase difference with time of $e^{i\delta_1\tau/\hbar}$, where τ is the time delay between the two emitted photons (keeping in mind that exciton and biexciton photons with different polarizations have different energies because of the FSS). Thus the two-photon polarization state produced by the biexciton–exciton cascade is given by

$$|\Psi\rangle = \frac{1}{\sqrt{2}}(|H_{XX}H_X\rangle + e^{i\delta_1\tau/\hbar}|V_{XX}V_X\rangle). \tag{7.26}$$

This expression is very similar to the middle term of Eq. (7.25) (rectilinear polarization basis), except for the additional phase factor. This means that the emitted photons of the biexciton–exciton cascade are also entangled, except that in this case the entangled state depends upon the delay in emission time τ between the two photons.

For the further discussion, we have to distinguish between time-correlated single-photon counting experiments, which allow us to measure the time delay between the two photons to be recorded for each pair, and time-integrated measurements.

In time-integrated measurements, we integrate over all possible delay times τ between the two photons, and therefore the total QD emission is used in an experiment or for a specific application. Now the QD will emit photon pairs with a range of temporal spacings in the order of the biexciton and exciton radiative lifetimes. The integration over all events will randomize the phase between the two terms in Eq. (7.26), effectively destroying the polarization entanglement between the two photons. It is important to remark that if the exciton fine-structure splitting δ_1 is nonzero but smaller than the homogeneous linewidth of the exciton, then the phase relationship between the two terms remains well defined, and thus the polarization entanglement is preserved. In the energy picture, this means that the H and V polarized photons cannot be distinguished with respect to their energy, and thus the two possible decay paths of the biexciton–exciton cascade are no longer distinguishable without measuring the polarization of the respective photons.

In time-resolved measurements the time delay between the photons can be accurately measured for each photon pair. This means in practice that the time resolution has to be much faster than the beating period of the phase. In this case, we can in principle use each generated state, which has been demonstrated in recent time-gated experiments, even in case of $\delta_1 \neq 0$. In this case, we can observe a sinusoidal variation in the fidelity with the maximally entangled state $|\Psi^+\rangle$. In practice, this complicates the analysis. This is why different methods have been developed to erase the fine-structure splitting, which will be discussed later in Chapter 9.

The radiative biexciton–exciton cascade offers an additional possibility to generate entangled photon pairs in another degree of freedom, namely the time degree of freedom that uses the so-called time-bin entanglement with two different "time-bins" (see also Chapter 5). Time-bin-entangled photon pairs are insensitive to the polarization non-degeneracy, and therefore no stringent requirements on excitonic fine-structure splittings are present. The basic principle of the ideal scheme is the following. The QD interacts with two pump pulses that are able to resonantly excite the biexcitonic state ($|XX\rangle$). It is important that the two pump pulses have to be in a coherent superposition state, which means that they have a well-defined delay and relative phase ϕ. Their phase difference has to be carried over to the phase difference between the emitted photon pair. The excitation pulses define the time bins "early" (E) and "late" (L). The first pulse excites the biexciton with probability p_1, and the second pulse with probability p_2. The time difference between the two excitation pulses has to be larger than the coherence time of the emitted photons from the biexciton–exciton cascade which is in the order of 1 ns. This specific excitation scheme results in either a biexciton–exciton photon pair emitted after the early or late excitation. The entangled two-photon state is then given by

$$|\Psi\rangle = \sqrt{p_1}|E_{XX}E_X\rangle + e^{i\phi}\sqrt{(1-p_1)p_2}|L_{XX}L_X\rangle, \tag{7.27}$$

where ϕ is the relative phase between the two excitation pulses. A maximally entangled pulse is achieved for $p_1 = 1/2$ and $p_2 = 1$ and is given by (see Section 5.2.9 and [126], Chapter 8)

$$|\Psi\rangle = \frac{1}{\sqrt{2}}(|E_{XX}E_X\rangle + e^{i\phi}|L_{XX}L_X\rangle). \tag{7.28}$$

From this form of the wavefunction, which includes the phase factor $e^{i\phi}$, it also becomes clear that a coherent excitation process is necessary. If the pump process is not phase preserving, as is the case for nonresonant pumping schemes, then ϕ will not be the phase between the pump pulses but some random phase in each emission cycle. This sums up in an overall mixed state destroying the observation of entanglement.

Two possible coherent pumping schemes of the biexciton are discussed in the following (and at the end of Section 5.2.9).

In the first scheme the biexciton is coherently excited from the ground state $|0\rangle$ using a two-photon excitation scheme (see Fig. 7.9(a)). This scheme has been already experimentally demonstrated and is discussed in somewhat more detail in Section 10.2.3.4. Here the laser is detuned from the $|X\rangle$-state and only excites the quantum dot resonantly over a virtual state into the $|XX\rangle$-state. The following photon pair generation scheme uses only one recombination pathway of the biexciton–exciton cascade to have the photons in a well-defined polarization mode. This scheme with the direct excitation from the ground state has some drawback since it does not suppress four photon emission events caused by double excitation, i. e., excitation of both early and late time bins. Such a photon cascade in both time bins is called a four-photon event. This is because the entanglement generation of the time bins depends on a probabilistic generation of one photon cascade either in the E or in the L time bin. To avoid for photon events in practice, low excitation probabilities have to be used, which prevents on-demand emission of time-bin-entangled photons. Yet the general possibility of generating time-bin entanglement has been proven via this technique.

The second scheme avoids the problem of double excitation by using a coherent excitation from a metastable state $|M\rangle$ (see Fig. 7.9(b)). In a first step the quantum dot has to be excited into a metastable state, and in a second step, further excited to the biexciton state with two consecutive excitation pulses (with $p_1 = 1/2$ and $p_2 = 1$). The scheme again uses only one recombination pathway of the biexciton–exciton cascade.

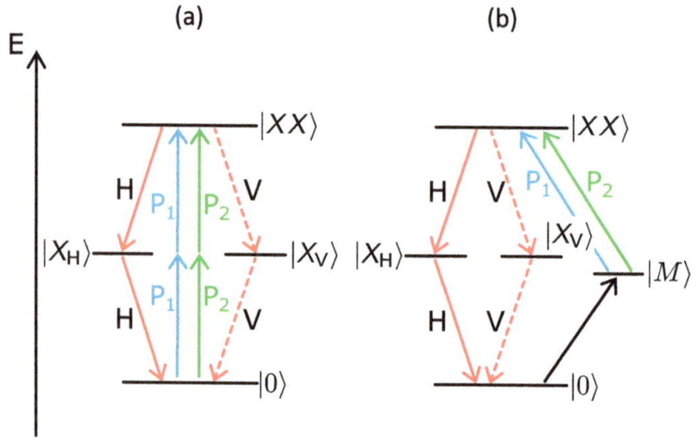

Figure 7.9: Schematic excitation schemes to generate time-bin-entangled photon pairs from the biexciton–exciton cascade. Only one recombination pathway of the biexciton–exciton cascade is used in both schemes. (a) The scheme is based on a resonant two-photon excitation of the biexcitonic state and the subsequent radiative biexciton–exciton cascade. (b) The excitation of the biexciton is performed in a two-step process via a metastable state. The excitation probabilities are described by p_1 and p_2 (for details, see text).

This scheme requires the optical addressability of a metastable state. Dark excitonic states or an off-resonant bright exciton in a photonic band-gap structure have been proposed. The drawback here is the difficulty of optically exciting both states, and practical demonstrations are still lacking.

As seen above, quantum information can be encoded in different photon degrees of freedom, e. g., polarization and time, each having its pros and cons. The polarization encoding scheme distinguishes itself by low complexity. The polarization of a photon can be easily manipulated using retarders (waveplates) and measured by polarizers. However, for fiber-based applications, noise in the fiber and in the environment could spoil the polarization state and therefore limit the success rate of this method. Especially, for long distance communication trough optical fibers, time-bin encoding has some advantage. The two time-bin-entangled photons are only some nanoseconds apart. Thus only changes in the environment that are faster than their temporal separation will lead to a degradation of the success rate, i. e., the fluctuations have to be in the GHz range.

It is important to note that QDs have also been used to generate states that exhibit entanglement in more than one degree of freedom, so-called hyperentangled states [154]. A hyperentangled state written in the polarization and time-bin degrees of freedom reads

$$|\Psi\rangle = \frac{1}{2}\left(|H_{XX}H_X\rangle + |V_{XX}V_X\rangle\right) \otimes \left(|E_{XX}E_X\rangle + |L_{XX}L_X\rangle\right). \tag{7.29}$$

These states represent a valuable tool to eventually reduce the experimental requirements and resource overheads for the successful implementation of quantum information protocols. For example, they allow complete Bell state measurements using only linear optics. However, the drawback is that they require more complex generation and detection schemes as outlined in [126] (Chapter 8).

7.3.3 Limitations on photon indistinguishability from cascaded emissions

It is important to note that the radiative biexciton–exciton cascade leads to a time-ordering of the emission of biexcitonic and excitonic photons due to the three-level ladder system. This results in a temporal correlation between the two photons. Therefore the individual cascades of the biexciton–exciton cascade, which are denoted above, e. g., as $|E_{XX}E_X\rangle$, $|L_{XX}L_X\rangle$, $|H_{XX}H_X\rangle$, or $|V_{XX}V_X\rangle$, are not in a product state as suggested by the notation, but are themselves entangled states. In the following, we can assume that the biexcitonic state is excited at $t = 0$, e. g., with a short laser pulse (two-photon excitation), and the biexciton state then decays via the exciton state to the ground state. The two-photon wavefunction of such a photonic cascade is described by the photon pair creation operator (see Eq. (4.33)) and reads

$$|\Psi_{XX,X}\rangle = P^{\dagger}_{\beta,XX,X}|0\rangle = \int_{t=0}^{\infty} \mathrm{d}t \int_{t'=t}^{\infty} \mathrm{d}t' \beta(t,t')\hat{a}^{\dagger}_{XX}(t)\hat{a}^{\dagger}_{X}(t')|0\rangle, \tag{7.30}$$

where $\hat{a}^{\dagger}_{XX}(t)$ and $\hat{a}^{\dagger}_{X}(t')$ are the time-domain creation operators that create the biexcitonic and excitonic photons in their respective modes, and $\beta(t,t')$ is the joint two-photon wavepacket of the emitted photon pair. It is given by

$$\beta(t,t') = \sqrt{\gamma_{XX}\gamma_X} e^{-i\omega_{XX}t} e^{-i\omega_X t'} e^{-\gamma_{XX}t/2} e^{-\gamma_X(t'-t)/2}, \tag{7.31}$$

where ω_{XX} and ω_X are the angular frequencies of the biexciton and exciton transitions, and γ_{XX} and γ_X are the decay rates of the biexciton and exciton states, respectively. The two-photon wavefunction is entangled due to the presence of the factor $e^{-\gamma_X(t'-t)/2}$, which describes the temporal correlation of the emission events.

This entanglement is not problematic for applications where only a single pair is used at any given time, such as quantum key distributions, but it becomes problematic for experiments and applications that rely on the indistinguishability I of the photons, such as entanglement swapping. The single photon emitted from the biexciton (exciton) state can be calculated by tracing out the exciton (biexciton) state from the pure two-photon state. Since the two-photon state is entangled, the individual exciton and biexciton single-photon states are mixed states, which can be described by density matrices ρ_{XX} and ρ_X. The separability of the emitted photon states can be quantified by its trace purity ($\mathrm{Tr}_{XX,X}[\rho^2_{XX,X}] = I$), which quantifies the indistinguishability of the emitted photons ([175, 184]):

$$I = \gamma_{XX}/(\gamma_{XX} + \gamma_X). \tag{7.32}$$

Obviously, the indistinguishability of both emitted photons is limited by their decay rate ratio. For QDs, the radiative recombination rate of the biexciton is typically twice that of the exciton, $\gamma_{XX} = 2\gamma_X$, which limits the indistinguishability of both photons to ~0.66. In Fig. 7.10 the indistinguishability of both biexciton and exciton transitions in dependence of their decay rate ratio is plotted. It is obvious that the indistinguishability of the photons can be raised by making the recombination rate of the biexciton much faster than that of the exciton. This could be implemented by embedding the QD into nanophotonic resonators in a way to selectively enhance the biexcitonic decay more than the excitonic one by making use of an asymmetric Purcell effect. Such an asymmetric Purcell effect could be implemented, e. g., in a circular Bragg grating cavity, which possesses a relatively broad spectral cavity resonance (meV range) and could enhance both transitions in an asymmetric way. An advantageous alignment comprises the biexciton line aligned close to the center of the cavity mode (maximum enhancement) and the exciton line therefore detuned by the biexciton binding energy (less enhancement; see Chapter 6, in particular, Eq. (6.68)).

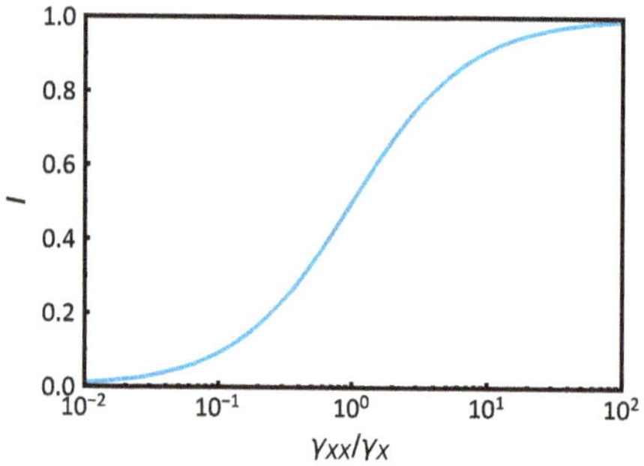

Figure 7.10: Indistinguishability of both biexciton and exciton transitions in dependence of their decay rate ratio.

Equation (7.32) represents an upper bound of the indistinguishability since in practice additional dephasing processes could further decrease I for both photons. Typically, the visibility of the exciton emission is lower than that of the biexciton emission due to the longer recombination lifetime, and therefore the biexciton state is less affected by the same dephasing rate (see Chapter 11).

7.3.4 Limitations on photon purity from the finite pulse length of a resonant excitation laser

The single-photon purity and indistinguishability critically depend on the type of optical or electrical excitation, which will be discussed in detail in Chapter 10. Limitations are caused by environmental effects and imperfections of the excitation process. In this section, we will highlight that even a perfect two-level system excited by a resonant laser pulse can affect the single-photon purity of the pulsed emission. In the following, we will show the effect of the finite duration of a resonant excitation π-pulse on the degree of photon antibunching, i. e., on the single-photon purity of two- and three-level systems. The fundamental limitation results from the emission of a photon during the presence of the excitation pulse, which allows a reexcitation of the two-level system after a first single-photon emission process, therefore allowing a subsequent second emission process, finally leading to multiphoton emission during the pulse. In this sense, it is an intrinsic limitation of the idealized two-level system and the finite excitation pulse width.

These effects have been theoretically [32, 45] and experimentally [32] studied in semiconductor quantum dots. Figure 7.11(b) displays measured photon autocorrelation data of on-demand triggered resonance fluorescence photons from a single QD at different excitation rates and with an excitation pulse width of 100 ps. The pulsed autocorrelation function shows strongly reduced areas of the zero-delay peaks indicating dominating single-photon emission. A zoom-in view of the zero-delay peak is shown in Figs. 7.11(c,d) and 7.11(e), displaying their dependence on the excitation pulse width. Here $g^2(\tau)$ represents the autocorrelation function of the continuous time delay τ, whereas $G^2(\tau_n)$ denotes the pulsed-mode autocorrelation function of the discretized time delay $\tau_n = nT$ obtained by integrating the nth pulse in $g^2(\tau)$, where $T = 1/f$ is the pulse period. In Fig. 7.11(e) a clear increase of $G^2(0)$ is observed with increasing excitation pulse width. This is caused by the above discussed reexcitation process of the QD, where the reexcitation probability increases with increasing excitation pulse width. Interestingly, the $g^{(2)}(\tau)$ values show a dip around $\tau = 0$ and are always zero at $\tau = 0$ for the deconvolved values. This means that more than one photon can be emitted during the 100-ps-long excitation pulse, but not at the same time as the first emission process.

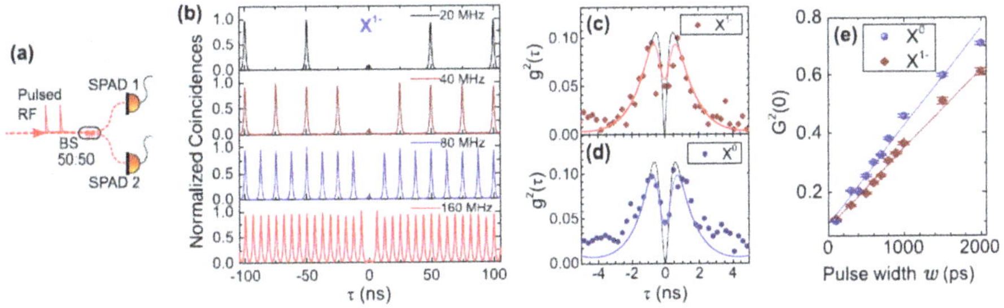

Figure 7.11: Second-order photon correlation measurements of a resonantly excited single QD. (a) Sketch of the HBT-setup. (b) Second-order photon correlation measurements for different excitation rates. (c), (d) Zoomed-in views of the time-zero peaks. (e) $G^2(0)$ as a function of excitation pulses width under π-pulse excitation. Reprinted with permission from [32]. Copyright (2023) the Optical Society.

In QDs, these unavoidable reexcitation processes can be significantly reduced by exploiting the more complex four-level system of the biexciton–exciton cascade [66]. Using a scheme based on resonant two-photon excitation of the biexciton significantly suppresses the reexcitation processes, therefore improving the achievable degree of photon antibunching. After the biexcitonic emission of the first photon, the system is found in the intermediate X state. For a two-photon excitation process of the biexciton (see Fig. 7.9), the reexcitation process of the biexciton is strongly suppressed, because the laser is far detuned from the biexciton transition ($|XX\rangle \rightarrow |X\rangle$) by half the biexciton binding energy (~ few meV). A reexcitation of the biexciton can only happen after the full radiative biexciton–exciton cascade to the ground state ($|XX\rangle \rightarrow |X\rangle \rightarrow |0\rangle$) has occurred. Thus the probability of the emission of a second biexcitonic photon is reduced accordingly. Figure 7.12 displays a comparison of the degree of second-order coherence as a function of the pulse length for a resonantly driven two-level system (black) and a two-photon excitation of the biexciton (red). For both cases, the values increase with increasing pulse length and finally approach one for very long pulses. For all pulse lengths, the values obtained from the two-photon excitation are smaller than for the resonantly excited two-level system. An analysis of the data reveals a scaling behavior of approximately $g^{(2)}(0) \sim \Delta t\, \gamma$ for the resonantly driven two-level system and $g^{(2)}(0) \sim (\Delta t\, \gamma)^2$ for the two-photon excitation scheme, with Δt the pulse length of the excitation laser and γ the decay rate of the investigated transition. For very short pulses, the achieved improvement amounts several orders of magnitude. The so far best reported $g^{(2)}(0)$ value from a QD transition under resonant two-photon excitation is ~$(7.5 \pm 1.6) \times 10^{-5}$ [179], a value hardly achievable from any other single quantum emitter such as single atoms, ions, molecules, or color centers in diamond. Nevertheless, it is important to note that the indistinguishability of the emitted photons from the biexciton–exciton cascade ($|XX\rangle \rightarrow |X\rangle \rightarrow |0\rangle$) is limited by their decay rate ratio as discussed in the previous section.

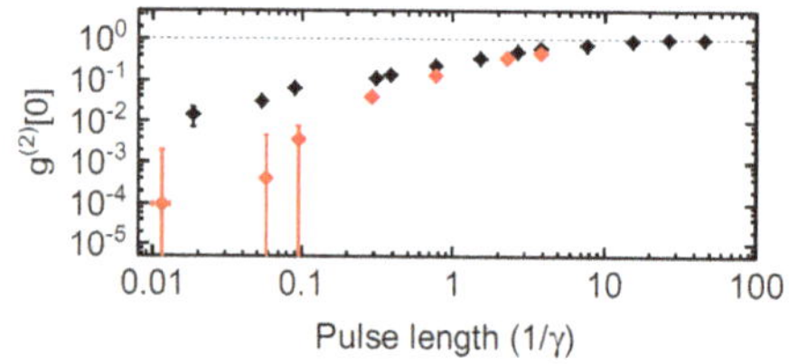

Figure 7.12: Measured values of $g^2(0)$ as a function of the pulse length for a resonantly driven two-level system (black) and two-photon excitation of the biexciton (red). The Poissonian level is indicated by the dashed line. Image reprinted with permission of Springer Nature from [66].

7.3.5 Trion and the generation of spin-photon entanglement

The simplest charged exciton configuration in a QD is either composed of two electrons and one hole, the negative trion state (X^-), or by two holes and one electron, the positive trion state (X^+), where all charge carriers are in their respective s-shells (see Fig. 7.13). Due to the Coulomb correlation between three charge carriers, the trion states are split from the neutral exciton line by a binding energy of several meV. Both positive and negative shifts have been reported. The sign of the shift depends on whether the binding energy with the additional carrier dominates over the repulsion energy between carriers with equal charge or vice versa.

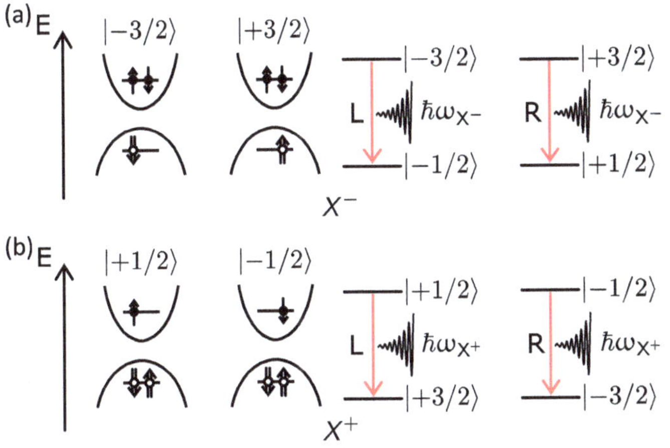

Figure 7.13: (a) Left: Schematic single-particle picture of a negative trion ground state in a quantum dot. All states are labeled according to their angular momentum projection, where small arrows indicate the involved possible spin states of the electrons ($s_z = \pm 1/2$) and the holes ($J_z = \pm 3/2$). Right: Excitonic picture, indicating both possible spin states of the X^- state ($J_z = \pm 3/2$) prior to a radiative transition of one electron–hole pair (long red arrow), which results in the emission of a single photon leaving a single electron in the QD with two possible spin states ($s_z = \pm 1/2$). (b) Left: Schematic single-particle picture of a positive trion ground state in a quantum dot. All states are labeled according to their angular momentum projection where small arrows indicate the involved possible spin states of the electrons ($s_z = \pm 1/2$) and holes ($J_z = \pm 3/2$). Right: Excitonic picture, indicating both possible spin states of the X^+ state ($J_z = \pm 1/2$) prior to a radiative transition of one electron–hole pair (long red arrow), which results in the emission of a single photon leaving a single hole in the QD with two possible spin states ($j_z = \pm 3/2$).

In the following, the negative trion is exemplarily discussed. In the X^- state the two s-shell electrons must have opposite spins due to the Pauli exclusion principle, whereas both spin orientations ($j_z = \pm 3/2$) for the hole are allowed. Consequently, the electronic configuration of the electrons is a spin-singlet ($S = 0$). Therefore these states are not subject to exchange interaction, and the same is true for the single spin of the hole in the ground state (Kramers degeneracy). Thus the X^- is double degenerate, and the trion transitions show no fine-structure splitting like it can be observed for the exciton transitions. The trion transitions for opposite spins have orthogonal circular polarizations (R, L), which can be exploited for spin-readout of the system. The same discussion is analogously valid for the X^+ state. The radiative transition of the trion state can be used to generate on-demand single-photon states, which can be applied in single-photon sources (see Chapter 5).

It is important to note that after the radiative recombination of an electron–hole pair of the trion state, the spin of the remaining charge particle (electron or hole) can be utilized for more complex and powerful quantum-light generation schemes, e. g., required for quantum network applications. Quantum network architectures typically consist of quantum memories to store quantum information, and the memories are interconnected by photons to exchange quantum information. Realizations of such networks require a quantum interface between stationary (matter) qubits and "flying" (photonic) qubits. Now the trion state allows the generation of entanglement between a stationary and a "flying" quantum system. The flying quantum system is represented by the emitted photon of the negative (X^-) or positive trion state (X^+), whereas the stationary quantum bit is represented by the remaining electron or hole spin, respectively.

To understand the basic idea, let us first discuss the energy level scheme of the negative trion state in a magnetic field directed perpendicularly to the QD growth direction (Voigt geometry). The application of an in-plane magnetic field induces a Zeeman splitting between the trion states and electron states, and it also modifies the system eigenstates and thus the optical selection rules (see Fig. 7.14). For high enough magnetic fields (B ~ 1T), each trion state is optically coupled to each of the two ground states, i. e., to both of the electron spin states in a Λ-configuration. In the following, we will consider the Λ-system of the trion state with the hole in the $|\Uparrow\rangle$-state (see Fig. 7.14(b)). Spontaneous emission of an H- or V-polarized photon at frequency $\omega + \delta\omega$ or, respectively, ω from the trion state brings the quantum dot back into the state $|\downarrow\rangle$ or, respectively, $|\uparrow\rangle$. Both transitions occur at the same rate, and their frequency difference $\delta\omega$ is given by the Zeeman splitting between the respective electron spin states $\Delta E^e = g_e \mu_B B$, where μ_B is the Bohr magneton, and g_e is the Landé g-factor for the electron in the QD.

The basic principle of the spin-photon entanglement is the following. The spontaneous emission decay in a Λ-system leads to entanglement between the emitted photon and the electron spin:

$$|\Psi\rangle = \frac{1}{\sqrt{2}}\big(i|\uparrow\rangle|H;\omega + \delta\omega\rangle + |\downarrow\rangle|V;\omega\rangle\big). \tag{7.33}$$

As we can see from Eq. (7.33), the spin is entangled with two photonic degrees of freedom, the polarization and the frequency of the photon. This would allow the knowledge of one photonic degree of freedom, e. g., polarization, by measuring the corresponding other degree of freedom, i. e., frequency and vice versa. To avoid this so-called "which-path

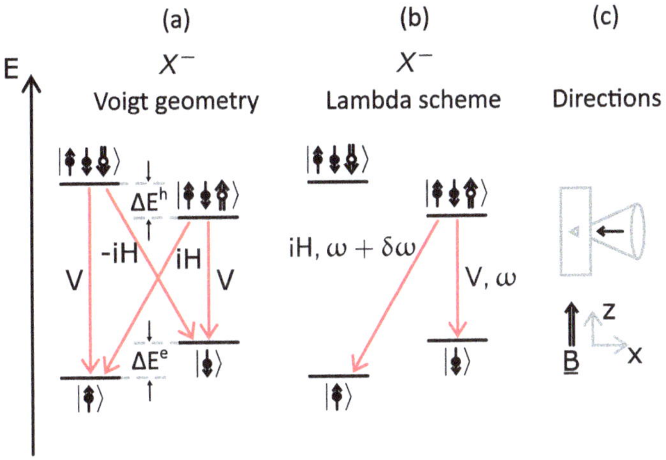

Figure 7.14: (a) Energy level structure of a negatively charged quantum dot in a magnetic field directed in plane (Voigt geometry), i. e., perpendicular to the growth direction z, with the respective Zeeman splittings. Small arrows indicate the involved possible spin states of the electrons (↑ or ↓) and the holes (⇑ or ⇓) denoting spins parallel and antiparallel to the magnetic field direction. The system has four linearly polarized transitions with the diagonal transitions carrying a $\pi/2$ phase factor [126]. V and H refer to linear polarizations, either perpendicular (V) or parallel (H) to the magnetic field. (b) Lambda scheme, which is used to discuss the generation of an entangled spin-photon state. Both radiative decays of the trion state are displayed with their corresponding polarization. $\delta\omega$ corresponds to the Zeeman splitting of the respective electron spin states. (c) Sketch of the sample and the respective directions of the magnetic field and optical axis.

information", one degree has to be erased to achieve an entangled spin-photon state. Possible eraser procedures will be discussed below. As the two trion transitions are energetically split by $\delta\omega$, they will experience a phase difference with time of $\exp i\delta\omega(t-t_g)$, where t_g is the generation time of the entangled state. The corresponding wavefunction is

$$|\Psi\rangle = \frac{1}{\sqrt{2}}(i|\uparrow\rangle|H;\omega+\delta\omega\rangle e^{i\delta\omega(t-t_g)} + |\downarrow\rangle|V;\omega\rangle). \tag{7.34}$$

There exist different methods to erase one of the two photonic degrees of freedom, frequency or polarization. For example, eraser of the center frequency degree of freedom has been obtained by frequency down-conversion using short pulses [34]. In this process the former two frequency-distinguishable photons ($\omega+\delta\omega$, ω) are so much frequency broadened by the down-conversion process that both converted photons are no longer distinguishable with respect to their center frequency. Therefore the resulting state becomes a spin-photon entangled state:

$$|\Psi\rangle = \frac{1}{\sqrt{2}}(i|\uparrow\rangle|H\rangle e^{i\delta\omega(t-t_g)} + |\downarrow\rangle|V\rangle). \tag{7.35}$$

A further method relies on time filtering using a fast detector with a time window faster than the beating period. After such a procedure, we also cannot distinguish the center frequencies of the two photons.

Furthermore, it is also possible to erase the polarization information using a polarizer at 45°, projecting the photon on $|H+V\rangle$ or $|H-V\rangle$ [53]. This eliminates the correlation with polarization, leaving an entangled state of photon color (frequency) and electron spin.

Note that the entanglement between the photon (flying qubit) and the spin (stationary qubit) can only persist on the timescale of the spin decoherence time T_2^{Spin} of the remaining electron or hole in the QD. So far the best reported values for T_2^{Spin} are in the order of microseconds. As a consequence, the photon can only travel about a kilometer ($L = cT_2$, c = speed of light) during this time, which limits the intended applications such as quantum communication or distributed quantum computation on this length scale.

7.3.6 Trion and the generation of a photon cluster state

The trion states can also be utilized to generate even more complex states, i. e., multiphoton entangled states like the so-called photonic graph states or photonic cluster states. Such states have been proposed for measurement-based quantum computation schemes and all-photonic quantum repeaters. The common theme of these approaches is that the entanglement structure between the photons allows implementing multiqubit logical gates by performing only single-qubit gates and measurements, instead of performing more complex quantum logical gates between single photons.

The basic idea for the generation of the photonic cluster states is to entangle the electron or the hole spin with a photon degree of freedom, e. g., polarization or the time of photon generation process (time-bin), and to repeat a series of basic operations such as spin rotations of the ground state electrons (or holes) and excitation and emission cycles of the radiative trion states.

In the following, we will discuss the photon cluster state generation process originally proposed by Lindner and Rudolph [114]. A QD with a single electron in the ground state where the electron possesses a degenerate spin-1/2 state manifold is considered (see Fig. 7.15). The basis denotes the spin projection along the z-axis. The excited states are negative trion states, and they are connected to the ground states by optical transitions governed by the selection rules. Thus only photons with right or left circular polarization are emitted for the corresponding spin configurations (see Fig. 7.15(b)). Only photons propagating along the z-axis are considered. Therefore, if one starts with the initial state $|\uparrow\rangle$ ($|\downarrow\rangle$), then an excitation to the state $|\uparrow\downarrow\Uparrow\rangle$ ($|\downarrow\uparrow\Downarrow\rangle$) followed by a radiative decay results in the emission of a right (left)-circular polarized photon $|R\rangle$ ($|L\rangle$) and leaves the QD in the state $|\uparrow\rangle$ ($|\downarrow\rangle$). Next, let us consider the superposition state $|\uparrow\rangle + |\downarrow\rangle$ as the initial state and excite the QD with a coherent excitation pulse with a linear polarization along the x-direction. Such a pulse will couple equally to both optical transitions

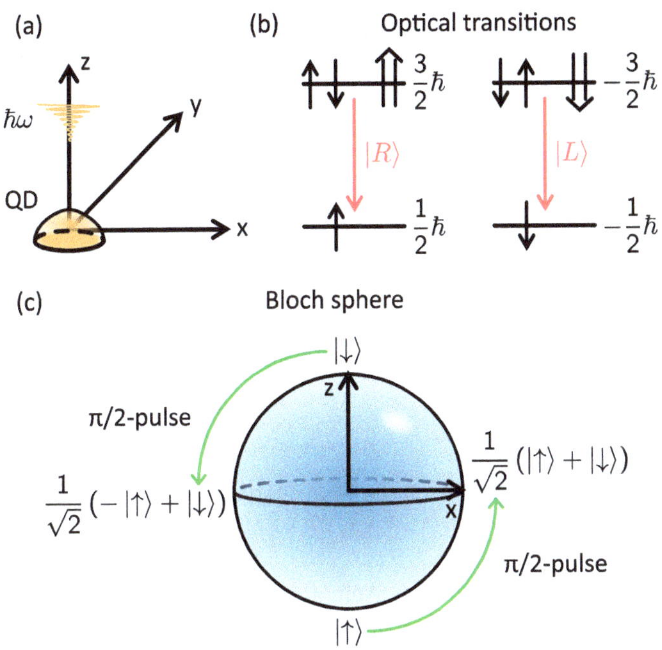

Figure 7.15: (a) Schematic picture of a QD located in the *xy*-plane (*z*-axis is the QD growth direction) together with single-photon emission along the *x*-direction. (b) Optical transitions and energy level structure of a negatively charged QD with a single electron in the ground state and a negative trion in the excited state (double degenerate states). Small arrows indicate the involved possible spin states of the electrons (↑ or ↓) and holes (⇑ or ⇓). The excited states have $J_z = \pm 3/2\hbar$ and the ground states have $J_z = \pm 1/2\hbar$, and thus only the single-photon transitions with right ($|R\rangle$) and left circular polarization ($|L\rangle$) are allowed. (c) Bloch sphere for a spin and spin transition for a $\pi/2$-pulse.

since the x-polarization can be regarded as a superposition of right and left circular polarized lights. Therefore both described processes happen in superposition, which will lead to an entangled state, i. e., the joint entangled electron–photon state is given by

$$|\Psi\rangle = \frac{1}{\sqrt{2}}(|\uparrow\rangle|R\rangle + |\downarrow\rangle|L\rangle) = \frac{1}{\sqrt{2}}(|\uparrow\rangle|R_1\rangle + |\downarrow\rangle|L_1\rangle), \tag{7.36}$$

where the subscript “1” denotes the first generated photon of the cluster state. In the next step, we apply a $\pi/2$-rotation on the spin around the y-axis. This rotation can be easily performed, e. g., by applying a magnetic field B (few tens of mT–100 mT) along the y-axis. The spin precession time is given by the g_e-factor of the electron in the QD and the strength of the magnetic field ($\omega_B = g_e\mu_B B/\hbar$). After the rotation, the state evolves to (without normalization and without considering a special method for spin rotation)

$$|\Psi\rangle = (|\uparrow\rangle + |\downarrow\rangle)|R_1\rangle + (-|\uparrow\rangle + |\downarrow\rangle)|L_1\rangle. \tag{7.37}$$

A second excitation pulse followed by the emission of a second photon will result in the following state:

$$
\begin{aligned}
|\Psi\rangle &= (|\uparrow\rangle|R_2\rangle + |\downarrow\rangle|L_2\rangle)|R_1\rangle + (-|\uparrow\rangle|R_2\rangle + |\downarrow\rangle|L_2\rangle)|L_1\rangle \\
&= |\uparrow\rangle|R_2\rangle|R_1\rangle + |\downarrow\rangle|L_2\rangle|R_1\rangle - |\uparrow\rangle|R_2\rangle|L_1\rangle + |\downarrow\rangle|L_2\rangle|L_1\rangle.
\end{aligned} \tag{7.38}
$$

Applying another $\pi/2$-rotation on the spin around the y-axis results in

$$
\begin{aligned}
|\Psi\rangle = {} & (|\uparrow\rangle + |\downarrow\rangle)|R_2\rangle|R_1\rangle + (-|\uparrow\rangle + |\downarrow\rangle)|L_2\rangle|R_1\rangle \\
& - (|\uparrow\rangle + |\downarrow\rangle)|R_2\rangle|L_1\rangle + (-|\uparrow\rangle + |\downarrow\rangle)|L_2\rangle|L_1\rangle,
\end{aligned} \tag{7.39}
$$

which is equal to

$$
\begin{aligned}
|\Psi\rangle = {} & |\uparrow\rangle|R_2\rangle|R_1\rangle + |\downarrow\rangle|R_2\rangle|R_1\rangle - |\uparrow\rangle|L_2\rangle|R_1\rangle + |\downarrow\rangle|L_2\rangle|R_1\rangle \\
& - |\uparrow\rangle|R_2\rangle|L_1\rangle - |\downarrow\rangle|R_2\rangle|L_1\rangle - |\uparrow\rangle|L_2\rangle|L_1\rangle + |\downarrow\rangle|L_2\rangle|L_1\rangle.
\end{aligned} \tag{7.40}
$$

This is a three-qubit linear cluster state with one spin and two photon states. Repeating the processes of excitation and spontaneous emission followed by $\pi/2$-rotation will produce a third photon such that the electron and the three photons are in a four-qubit linear cluster state. Further repeating this procedure will produce a continuous chain of photons in an entangled linear cluster state. A practical limitation of the length of the photonic cluster state (number of photons) is also given by the spin coherence time T_2^{Spin}, as already discussed before for the spin-photon entanglement.

The discussion of detailed implementations goes well beyond this textbook, and the special scheme depends on the favored final state and can be found in the original literature. A first proof-of-principle demonstration of a three-photon cluster state based on a complex scheme on a dark exciton state has been reported by Schwartz et al. [176]. Recent realizations used the hole spin in a QD as an entanglement tool, which also allowed high indistinguishability of the generated photons [28, 29]. The generation of even more complex graph states, i. e., two-dimensional cluster states, has been recently proposed by employing coupled quantum emitters [198]. This research field is rapidly growing, and the practical generation of new multiphoton-entangled states with quantum emitters will open new routes for quantum communication and quantum computing.

7.4 Summary

- In a quantum dot (0D-system) the electrons (holes) are confined in all three directions at a scale smaller than their De Broglie wavelength.
- The nomenclature of the discrete electronic states in QDs is similar to those of atomic shells, i. e., $s-, p-, d-, \ldots$ shells assign the lowest three electronic states in raising order.

- The conduction band ground states consist of two degenerate states $|s, s_z\rangle = |\frac{1}{2}, \pm\frac{1}{2}\rangle$. The valence band ground states posses six possible hole states, the heavy hole states $|j, j_z\rangle = |\frac{3}{2}, \pm\frac{3}{2}\rangle$ as the highest energy valance band, the light hole states $|j, j_z\rangle = |\frac{3}{2}, \pm\frac{1}{2}\rangle$, which are lower in energy due to the built-in strain by at least several meV, and the split-off states $|j, j_z\rangle = |\frac{1}{2}, \pm\frac{1}{2}\rangle$, which are further split-off in energy.
- The allowed optical interband transitions correspond to the radiative recombination of electron–hole pairs, where electron and holes occupy levels with the same shell index ($\Delta s = 0$), i. e., s-shell to s-shell or p-shell to p-shell transitions are thus allowed.
- Typical oscillator strength of an exciton, e. g., in InAs/GaAs QD, is $f \sim 10$, which results in a typical exciton radiative lifetime of ~1 ns.
- The neutral exciton (X) possesses four different excitonic configurations (noninteracting spins) with total angular momentum projections of $M_J = \pm1, \pm2$, i. e., two dark states ($|+2\rangle$, $|-2\rangle$) and two bright states ($|+1\rangle$, $|-1\rangle$). The emission of the bright state $|+1\rangle$ ($|-1\rangle$) results in the emission of a circular right R- (left L-)polarized photon.
- The electron-hole exchange interaction in real QDs couples their different spin configurations. The initial fourfold degeneracy for noninteracting spins is therefore lifted. This leads to optically allowed "bright" ($|\pm1\rangle$) and forbidden "dark" exciton states ($|X_{\pm}^{\text{dark}}\rangle$). The bright states are further split into a doublet ($|X_{H,V}\rangle$) under the conditions of lowered in-plane symmetry. Accordingly, the radiative transition of the state $|X_H\rangle$ ($|X_V\rangle$) results in the emission of a linear H- (V-)polarized photon. The H- and V-photons differ in energy according to the so-called fine-structure splitting δ_1.
- The neutral biexciton (XX) consists of two electron–hole pairs and forms a spin-singlet. Typical biexciton binding energies are in the range of several meV (positive and negative values are possible). The biexciton decays radiatively via the biexciton–exciton cascade by emitting a biexcitonic photon followed by an excitonic photon.
- For vanishing fine-structure splitting the two-photon state after radiative decay of XX is given by a Bell state, e. g., by $|\Psi^+\rangle = \frac{1}{\sqrt{2}}(|H_{XX}H_X\rangle + |V_{XX}V_X\rangle)$, and the photons are polarization entangled. In case of nonvanishing fine-structure splitting the two-photon state receives an additional phase factor between the two components. Since in a time-integrated measurement the phase factor randomizes, the polarization-entanglement is lost.
- The radiative biexciton–exciton cascade also offers the possibility to generate time-bin entangled photon pairs. The maximally entangled state is given by $|\Psi\rangle = \frac{1}{\sqrt{2}}(|E_{XX}E_X\rangle + e^{i\phi}|L_{XX}L_X\rangle)$ with E (L) the early (late) time bins and ϕ the relative phase between the pairs.
- The radiative biexciton–exciton cascade leads to a time-ordering of the emission of biexcitonic and excitonic photons, which results in a temporal correlation between them. Consequently, the indistinguishability I of both emitted photons is limited by their decay rate ratio according to $I = \gamma_{XX}/(\gamma_{XX} + \gamma_X)$.

- The trion states (X^-, X^+) are split from the neutral exciton line by a binding energy of several meV (positive and negative values are possible). The trion state is double degenerate, and the trion transitions for opposite spins have orthogonal circular polarizations with no fine-structure splitting.
- The trion state allows for the generation of entanglement between a stationary (spin) and a flying (photon) quantum system (qubits), a valuable resource for quantum networks.
- The trion state can also be utilized to generate multiphoton-entangled states, so-called photonic cluster states.

8 Environmental influences on the physical properties of quantum dots

8.1 Introduction

So far, we have discussed the physical properties of QDs as they were isolated objects, which is commonly called the artificial atom model. This model is a valuable tool to understand the basic physical properties of the QDs as discussed in the previous sections. However, it has its natural limits, and useful extensions are discussed within this chapter.

The electronic excitations of QDs are coupled to vibrational modes, i. e., phonons of the host material, and can also experience interaction with electric and magnetic fields caused by fluctuating charges and magnetic moments in their solid-state environment. These effects lead to dephasing, spectral diffusion, and therefore to linewidth broadening, which can also seriously limit their usability as quantum light sources in quantum photonic applications. However, a detailed physical understanding of all these effects allows us to predominantly avoid the thereby connected drawbacks. It is important to note that QD single-photon sources have demonstrated close to ideal single-photon emission at cryogenic temperatures, exploiting photonic structures and optimized resonant optical excitation schemes.

In the mostly used Stranski–Krastanov growth mode (see Section 12.2), QDs are grown on a 2D semiconductor layer, the so-called wetting layer (WL), and capped by a 3D semiconductor, the barrier layer. Figure 8.1 shows the modified potential landscape, now including the wetting layer. The wetting layer plays an important role for the carrier confinement and has to be considered especially under nonresonant excitation schemes. Moreover, additional optical interband transitions must be considered between the electron and hole states of the QD, the wetting and barrier layers [206]. These transitions, which involve one bound state and one delocalized state, are a consequence of the joint nature of the valence-to-conduction density of states in QDs. They also play a role on the broadening of quantum dot emission lines.

In the following, we will discuss the influence of phonon effects and fluctuating electric and magnetic fields on the photon properties.

8.2 Phonon effects

An epitaxially grown QD is embedded in a solid-state matrix, and therefore its electronic excitations are also coupled to vibrational modes of the host matrix. Both acoustic and optical phonon couplings are present in III–V semiconductor quantum dots. The interaction with phonons plays an important role for understanding the physics of semiconductor quantum dots and distinctly determines the performance of semiconductor quantum light sources. Phonons significantly contribute to carrier capture processes from

https://doi.org/10.1515/9783110703412-008

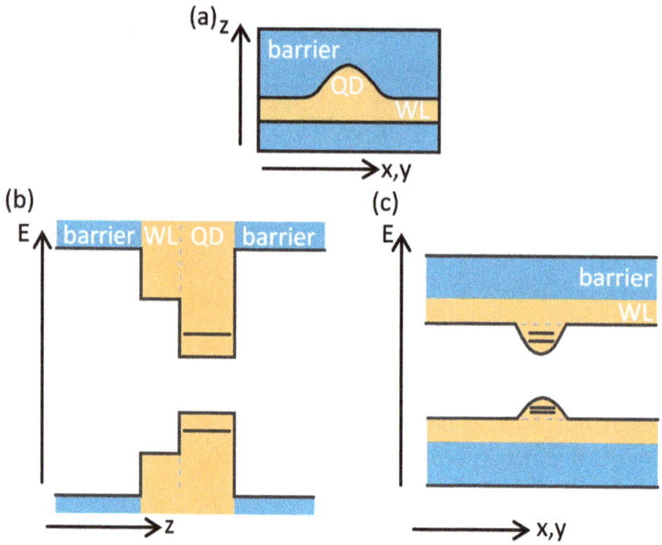

Figure 8.1: (a) Illustration of an epitaxially fabricated QD grown in the Stranski–Krastanov growth mode. (b) Scheme of the bandgap energy along the growth axis z. (c) A schematic representation of the energy levels arising from the harmonic oscillator-like potential in the growth plane (x, y) of the QD.

excited barrier and wetting layer states, and they are also relevant for carrier relaxation processes from higher excited states (p- and d-shells) to the QD ground state (s-shell). Moreover, their interaction with the electronic excitations (excitons, trions, biexcitons, etc.) determine the spectrum and linewidth of the respective optical transitions. Phonon scattering leads to dephasing in the QD system and thus reduces the coherence of the emitted photons. However, phonons can also be used in a beneficial way to make the excited state preparation of excitons and biexcitons more efficient, robust, and flexible. Phonon-induced carrier relaxation from excited states and excited state preparation via phonons will be discussed below and in Chapter 10.

This coupling of the exciton to either optical phonons (longitudinal optical (LO)) or acoustic phonons (longitudinal acoustic (LA) and transversal acoustic (TA)) can be due to different mechanisms with distinct origins. For all types of mechanisms, the coupling matrix elements can be factorized in a constant term, which accounts for the specific coupling mechanism, and another contribution term. This can be calculated starting from the wave function of the electron and hole in the quantum dot potential according to [103]

$$F_q^{e,h} = \int d^3r \left|\Psi^{e,h}(\vec{r})\right|^2 \exp(i\vec{q}\,\vec{r}), \tag{8.1}$$

where $F_q^{e,h}$ is the form factor, $\vec{q}$ is the phonon wave vector, and $\Psi^{e,h}(\vec{r})$ are the wavefunctions of the electron and hole inside the QD potential. The form factor determines

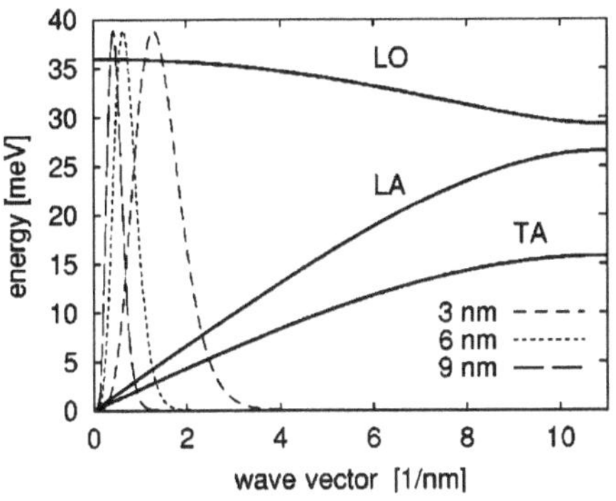

Figure 8.2: Dispersion relations of the LO, LA, and TA phonons together with normalized effective form factors (see Eq. (8.1)) describing the coupling of quantum dots with three different sizes to the various phonon modes. Reprinted with permission from [103]. Copyright (2023) by the American Physical Society.

the region in the phonon dispersion $\vec{q}$ to which the dot is effectively coupled. It is very instructive to plot both the phonon dispersion relations for LO, LA, and TA and the form factor for different QD sizes in one graph (see Fig. 8.2). In polar semiconductors the polarization of the lattice induced by the different charge contributions of the electron and hole in the QD is responsible for the LO-phonon exciton coupling. It is accounted for by the usual Fröhlich-coupling,[1] and it is sensitive to the specific wavefunctions of the carriers. In the particular case of identical electron and hole wavefunctions, LO phonon features would not be observable. Nevertheless, the Stark effect, shape asymmetries, and compositional nonuniformities of the QD that reduce the electron–hole wavefunction overlap, and all can enhance the dipole moment and, as a consequence, the strength of the coupling. For larger QDs with diameters d >10 nm, the LO phonon energy is almost constant for the wavevector range covered by the effective form factor. Therefore in absorption and PL-spectra, sharp peaks at single or multiple LO-phonon energy detunings (e. g., 36 meV for GaAs) from the zero-phonon line (ZPL) are observed. For smaller dots, the dispersion of the LO phonons becomes important, leading to a broadening of the LO phonon features with decreasing QD size. However, the ratio of the intensity of the first LO phonon satellite to the ZPL in absorption and PL is usually very low (~0.001) at low temperatures.

The acoustic phonons exhibit a continuous dispersion starting at zero frequency as depicted in Fig. 8.2. Therefore continuous acoustic phonon sidebands (PSB) appear in the spectra, which approach the ZPL. As we can see from Fig. 8.2, the assumption of a linear dispersion of both LA and TA is well satisfied for typical QD sizes.

1 In polar semiconductors a long-wavelength LO phonon leads to uniform displacements of the charged atoms within the unit cell. These relative displacements of the oppositely charged atoms generate a macroscopic electric field. An electron/hole can then interact with this field, which is known as the Fröhlich interaction.

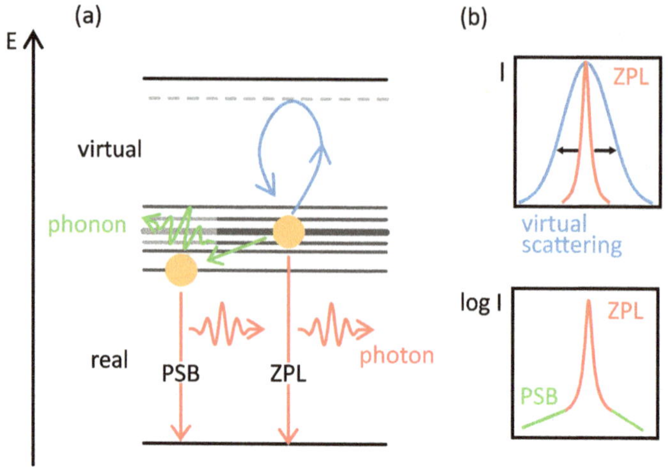

Figure 8.3: (a) Schematic figure of real and virtual phonon-induced processes. (b) The exciton can decay either purely radiative, which leads to a narrow zero-phonon line (ZPL) or may emit or absorb a phonon during the decay. This leads to an additional broad phonon sideband (PSB) in the emission spectrum. Virtual scattering of the exciton with thermal phonons leads to broadening of the ZPL due to pure dephasing.

For the following discussion, we can assume acoustic phonon coupling and a fully relaxed electronic excitation (e. g., exciton) in the s-shell of the QD. Two phonon-induced processes affect the optical emission properties of an excited state of a QD (see Fig. 8.3). First, emission and absorption of phonons during exciton recombination lead to phonon sideband (PSB) emission. When the QD relaxes from its excited state to its ground state, a phonon can be simultaneously emitted, thereby leading to the generation of a photon with energy lower than that emitted by the pure radiative decay (ZPL). Similarly, if the temperature is high enough that phonon modes are thermally occupied, then a phonon can be absorbed, resulting in the emission of a blue-detuned photon. At low temperatures, the acoustic phonon sidebands are asymmetric due to unequal probabilities of phonon absorption and emission. For both processes, the precise energy of the photon is not known, since it depends on the energy of the phonon, which is lost to the environment or absorbed from it. Consequently, the photons produced by these phonon-mediated transitions are inherently distinguishable. Second, broadening of the zero-phonon line (ZPL) can be observed because scattering of thermal phonons leads to virtual transitions to higher-lying excitonic states (see Fig. 8.3). Although these scattering processes do not change the population of the excited state, they can induce a dephasing of the QD dipole moment during the emission process, which also reduces the photon indistinguishability.

We will now discuss the impact of phonons at low temperature with a special focus on their implications on the brightness, purity, and indistinguishability of quantum light sources (see Chapter 5). At low temperatures, we observe asymmetric phonon sidebands due to the low thermal occupation of phonon modes. Figure 8.4 shows a low-

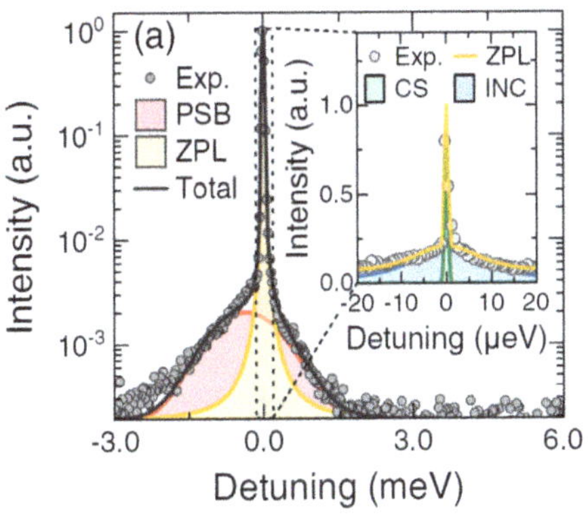

Figure 8.4: Resonant light scattering spectrum from a single quantum dot showing the zero-phonon line (ZPL) and phonon sidebands (PSB). Gray circles, experiment; yellow, fit to ZPL; red, fit to PSB; black line, total fit. Inset: High-resolution spectrum. Gray circles, experiment; green, fit to coherent scattering; blue, fit to incoherent resonance fluorescence; yellow, total fitted profile. Reprinted with permission from [19]. Copyright (2023) by the American Physical Society.

temperature resonant light scattering spectrum of a QD. The spectrum is dominated by the ZPL, but coupling to vibrational modes also leads to a broad and very weak PSB in the spectrum. The probability of a photon scattering event into the PSB is given by the square of the Franck–Condon factor B.[2] At low temperatures, typically, more than 90 % of the photons are funneled into the ZPL, which distinguishes semiconductor QDs from other single quantum emitters, e. g., from color centers in diamond, where typically a large fraction of photons are funneled into the PSB. Since the photons that are funneled into the PSB are distinguishable, this fraction cannot be used for quantum photonic applications, such as optical quantum computing and quantum repeaters, which require highly indistinguishable photons. Thus spectral filtering of the PSB becomes necessary, which will limit the achievable brightness of the quantum light source. Since at low temperatures $T < 5\,\mathrm{K}$, the occupation of the phonon modes is low, virtual transitions to higher excited states are negligible. Therefore the ZPL linewidth can be close to be Fourier limited, i. e., it is only lifetime broadened, which leads to photon emission into the ZPL with a high degree of indistinguishability.

It is important to note that the adverse effects of phonons on the performance of quantum light sources can be significantly reduced by embedding the QD into a waveguide or a microcavity and utilizing the Purcell effect [35]. A large coupling to a high-quality factor cavity can simultaneously reduce the effect of pure dephasing on the ZPL line and efficiently redirect the phonon sidebands in the zero-phonon line. Figure 8.5 clearly shows the positive cavity effect on the photon indistinguishability. The QD-cavity device was fabricated by a deterministic lithography process (see Chapter 12), thereby

2 In the literature, the letter B is typically used for the Franck–Condon factor.

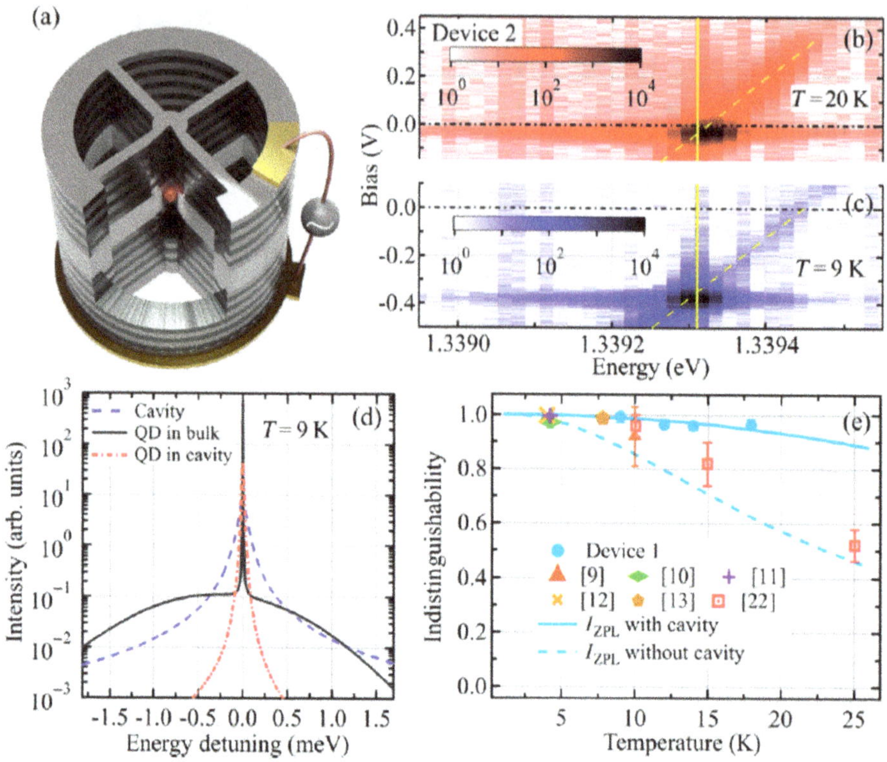

Figure 8.5: (a) Schematic sketch of an electrically tunable QD pillar cavity. (b), (c) Resonant fluorescence maps as functions of energy and bias voltages at temperatures 20 and 9 K, respectively. The cavity and QD spectral positions are indicated with solid and dashed yellow lines, respectively. (d) Comparison of the calculated emission spectra of a QD in a bulk photonic environment (black solid line) and coupled resonantly to a cavity (red dashed dotted line) at 9 K. The cavity spectrum is also indicated (blue dashed line). (e) Indistinguishability of the ZPL as a function of temperature. The measurements (device 1) are shown in blue circles. Calculations for the device 1 (solid line) and without cavity-QED effects (dashed line) are shown. The measurements reporting high indistinguishability of the ZPL and the temperature dependence in the absence of the Purcell effect are also indicated. Reprinted with permission from [58]. Copyright (2023) by the American Physical Society.

ensuring an optimized spatial overlap of the QD with the field maximum of the optical mode. The spectral matching of the QD at different temperatures can be achieved by applying electric fields and utilizing the QCSE. Figures 8.5(b,c) show the above-discussed ZPL on top of a broad phonon sideband emission. Furthermore, Fig. 8.5(d) highlights the influence of the cavity on the spectrum for a bulklike emission (i. e., without cavity) and for a QD in cavity, based on nonequilibrium Green's function calculations [58]. In the bulk case the ZPL sits on the broad PSB. Two main effects become evident when the QD is coupled to the cavity mode. First, the linewidth of the ZPL broadens due to the Purcell effect on the radiative emission. Second, the PSBs that fall outside the cavity spectrum are strongly suppressed. Consequently, the PSB emission is redirected toward the ZPL. The authors calculate the fraction emitted in the ZPL to increase from 81 % in bulk to 98 % in the cavity. At the same time the photon indistinguishability is considerably increased as we can see in Fig. 8.5(e).

Finally, it is important to note that phonon scattering inhibits simultaneous near-unity efficiency and indistinguishability in semiconductor single-photon sources. The trade-off has been extensively discussed in the literature. However, in a Purcell-enhanced source with near-unity indistinguishability (>99 %), an efficiency of approximately 96 % is reachable for realistic parameters [86]. Furthermore, it is important to remark that suppressing phonon-assisted photon emission in polarization-entangled photon pairs sources is even more challenging. It requires a polarization-degenerate Purcell enhancement of the exciton and biexciton lines together with PSB suppression. For example, due to the spectrally broad cavity resonance a bullseye cavity can support Purcell enhancement of both exciton and biexciton emission lines but hence fails in PSB suppression. Photonic molecules or double cavities (see Chapter 14) exhibiting two spectrally sharp cavity resonances have the potential to fulfill both aspects but rely on highly challenging photonic engineering. Exciton and biexciton transitions have to be simultaneously in resonance with one of the photonic modes at a time.

With increasing temperature, the Franck–Condon factor B quickly increases, which leads to increasing PSB emission and simultaneously reducing and broadening of the ZPL. Figure 8.6 displays μ-PL spectra (see Section 5.2.1) of an individual exciton emission line emitted from a CdSe QD for various temperatures. Each PL spectrum is normalized to the integrated intensity at 4.5 K. At low temperature a ZPL with a weak PSB is visible. With increasing temperature, we observe the expected redshift and broadening of the ZPL. Furthermore, asymmetric PSBs progressively develop in a range of a few meV. Above 70 K the PSBs become symmetric and tend to completely dominate the exciton emission line. The spectral linewidth increases to 7 meV at 100 K. Since the photons emitted into the PSBs are not indistinguishable, it becomes clear that semiconductor quantum dots cannot be used as sources of indistinguishable single photons at high temperatures. However, for applications, such as quantum cryptography (e. g., BB84 protocol) and random number generators, where photon indistinguishability is not required, QDs may serve as on-demand single-photon sources also at elevated temperatures.

Coupling with optical phonons is an important additional mechanism to linewidth broadening and therefore to decoherence of excitonic excitations in QDs. The overall temperature dependence of the ZPL of the exciton linewidth Δ can be approximated by [92]

$$\Delta(T) = \Delta_0 + aT + b \times \exp\left(-\frac{\Delta E}{k_B T}\right), \tag{8.2}$$

where Δ_0 is the linewidth at 0 K, and it is ideally the natural linewidth determined by the radiative limit. The linear increase, which dominates at low temperatures, resembles to

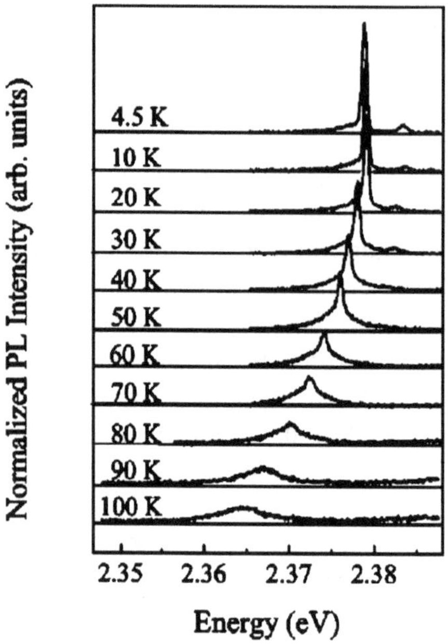

Figure 8.6: μ-PL spectra of a CdSe QD for different temperatures. The spectra are normalized to the integrated intensity at 4.5 K and shifted vertically for better representation. Reprinted from [180] with the permission of AIP Publishing.

acoustical phonons. The exponential rise is attributed to optical phonons, which become increasingly excited at higher temperatures, ΔE is the activation energy, which is in the range of ~25–27 meV for the InAs/GaAs material system, and a and b are characteristic constants related to phonon coupling.

8.3 Linewidth broadening, dephasing, and spectral diffusion

Quantum dots are embedded into their semiconductor host material and have to be treated as open quantum systems in interaction with their environment of lattice vibrations, charge carriers, and magnetic moments (spins) as well as coupling to the relevant photons modes, e. g., to free space-, waveguide-, or cavity modes. An adequate description and discussion, especially for dissipative processes such as carrier–carrier and carrier–phonon interaction are typically described and modeled by sophisticated many-particle theories, which go well beyond the level of this introductory textbook. Very detailed reviews can be found, e. g., in [126] in the chapters of Frank Jahnke and Stephen Hughes and their coworkers and references therein.

Dephasing in semiconductor quantum dots can be discussed in different contexts. A single quantum dot can be used both for generation of quantum light and for an interface between light ("flying qubits") and "stationary" qubits, e. g., a spin of a charge carrier confined in a QD. In the first case the dephasing processes inside the QD affect the coherence of the generated quantum light (see experimental methods in Chapter 5). Therefore the physical properties of the emitter and the whole excitation and recombination processes have to be taken into account for a complete description of single-photon pulses. Phenomena like phonon-assisted relaxation processes, electron–phonon scattering, and carrier–carrier scattering are some of the prominent dephasing mechanisms, which can affect the coherence of generated quantum light from the QD. In the second case, dephasing deteriorates the coherent evolution of the stationary qubit (spin of the charge carrier). This can also be relevant for quantum light generation, e. g., for the implementation of spin-photon entanglement and generation of multiphoton-entangled states, so-called photon cluster states (see Sections 7.3.5 and 7.3.6). In this book, we discuss in more detail the dephasing processes affecting the photon pulses.

A damped two-level system (TLS) that is in interaction with its environment can provide important intuitive insight into the physics of relaxation and dephasing of semiconductor quantum dots. In this context the empty QD corresponds to the TLS ground state, whereas, e. g., a single exciton (e. g., neutral exciton or trion) characterizes the TLS-excited state. The discussion of dephasing will be based on such a simplified model.

For a TLS, two mechanisms of dissipation are possible. First, it can decay spontaneously, e. g., via a radiative transition from the excited to the ground state, the corresponding spontaneous decay rate is $\Gamma_1 = 1/T_1$ with T_1 the lifetime of the upper level. Nonradiative processes can also contribute to T_1. Examples are nonradiative recombination via deep levels or tunneling of electrons and holes out of a QD in an electric field. Second, the time evolution of the coherence is damped by the rate $\Gamma_2 = 1/T_2 = \Gamma_1/2+\Gamma_{\mathrm{deph}}$ with the pure dephasing time $T_2^* = 1/\Gamma_{\mathrm{deph}}$; T_2^* describes the characteristic pure dephasing that leaves the population of the TLS unchanged but disturbs its phase. The times are therefore connected by (see also Section 5.2.6):

$$\frac{1}{T_2} = \frac{1}{2T_1} + \frac{1}{T_2^*}. \tag{8.3}$$

For the case of negligible pure dephasing and absent non-radiative recombination ($T_1 = \tau_{\mathrm{rad}}$), the coherence can maximally reach the so-called Fourier-transform limit yielding $T_2/(2T_1) = 1$. Here the decay of excitation (e. g., exciton or trion) leads to the loss of phase due to population change in the QD. In the case of pure radiative recombination, the homogeneous linewidth of the transition corresponds to the natural linewidth, and it is directly linked to the radiative emission rate. This regime is highly desirable for most quantum photonic applications using single- and entangled photon sources, and it has been nearly reached in some rare studies, where electrically gated quantum dots of highest crystal quality have been studied at low temperature.

In general, QD-confined charge carriers experience carrier–phonon interactions, carrier–carrier scattering, and nuclear spin noise. Therefore the coherence of the QD is not only limited by the spontaneous emission process, but also by coupling to the surrounding of the QD. We can basically distinguish three different processes (see Fig. 8.7):

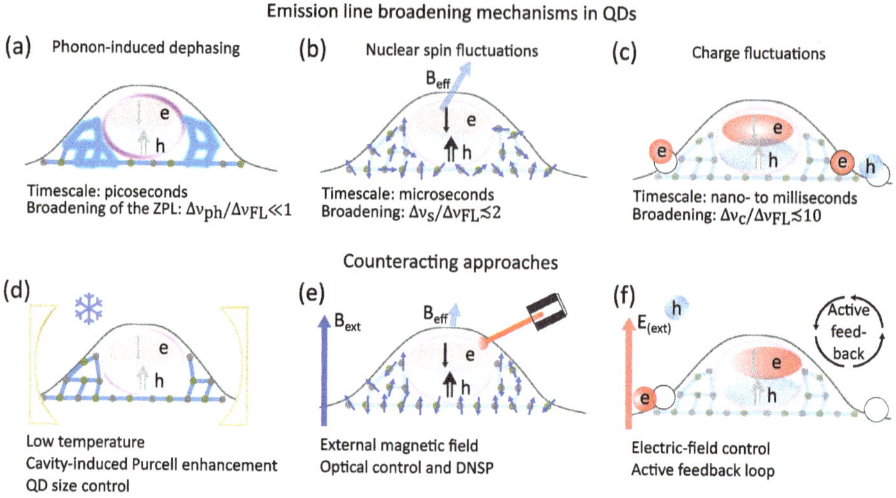

Figure 8.7: (a)–(f) Illustration of different line broadening mechanisms in QDs. Schematic sketch of an electron–hole pair confined in a self-assembled QD under environmental coupling of (a) lattice vibrations, (b) nuclear spins constituting an effective magnetic field B_{eff} refereed to as the Overhauser field, and (c) local charges. Their fluctuations induce dephasing and emission line broadening over the indicated timescales. $\Delta_i/\Delta_{\mathrm{FL}}$ is the ratio between common environment-induced broadening and the Fourier-limited homogeneous linewidth. (d)–(f) Illustration of active and passive counteracting approaches including (d) a quantum dot in a microcavity at cryogenic temperature, (e) an external magnetic field B_{ext} and self-locking effects through dynamic nuclear spin polarization (DNSP), and (f) an external electric field E_{ext} and active feedback control [215]. Reprinted from [215] with the permission of AIP Publishing.

- Phonon-induced dephasing processes depend strongly on the temperature, quantum dot size, and structure embedding the QD (see Section 8.2). It is important to note that coupling to phonons can never be completely quenched, as even at $T = 0$ K, phonon emission is still possible [86], which limits photon indistinguishability to high nineties. The linear exciton–photon coupling causes PSBs and leads to dephasing in a few picoseconds. For applications, these photons can be filtered out from the emitted photon stream at the expense of photon collection efficiency. The ZPL is coupled quadratically to phonons, causing pure dephasing of the photonic wavepacket. The impact of both PSB and pure dephasing on two-photon interference can be drastically reduced in narrowband microcavities, where Purcell enhancement increases the radiative decay rate (see discussion in Section 8.2). A detailed discussion of electron–phonon coupling is given, e. g., in the chapter of K. Roy-Choudhury and S. Hughes in [126].
- Spin noise arising from fluctuations in the nuclear spins of the host material leads to linewidth changes, and it can finally limit photon indistinguishability. The hyperfine interaction of the charge carrier spins with typically 10^5 nuclear spins of the atoms forming the QD results in a fluctuating magnetic field, the Overhauser field (OH), with a distribution variance of several tens of mT. This field weakly affects

the generally present electron–hole exchange interaction for the exciton, whereas for charged excitons, the degeneracy of the ground and excited states is lifted due to Zeeman splitting (see Section 9.5). The OH field is static over the duration of the lifetime of QD transitions, but it fluctuates over time. In strain driven QDs the OH-field fluctuation is drastically enhanced in the presence of QD-confined charge carriers with an unpaired spin [215]. The electron or the hole spin can be an important resource for spin-photon entanglement-based applications (see Section 7.3.5). Here it is important to note that the electron spin interacts predominantly via the Fermi-contact interaction, which is much stronger than the heavy-hole spin interaction, which is of dipole nature. However, dynamic nuclear spin polarization (DNSP) schemes, which use complex echo-like sequences, allow us to protect the spin from environmental fluctuations. A detailed discussion of spin noise is given, e. g., in the chapter of R. Warburton in [126].

- Charge noise arises from occupation fluctuations of charges trapped in the vicinity of a QD and causes fluctuations in the local electric field. This leads to shifts in the optical transition energy of a quantum dot via the Stark effect on the nanosecond to millisecond timescale (see also Section 9.3) and results in a linewidth increase of a QD transition. The dynamics of charge noise strongly influences the photon indistinguishability and therefore the two-photon interference visibility (see Section 5.2.8). Charge fluctuations strongly depend both on the excitation scheme, i. e., optical (resonant or nonresonant excitation) or electrical schemes, and also on the material and surface quality of the respective nanostructure embedding the QD. The following sources of charge noise have been discussed: impurities due to intentional doping, trapped charges of residual doping in defects induced by the QDs strain field, and surface states at processed or etched surfaces. The influences can be reduced by employing resonant excitation schemes, where less charge carriers are created in the vicinity of the QD. Moreover, high-quality samples with less defects and applying surface passivation techniques to reduce surface states have proven to be beneficial. Furthermore, using electrical gating also proved advantageous to reduce charge fluctuations.

It is important to note that the finite radiative lifetime and the interactions with the phonon bath represent a homogeneous broadening mechanism with a Lorentzian lineshape in the frequency domain. In contrast, the fluctuating electrical and magnetic environment represents an inhomogeneous broadening mechanism, which results in a Gaussian lineshape in the frequency domain. These effects are slow with regard to the radiative lifetime and are usually termed *spectral diffusion* or *spectral wandering*. In general, pure dephasing (expressed by the T_2^*-time) include both homogeneous and inhomogeneous parts. In cases where both Lorentzian and Gaussian line broadening processes are active, the resulting lineshape is described by a Voigt profile (see also Section 5.2.7).

As we will discuss later (Chapter 10), pulsed resonant optical excitation of an exciton, trion, or biexciton is very beneficial for achieving a high photon indistinguishability. In this case the TLS is inverted by a so-called π-pulse from an excitation laser, which leads to a nearly coherent and deterministic generation of the excited state. However, loss of coherence is typically also observed under pure resonant pumping. The loss of coherence becomes obvious by the observation of damped Rabi oscillations as a function of applied laser power as shown in Fig. 8.8. The graphs show typical Rabi oscillations of the excited level population of a single self-assembled GaAs/AlGaAs quantum dot at 4 K, i. e., for an exciton under pure s-shell resonant excitation and for a biexciton for two-photon excitation. Four (eight) oscillatory periods are clearly visible for the exciton (biexciton), together with a clear damping for increasing pulse area (see also Section 6.2). More oscillations can be observed for the biexciton since additional spectral filtering of the excitation laser in addition to cross polarization (as employed in Fig. 8.8(a)) can be utilized. The damping at high-pulse areas is related to the increase of the excitation intensity, and it is therefore termed *excitation induced dephasing* (EID). The dominant EID arises from the interaction with phonons in which the dominant coupling is to longitudinal-acoustic (LA) phonons via deformation potential scattering. Furthermore, Ramsey et al. [156, 157] showed that for increasing temperature, the damping gets stronger, and the oscillation period also increases. It is important to note that coherent exciton (biexciton) generation for subsequent single-photon emission (photon pair generation) is typically carried out at pulse areas of ~π, where EID is small. State occupation probabilities for excitons, trions, and biexcitons above 90 % have been reported. This is also true for laser-driven qubit-control experiments of single excitations in QDs.

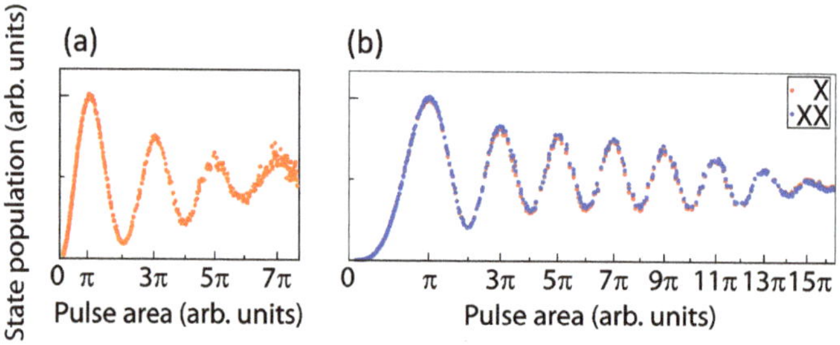

Figure 8.8: Excitation laser power-dependent Rabi oscillation up to pulse of (a) 7π under s-shell resonant excitation of an exciton and (b) 16π for two-photon excitation of a biexciton. Reprinted after Schöll et al. [175].

For most of the applications in photonic quantum technology, the two-photon interference of single photons on a beamsplitter plays an essential role. For indistinguishable photons, we can observe the so-called Hong–Ou–Mandel effect, i. e., two photons will never leave the beamsplitter on different output ports, but instead we will detect them simultaneously on one of the two output ports, e. g., with photon-number-resolving detectors (see Sections 4.4 and 5.2.8). Photon-dephasing processes will reduce the Hong–Ou–Mandel effect and will introduce errors in the quantum photonic protocols of the respective application. For example, to realize the so-called "Quantum Advantage" in a boson sampling quantum simulation algorithm, we require about 50 high-quality photons. For applications such as fault-tolerant photonic quantum computing, up to thousands and millions single-photon qubits can be foreseen. Two different modus operandi are utilized for the allocation of the necessary number of required photons. The photons are taken either from one demultiplexed *single* source or from different independent single-photon sources (see Fig. 8.9). In practice, this makes a huge difference due to the distinct time behavior of dephasing processes and because two different QDs will never emit the same photon wavepacket, which requires additional tuning possibilities (see Chap-

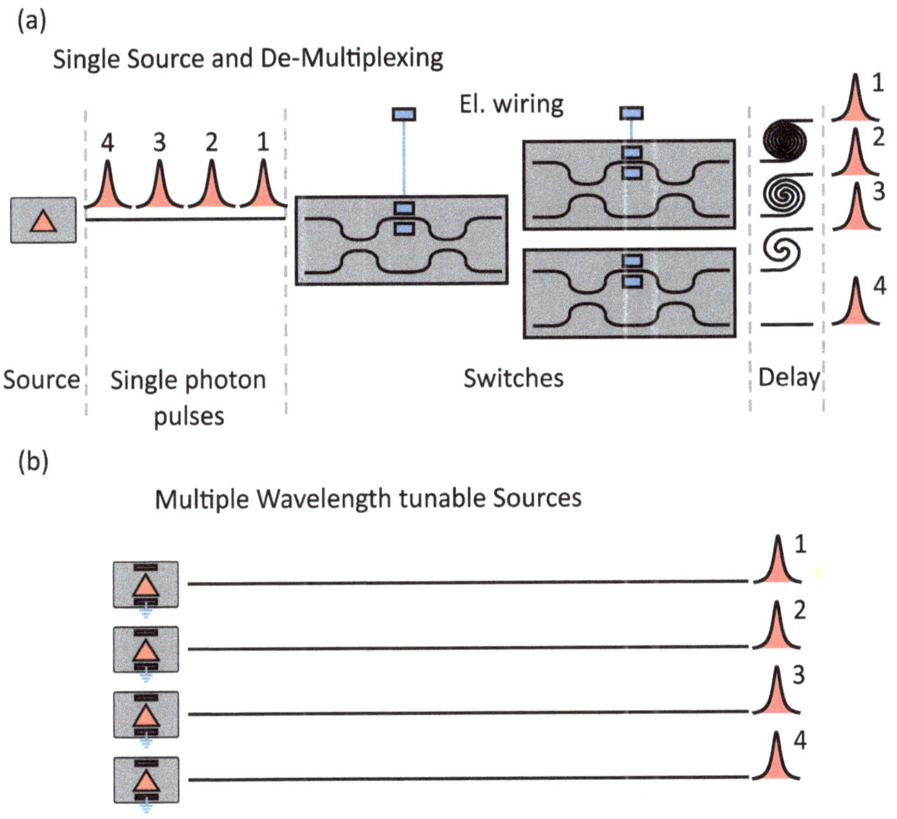

Figure 8.9: Illustration of (a) a demultiplexed single-photon source and (b) multiple wavelength-tunable single-photon sources for the synchronous preparation of single-photon pulses at the same wavelength. In (a) the deterministic stream of photons is routed to different spatial modes by cascading electro-optical switches. Delay lines will compensate the photons mutual delays. (b) Several wavelength-tunable sources are implemented. The wavelength tunability is necessary since QDs possess inherently different emission wavelengths due to the self-assembled growth.

ter 9). For a demultiplexed source, a detailed understanding of the temporal dynamics of the line broadening processes are important to select an appropriate time window before distinct dephasing starts to reduce the indistinguishability of subsequently emitted photons.

Photon correlation Fourier spectroscopy (PCFS) and slow-light photon-correlation spectroscopy are powerful tools (see Section 5.2.7) to study the temporal dynamics of charge and spin noise from nanoseconds up to the millisecond time range [174, 213]. These techniques allow us to study the emitter properties exactly under the same conditions used to generate single photons or photon pairs, e. g., after coherent π-pulse excitation. Figure 8.10 presents the so-called distribution of spectral shifts $p_\tau(\epsilon)$ of the exciton and biexciton emission of a GaAs/AlGaAs QD at 4 K measured by PCFS. It is important to note that $p_\tau(\epsilon)$ cannot be directly identified as the spectral distribution of the light emission, but it contains the desired information about the time-dependent spectral properties of the measured QD emission line [174]. In Figure 8.10(a–c), $p_\tau(\epsilon)$ is displayed for both X and XX transitions for three different time regimes. For the lowest accessible time

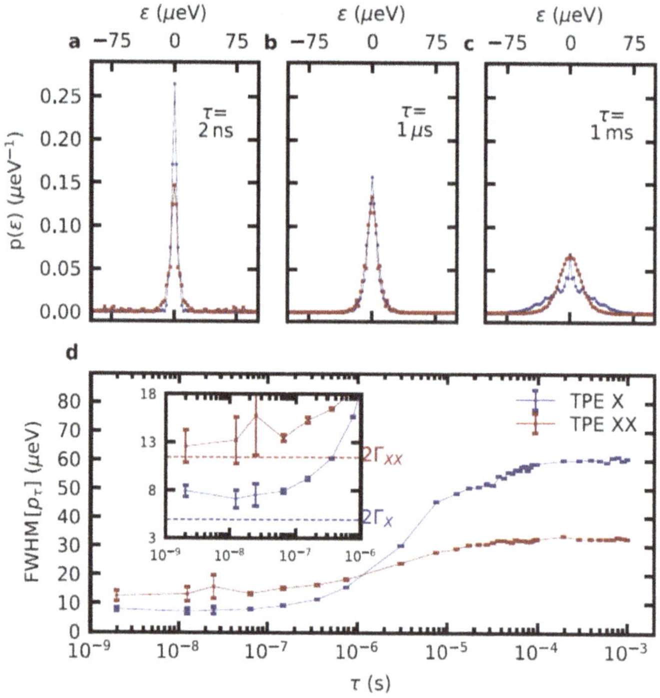

Figure 8.10: (a–c) Distribution of spectral shifts $p_\tau(\epsilon)$ of the exciton and biexciton emission of a GaAs QD at time delay τ. The measurement was performed on one fine-structure component of the respective transition. (d) Full width at half-maximum of $p_\tau(\epsilon)$ as a function of τ in the range from 2 ns to 1 ms. The dashed lines in the inset correspond to the X and XX Fourier limits, respectively. Reprinted with permission from [174]. Copyright (2023) the Optical Society.

delay (2 ns), X and XX emissions exhibit an FWHM close to the Fourier limit (see inset of (d)). At longer time delays (1 µs and 1 ms) the FWHM is substantially broader than at 2 ns delay. Figure 8.10(d) shows the FWHM of $p(\epsilon)$ for a time range τ between 2 ns and 1 ms. For small time delays between 2 and 100 ns, the linewidth is nearly constant followed by a distinct increase for longer time delays. Whereas for short time delays, the X line remains narrower than the XX line, a crossover of the linewidths is observed around 1 µs. The more dominant broadening of the X line compared to the XX line has been attributed to a higher polarizability of the X state in GaAs QDs, leading to a stronger coupling to environmental charges. Finally, at the millisecond time scale, both linewidths converge to constant but distinct values.

The processes that cause spectral diffusion and their magnitude vary for different material systems and from QD to QD. However, in many cases the stationary distribution of the spectral diffusion is represented by a Gaussian lineshape, often a consequence of the central limit theorem. This suggests a statistical description of the dynamics of spectral diffusion. Assuming that the QDs transition frequency shifts are caused by n independent two-state jump processes, the limit for large n converges to the so-called Ornstein–Uhlenbeck (OU) stochastic process, a stationary Markovian Gaussian noise process (see Section 5.2.7 and [212, 213]). Such a model description can be justified since QD confined charge carriers experience a hyperfine coupling to $\sim 10^5$ fluctuating nuclear spins and a large number of Stark shifts caused by trapped and released charges in its environment, which finally leads to the observation of a Gaussian lineshape. Several photon correlation and HOM measurements have been successfully modeled on the basis of the OU processes and therefore support this understanding of spectral diffusion processes [213, 214].

The results of the PCFS measurements discussed above suggest that harmful dephasing starts in this case around 100 ns after the generation of the XX and X states, and photon pairs with longer time delays between them would exhibit lower HOM visibility than at shorter time delays. The trend of decreasing HOM visibility with increasing time delay between the photons has been confirmed in this study and also by other studies, and basically limits the available number of photons with high indistinguishability of demultiplexed single-photon sources. In ultra-high quality QDs, a distinct increase in dephasing starts only around 1 ms (see, e. g., Chapter 9 of R. Warburton in [126]), which would in principle allow us to use ~10.000 photons for quantum photonic gate applications, assuming a Purcell enhanced radiative decay with a transition lifetime below 100 ps.

In contrast, for remote single-photon source applications, the fully integrated spectra (long time values) determine their quantum interference. It will be exciting to see in the future whether ultra-high quality QDs exhibiting Fourier transform-limited linewidth will be routinely fabricated and/or suppression of fluctuations by stabilization of the QD environment can be achieved for all relevant time scales.

8.4 Nonradiative recombination, carrier traps, and blinking

The brightness of a quantum light source is of uttermost importance for most of the applications in quantum information science (see Section 5.2.4). In the case of an ideal source, the brightness is equal to one. It requires that each trigger pulse generates a single photon pulse. The brightness of the source is also influenced by the radiative efficiency of the QD, which can deviate from unity. Possible reasons are nonradiative recombination processes, e. g., via defect states inside or close to the QD or fluorescence intermittency, an on-off intensity modulation (telegraph noise), also called *blinking*.

Blinking has long been observed in a variety of single quantum emitters, ranging from dye molecules, carbon nanotubes, color centers, defects in silicon carbides to nanocrystals and quantum dots. Semiconductor nanocrystals, also often referred to as quantum dots have to be distinguished from epitaxially grown quantum dots since different physical mechanisms are responsible for blinking. For the former, nonradiative Auger processes have been identified as important processes for blinking, which is considered to be of no or minor importance in epitaxially grown QDs. This is related due to their typically larger size and smoother potential-energy functions for charge carriers [31]. This conclusion is signified by the bright biexciton emissions routinely measured in QD PL.

We can basically distinguish the following main processes for blinking in epitaxially grown QDs:

- Nonresonant excitation of charge carriers in the surrounding barrier of the QD and the subsequent independent capture process of single carriers into the QD results in a random population of neutral and charged states in the QD. We can consider this as a fast blinking, e. g., of the neutral exciton line, which is randomly rendered dark by the statistical occupation of the QD with an additional charge, which would result in the subsequent emission of a charged exciton line (trion line). The time scale of the blinking is related to the time between carrier capture events. Under resonant excitation, charge carriers are generated in pairs, which avoids this reason for blinking.
- A charge carrier trap in close vicinity of a QD, e. g., a deep defect, a surface state, or an impurity, can individually influence the ratio between the differently charged states in a QD. Here the dynamics is determined by individual carrier capture rates, i. e., time constants τ_{in} and τ_{out}, for the relaxation (tunnel process) of an electron/hole from the trap to the dot and vice versa (see the following).
- If neutral exciton recombination is used for the generation of single photons, then a spin-flip process to the dark exciton state and back leads to blinking (see Section 7.3). This process can also affect the biexciton population probability under certain conditions and can also be relevant for resonant pumping schemes. Using trion recombination for the generation of single-photon states avoids this possible drawback since trions do not exhibit a natural dark state.

The presence of blinking leads to characteristic signatures in the population transients (bi- or multiexponential decays) and intensity autocorrelation transients (bunching), which can be conveniently measured by time-correlated single-photon counting and photon correlation measurements, respectively (see Sections 5.2.3 and 5.2.5).

First, we will present a simple formula, which allows the estimation of the on- and off-times from the bunching signature in photon autocorrelation measurements under pulsed excitation conditions. The underlying model assumes that the dot randomly blinks between two conditions, a fully functioning condition and a "dark" condition, in which photons are not observed [166]. The model gives the mth normalized correlation peak h_m to

$$h_{m\neq 0} = 1 + \frac{\tau_{\text{off}}}{\tau_{\text{on}}} e^{-(1/\tau_{\text{off}}+1/\tau_{\text{on}})[mT]}, \tag{8.4}$$

where T is the excitation laser repetition period. Fitting this formula to the experimental data gives the values for τ_{off} and τ_{on}.

Second, results from a more sophisticated model will be shortly presented, where the interaction of the dot with electron and hole traps is discussed, and the calculations are based on a Monte Carlo algorithm to model incoherent population dynamics of individually assembled QD systems. It simulates the evolution of individual carriers under consideration of transition rates and population limits (including Pauli blocking). If not stated otherwise, a typical e-h pair recombination time of 1 ns and carrier capture and relaxation times on the order of 10 ps are assumed in the following. Further details can be found in [96]. The discussion will start with a single charge carrier trap in interaction with a single QD following in extracts Chapter 4 of [96] (see Fig. 8.11(a)). If we assume a geminate capture process of electrons and holes and an initially occupied trap, then the state of the QD can completely switch from a neutral to a negatively charged configura-

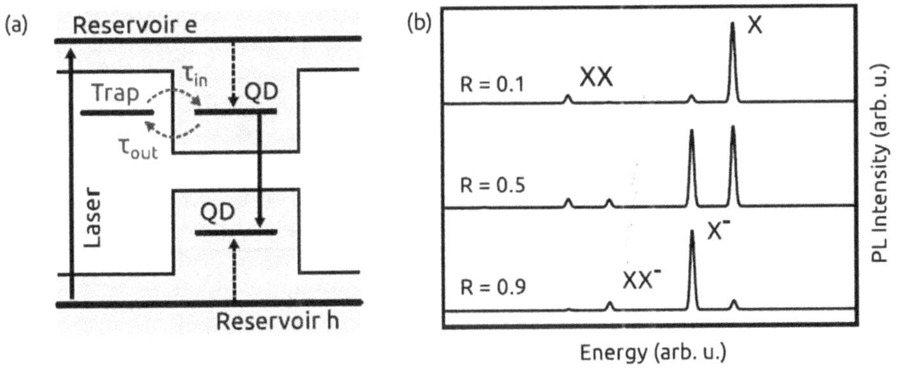

Figure 8.11: (a) A single charge carrier trap in interaction with the electron s-shell of a QD. (b) Charge state switching for different tunneling ratios $R = \tau_{\text{out}}/\tau_{\text{in}}$ and an initially occupied trap. Reprinted after Kettler [96].

tion, e. g., from an exciton to a negatively charged exciton (trion), provided that the relaxation from trap to dot prevails. The ratio of the tunneling rates $R = R_{in}/R_{out} = \tau_{out}/\tau_{in}$ directly translates to the relative line intensities (see Fig. 8.11(b)). A change from a geminate (e. g., by resonant excitation) to a separate charge carrier capture (e. g., by above band excitation) may strongly influence the observed charge state. A more elaborate discussion for separate carrier capture also with enhanced capture probability for carriers of opposite charge can be found in [96].

However, competition of different charge states leads to blinking, and a bunching signature appears in the intensity autocorrelation function. Figure 8.12 shows simulations of photon autocorrelations for the exciton transition of the system discussed above in the case of geminate carrier capture under continuous wave excitation. With increasing tunneling ratio R, the bunching strength, i. e., the height of the bunching peak, increases, and the bunching time constant $\tau_{bunching}$ decreases (Fig. 8.12(a)). The bunching time constant is determined by both tunneling times and is given by $1/\tau_{bunching}$ =

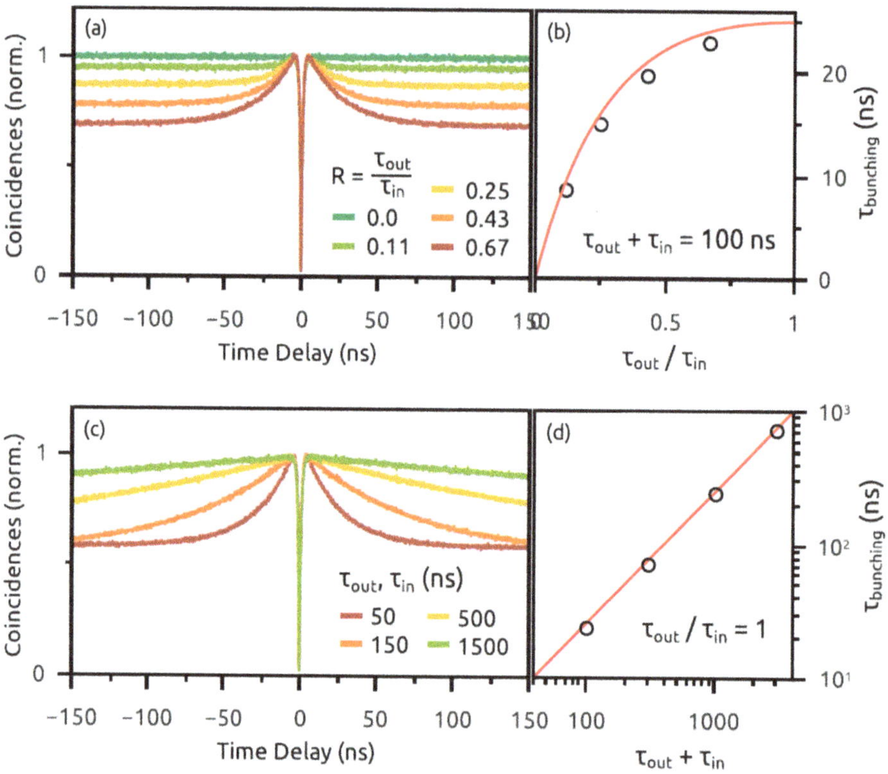

Figure 8.12: Influence of a charge carrier trap in interaction with a QD. Charging and uncharging time constants control the blinking behavior. (a) Exciton autocorrelation with increasing tunneling ratio R. (b) Related bunching time constant. (c) Exciton intensity autocorrelation with fixed ratio R but increasing time constants. (d) Related bunching time constant. Reprinted after Kettler [96].

$1/\tau_{in}+1/\tau_{out}$. If the tunneling ratio is kept constant, and τ_{in} and τ_{out} are equally increased, then the bunching peak is temporally broadened (Fig. 8.12(c,d)). Obviously, the bunching strength and the respective bunching time constant are closely connected to the dynamics of the charge carrier exchange between the trap and QD.

The presence of a charge carrier trap nearby a QD can also lead to a characteristic signature in the population transients and intensity autocorrelations under pulsed excitation conditions. After a nonresonant excitation pulse, charge carriers are captured into the QD and into the trap via nonradiative relaxation processes (see Chapter 10 for details on the excitation process). A trap may keep its population substantially longer than the carrier reservoirs of barrier and wetting layer, depending on the nature and position of the trap. After a first radiative recombination process caused by the charge carrier population inside the QD, the population of the trap can then "refill" the QD, leading to a secondary recombination, which is delayed in time. This process destroys the purity of single-photon emission and therefore represents a serious drawback for a pulsed single-photon source. In the following the population transients and intensity autocorrelations of the exciton in a QD under the influence of an electron and a hole carrier trap will be discussed (see Fig. 8.13). The possibility to detect two radiative recombination processes from excitons inside the QD, the first originating from the direct capture of e-h pairs from the barrier, and the second by tunneling processes from the carrier traps, lead to a biexponential shape of the decay transient. The slower, secondary decay time reflects the typically longer relaxation time of a carrier stored in the trap. The ratio between the carrier capture times of QD and trap determines approximately the ratio between the two observed decay processes (Fig. 8.13(a), bottom). In the case of a very strong contribution from excitons stemming from the trap, even a rise time of the secondary transient can be observed. The intensity autocorrelation functions are likewise affected by the different carrier relaxation and recombination rates of the trap and QD (see Fig. 8.13(b)). For a relatively fast carrier capture processes from the trap to the QD ($\tau_{tl} = 2$ ns), a curious peak with an additional antibunching dip occurs at zero delay time. This corresponds to the situation where both QD and trap are initially populated by an e-h pair (exciton) after an excitation pulse. If the trap decays first, i. e., the electron and the hole stored in the trap are captured by the QD, then a biexciton forms inside the QD, and no coincidence event is registered. In the opposite situation, where the exciton of the QD decays first, it triggers the zero time delay with respect to the recombination of the second exciton, stemming from the trap. Hence the dynamical behavior of the trap is reflected in the shape of the coincidence peak at zero delay, exhibiting a rise and decay time corresponding to the trap relaxation constant. With an increasing trap relaxation time, the width of the zero delay peak broadens and finally completely vanishes ($\tau_{tl} = 10$ ns and 50 ns). This is because coincidences from carriers stemming originally from the trap are more and more distributed over a larger time span and eventually overlap with coincidences from subsequent excitation cycles. This leads to a quasi-cw background below the coincidence peaks. At zero time delay, no coincidences are visible

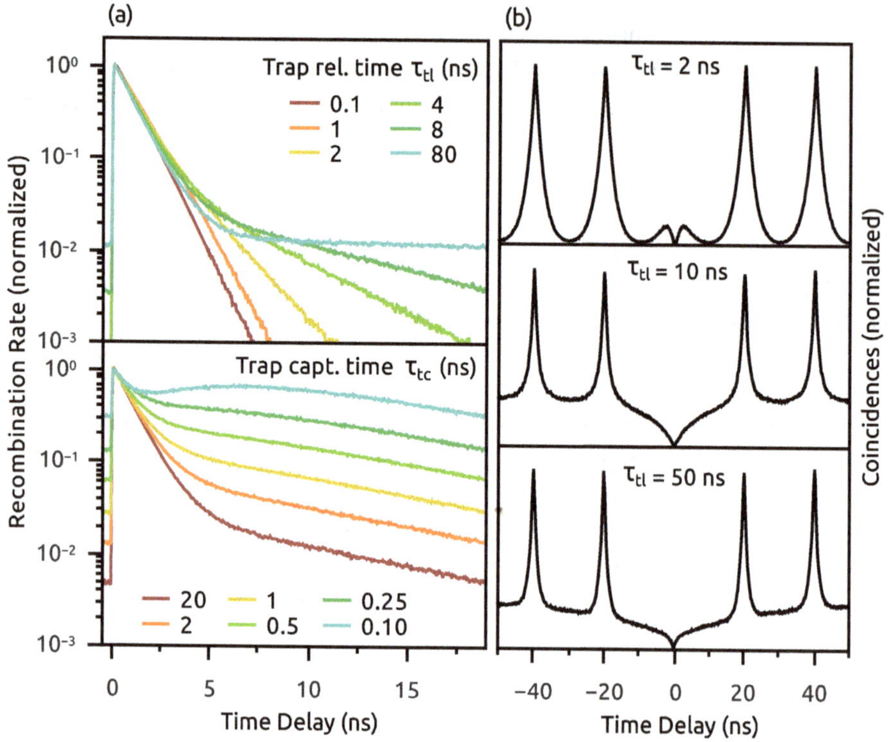

Figure 8.13: Influence of refilling of electrons and holes from charge carrier traps into a QD on the population dynamics and intensity correlation of QD exciton emission. (a) Exciton recombination transients with varying trap relaxation times τ_{tl} (top) and trap capture times τ_{tc} (bottom). (b) Exciton intensity autocorrelation histograms for different trap relaxation times. Reprinted after Kettler [96].

since the QD can only emit one photon at a time, and the visible antibunching dip forms on a timescale of the exciton radiative decay time.

The negative effects of charge carrier fluctuations in QDs can be considerably reduced or even avoided by applying different techniques. Using resonant laser excitation techniques (see Chapter 10), the statistical carrier capture processes into the QD are avoided. Thus preselected excitations, e. g., neutral exciton, trion, or biexciton states, can be exclusively generated. Nevertheless, nearby traps may inject statistically charge carriers into the QD on different timescales as discussed above. It has been shown that the application of an additional weak above-band laser may stabilize the charge environment around the QD. However, this may come at the price of increased dephasing processes. A more adequate and flexible method is to embed the QD into a vertical tunneling device, where a controlled QD charging process can be established; see Section 9.4. With charge-tunable structures, close to Fourier-limited lines have been observed in resonance fluorescence experiments for both InGaAs/GaAs (see Chapter 9 in [126]) and GaAs/Al(Ga)As [234] based QDs, and the emission of triggered indistinguishable photons in open cavities at ultrahigh repetition rates have been demonstrated [196].

The observation of biexponential transients for specific transitions can also result from the existence of optically dark QD states. For the neutral exciton in a QD, there exist two bright and two dark states, which are coupled by spin-flip processes (see section 7.3.1). As discussed above this process leads to blinking and can result in a biexponential transient. Furthermore, nonradiative recombination processes via deep defect states in the vicinity of QD provide alternative pathways for exciton recombination. Analyzing the dynamics, i. e., the intensity transients, provides information on radiative and nonradiative recombination rates, spin-flip times, and therefore also on the radiative quantum efficiency of the QD, as outlined in Chapter 5 from Tighineanu et al. in [126]. Spin-flip processes can be avoided by using the trion state for the generation of single-photon states due to the absence of a dark trion ground state.

Finally, it is important to point out that blinking can be completely avoided in high-crystal quality QDs (no deep defects) embedded in high-quality photonic cavity structures (no surface trap states) under pure resonant pumping (only geminate carrier generation) of trion states (no dark exciton states). Thus radiative quantum efficiencies close to one can be achieved with epitaxially grown QDs.

8.5 Nonresonant quantum dot-cavity coupling

The phenomenon of nonresonant QD-cavity coupling (NRC) may have an important impact on the operation and quality of light emission of QD-based quantum light sources. Cavity mode emission of QD-microcavity systems can be observed even under large spectral detuning between the quantum dot and the mode. The effect is referred to as "nonresonant quantum dot-cavity coupling." This behavior indicates a complex light-matter interaction in a semiconductor, which is not expected for a simple two-level system inside a cavity (e. g., an atom in a cavity). Several mechanisms can be responsible for NRC including i) phonon-assisted optical processes, ii) nonresonant coupling mediated by multiexciton transitions, and iii) Coulomb-assisted nonresonant coupling.

The discussion starts with phonon-assisted optical processes, which are mainly relevant for small detunings (<3 meV between the cavity mode resonance and the emitter transition frequency) and may be present even for truly resonant pumping schemes. A schematic in Fig. 8.14 shows the two simplest cases of cavity feeding by exciton recombination and phonon emission and absorption processes. If the cavity mode has a lower energy than the exciton inside the QD (Fig. 8.14(a)), then the cavity can be excited in a two-step quantum process by exciton recombination along with a phonon emission process. In the opposite situation, i. e., the cavity has a larger energy than the exciton (Fig. 8.14(b)), the cavity can be excited by exciton recombination together with a phonon absorption process. At low temperatures, phonon emission is more likely than phonon absorption processes. The adequate framework to describe the phonon-assisted energy

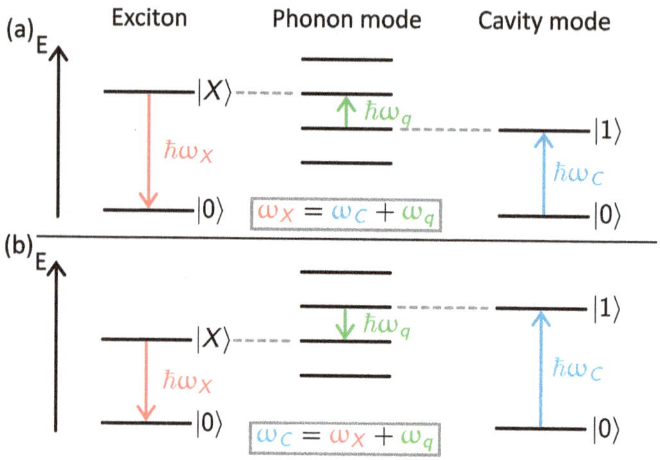

Figure 8.14: The cavity is excited by one photon ($|0\rangle \to |1\rangle$) through exciton recombination ($|X\rangle \to |0\rangle$): (a) generation of one phonon; (b) absorption of one phonon.

transfer processes is provided by a polaron picture, in which the exciton generates an electric field to which the lattice ions react by displacements of their oscillation centers. Different models have been used to quantify the feeding rates (see, e. g., Gies et al., Chapter 1, and K. Roy-Choudhury and S. Hughes, Chapter 2 of [126]). Roy-Choudhury and Hughes [162] have calculated these processes using a sophisticated polaron reservoir master equation (ME) method, and the results for some exemplary detunings are shown in Fig. 8.15. We can clearly see that the cavity can be excited more efficiently by phonon contributions where a cavity photon is created along with phonon emission.

Under nonresonant pumping conditions, e. g., into the barrier or wetting layer, many charge carriers (electrons and holes) are generated around the QD (see Chapter 10). As a result of the statistical carrier capture processes into the QD different charge configurations, i. e., excited multiexciton configurations, are formed during the relaxation processes into the respective ground state. Due to the different Coulomb correlation interaction, these configurations possess largely different multiexciton energies. As soon as one of these configuration energies is resonant with the cavity mode, Purcell enhanced emission into the modes becomes possible, and NRC is observed. This kind of cavity feeding has experimentally been observed for detunings between +10 meV and −45 meV. Figure 8.16 shows as an example a PL spectrum of a single QD coupled to a photonic crystal defect cavity together with a calculated energy-level diagram with the respective possible transitions [227]. In this case the cavity mode (CM) is tuned 15.3 meV to the red of the neutral exciton emission (X^0). The cavity mode was visible even if the excitation laser was kept well below the QD saturation level. The major QD lines stem from the neutral exciton X^0, the negatively X^{1-} and positively X^{1+} charged trions, and the biexciton transitions XX^0. In addition, a two orders of magnitude weaker broad single-QD background is observed. The part resonant with the cavity mode is Pur-

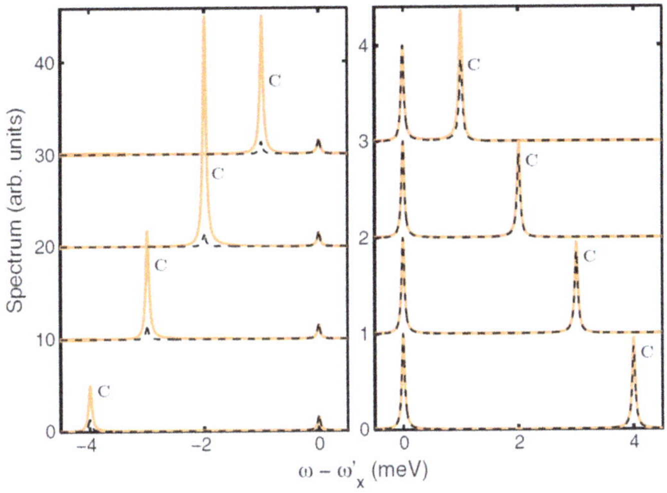

Figure 8.15: Nonresonant quantum-dot cavity coupling with phonons. Normalized linear emission spectra for a QD-cavity system without phonons (black dashed lines) and with phonons (solid orange lines) calculated at $T = 4$ K. The QD-cavity detunings are 1, 2, 3, 4 meV and −1, −2, −3, −4 meV (top to bottom) in left and right panels, respectively. The individual spectra are normalized by peak ZPL intensity, and data in bottom-left panel are multiplied by 1.5 for better visibility. The QD and cavity parameters are QD-cavity coupling rate $g = 100$ μeV, cavity decay rate $\gamma = 65$ μeV, pure dephasing rate $\gamma_d = 55$ μeV, and the radiative exciton decay rate $\gamma_0 = 5$ μeV [162]. Reprinted with permission from [162]. Copyright (2023) by the American Physical Society.

cell enhanced and is responsible for the off-resonant luminescence of the cavity mode. Winger and coworkers have calculated the full excitation spectrum of electron and hole states in the QD [227]. As a result, they obtained an excitation spectrum consisting of a series of QD manifolds ($n = 1, 2, 3$) separated approximately by the bandgap energy (see Fig. 8.16(b)). The narrow QD lines stem from states where electron and holes occupy the lowest possible energy levels (X^0, XX^0), whereas the respective higher lying levels stem from transitions for which, e. g., a carrier is excited to a p- or d-shell state. Based on the large number of excited multiexciton states, some of these states overlap with the cavity mode and give rise to the CM emission (see Fig. 8.16(c)). Similar excitation spectra exist also for positively and negatively charged trions, and their excited transitions also feed the CM. The NRC is signified in a superlinear pump-power dependence of the CM emission and proven by photon cross-correlation measurements between CM and fundamental QD transitions. Strong antibunching features are observed since CM and, e. g., X^0-transitions do not occur simultaneously.

Furthermore, it has been pointed out that in addition to the contributions from multiexciton states, delocalized states outside the QD can play a significant role for NRC. It has been shown that QD bound states ($n \geq 2$) are subject to strong hybridization with delocalized states [94]. The delocalized carriers in the wetting layer and/or barrier states can act as a thermal bath, which is able to compensate for the energy mismatch between

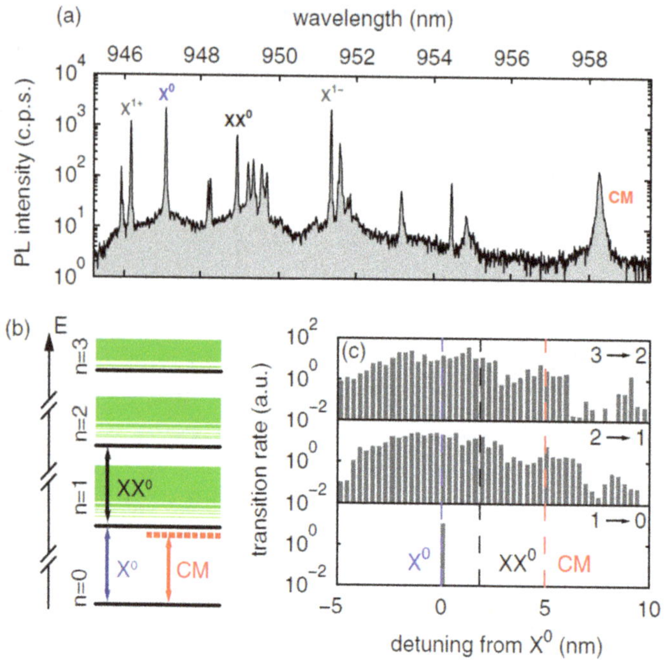

Figure 8.16: (a) PL spectrum of a single QD embedded into a photonic crystal cavity under nonresonant optical pumping (excitation wavelength $\lambda = 838$ nm) at ~4 K on a semilogarithmic scale. (b) Calculated energy-level diagram ($n \leq 3$) of a neutral exciton assuming a truncated parabolic in-plane confinement potential (energy separations between the manifolds not to scale). (c) Calculated transition rates between different manifolds. Reprinted with permission from [227]. Copyright (2023) by the American Physical Society.

the CM and the respective QD transitions via Auger-like processes. This process is even present in QDs that hold only one single electron and hole level. C. Gies et al. calculated the cavity-feeding rate and pointed out that for high carrier densities ($n > 10^{10}/\text{cm}^2$), these Coulomb-assisted processes prevail even at large detunings and lead to a significant cavity feeding (see Chapter 1 in [126]).

8.6 High-temperature operation of quantum dot single-photon sources

From the viewpoint of practical applications of single-photon sources, it is important to distinguish between applications that require two-photon interference processes on a beamsplitter (e. g., photonic quantum computing, quantum repeater) and those that do not rely on this quantum interference process (e. g., quantum key distribution). For the former, highly indistinguishable photons are indispensable, which require low-temperature operation (4 K) to avoid phonon-induced dephasing processes. For the latter, it would be advantageous to operate the single-photon source at elevated temperatures, achievable by nitrogen cooling, thermoelectric cooling, or ultimately without any cooling.

With increasing temperature, phonon-related effects like linewidth broadening and dephasing become more and more pronounced. Furthermore, thermally activated carrier loss processes from the quantum dot into the surrounding wetting layer or barriers set in. These thermal emission processes critically depend on the electronic confinement, i. e., barrier heights and energy level structure, i. e., the level spacing of the quantized states of the quantum dot, and they will limit the achievable brightness of the source at elevated temperatures. Figure 8.17(d) schematically illustrates the thermal emission process of carriers into the barrier. Many studies reported a characteristic thermal behavior of the single quantum dot intensity under nonresonant excitation conditions, i. e., optical excitation into the barrier. With increasing temperature, a first increase of the recombination rate at lower temperature followed by a more abrupt intensity drop at higher temperatures is observed. Figure 8.17(a–c) shows such an example for an InGaAs quantum dot emitting in the telecom O-Band. Wetting layer, p-shell and

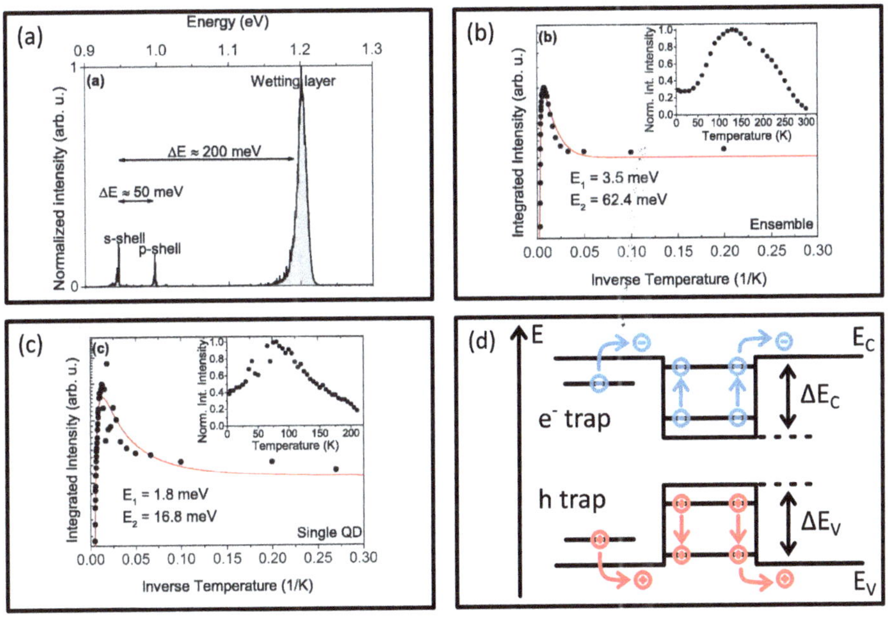

Figure 8.17: (a) Exemplary QD μ-PL spectrum including WL and QD p- and s-shells. Diagram of the temperature-dependent development of the integrated intensity (b) for a QD ensemble and (c) for a single QD. Black dots denote the measured data points, and red lines represent a fit to the data considering a thermal activated carrier reservoir (activation energy E_1) and a loss channel (activation energy E_2) (see text). Insets: Diagrams of the normalized, integrated intensities over temperature. Reprinted from [140] with the permission of AIP Publishing. (d) Schematic illustration of thermal emission of electrons and holes from electron and hole traps into the barrier (left part) and from the quantum dot into the barrier (right part).

s-shell emissions can be clearly observed. The typical energetic distance of the QD toward the wetting layer is around 200 meV, whereas typical shell spacings are between 10 and 60 meV for these QDs. Figure 8.17(b,c) shows the characteristics of the integrated intensity over the inversed temperature for the QD ensemble and a selected single QD, respectively. The different involved activation energies can be evaluated using a modified empirical Arrhenius function, which accounts both for an increase in the integrated intensity supplied by a finite carrier reservoir (e. g., by charge traps) and for a loss channel, e. g., the thermal emission of carriers out of the QD into the barrier. The temperature-dependent intensity $I(T)$ can be described as [140]

$$I(T) = \frac{I_0 + I_P \cdot [1 - 1/(1 + B_1 \cdot e^{-E_1/k_B T})]}{1 + B_2 \cdot e^{-E_2/k_B T}}, \tag{8.5}$$

where I_0 is the initial intensity, I_P is the reservoir intensity, E_1 and E_2 are the activation energies (see caption of Fig. 8.17), k_B is the Boltzmann constant, B_1 and B_2 are the coupling constants, and T is the temperature. With this relation, a satisfactory description is obtained regardless the underlying physical process. The fit of the data in Fig. 8.17(b,c) results in activation energies E_1 of several meV and E_2 of several tens of meV. The increase of the intensity could be caused by (i) the change in the temperature-dependent charge carrier mobility, (ii) the thermal activation of charge carrier traps originating of the wetting layer or of the strain reducing layer thickness variations located in the vicinity of the QDs, and (iii) the phonon-assisted overcoming of strain-induced barriers[3] surrounding the QD potential or a combination of the aforementioned effects. In the case of the single QD the activation energy responsible for the darkening of the QDs is in agreement with the typical shell spacing of the QDs, which seems to be the dominant loss channel for the charge carriers. It is interesting to note that for the ensemble, a higher activation energy (62.4 meV) is observed compared to the single QD (16.8 meV). This trend has been measured for many other QDs in different material systems. This indicates that a redistribution of charge carriers between different QDs takes place via the emission of carriers into the barrier and subsequent carrier recapture processes into other nearby QDs. Consequently, the emission of a single QD level could vanish, whereas the ensemble darkens at much higher temperatures.

For efficient high-temperature operation, several requirements have to be fulfilled. First, it becomes clear from the previous discussion that a large electronic confinement for both electrons and holes together with large energy-level spacings are necessary to prevent thermal emission of carriers into the wetting layer and/or barriers. Second, large biexciton binding energies are necessary for most of the used pumping schemes. Under nonresonant excitation conditions, typically, exciton and biexciton recombination lines are observed at high excitation powers, which are necessary for on-demand

3 A possible local strain field around a QD will modify the bandstructure and can introduce additional strain-induced barriers around the QD.

generation of photons. The same is true for resonant and phonon-assisted two-photon excitation schemes. The biexciton binding energy has to be larger than the linewidth of the corresponding transitions to prevent spectral overlapping of the exciton and biexciton lines at elevated temperatures. Third, the QD area density has to be kept low enough to prevent spectral overlap from neighboring spectrally broadened QD lines. Wide-bandgap semiconductors like II–VI and group-III nitride semiconductors offer large biexciton binding energies (20–30 meV). This is why single-photon emission at ambient temperatures of 300 K and also above have been realized with CdSe/ZnSe- [17] and GaN/AlGaN QDs [80]. The emission wavelength of these wide-bandgap QDs is in the visible and ultraviolet spectral ranges. Figure 8.18 shows exemplary photon autocorrelation measurements of a CdSe QD embedded into a ZnSe nanowire at 4 K, 220 K, and 300 K with the corresponding PL spectra. In the presented case, the intensity of the exciton recombination line is considerably weaker than the intensity of the biexciton line. This is because the bright exciton state can transfer very efficiently to the dark-exciton state, which is 4–9 meV lower in energy. Thus the photon statistics of the more intense biexciton line was studied up to room temperature. Clear single-photon emission ($g^{(2)}(0) < 0.5$) is observed until room temperature, however, with increasing $g^2(0)$-values for increasing temperature. This phenomenon is generally observed and most probably due to nonperfect spectral separation of different recombination lines from one and the same QD, or neighboring QDs, and to an unwanted background emission at higher temperatures. Finally, it is important to note that the excellent $g^2(0)$-values achieved at low temperatures (down to 10^{-4} level at 4 K) have not been achieved for high-temperature operation [5].

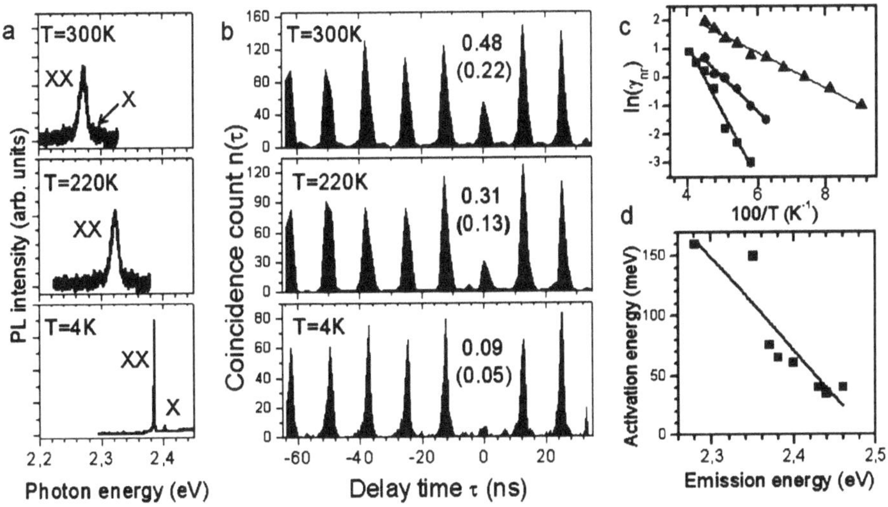

Figure 8.18: Emission spectra and photon autocorrelation data from the biexciton recombination for a CdSe QD in a ZnSe nanowire at various temperatures. Reprinted with permission from [17]. Copyright (2023) American Chemical Society.

8.7 Summary

- The electronic excitations in QDs are coupled to vibrational modes. Both acoustic and optical phonon couplings are present in compound semiconductor quantum dots (e. g., in InAs/GaAs- and CdSe/ZnSe-QDs)
- Emission and absorption of acoustic phonons during exciton recombination lead to phonon sideband (PSB) emission. The photons produced by these phonon-mediated processes are inherently distinguishable.
- Virtual transitions to higher-lying excitonic states through scattering of thermal acoustic phonons lead to broadening of the zero-phonon linewidth (ZPL). These processes lead to dephasing, which reduces the photon indistinguishability.
- A low temperatures (4 K), more than 90 % of the photons are funneled into the ZPL, which distinguishes semiconductor QDs from other single quantum emitters.
- The adverse effects of phonons can be significantly reduced by embedding the QD into a microcavity and utilizing the Purcell effect. A large coupling to a high-quality factor cavity can simultaneously reduce the effect of pure dephasing on the ZPL and efficiently redirect the PSB into the ZPL.
- With increasing temperature, the Franck–Condon factor B quickly increases, which leads to increasing PSB emission and a simultaneous reduction and broadening of the ZPL.
- LO phonon satellites to the ZPL in absorption and PL are usually very weak (~0.001) at low temperatures (~4 K). With increasing temperature, LO phonon scattering also contributes to linewidth broadening.
- For a damped two-level system, the spontaneous decay time T_1, the coherence time T_2, and the dephasing time T_2^* are connected by $1/T_2 = 1/(2T_1) + 1/T_2^*$. For negligible pure dephasing, the coherence can maximally reach the so-called Fourier transform limit yielding $T_2/(2T_1) = 1$.
- Dephasing can be caused by interaction with phonons, by spin noise arising from fluctuations in the nuclear spins of the host material, and by charge noise arising from occupation fluctuations of charges trapped in the vicinity of a QD. These processes lead to linewidth broadening.
- Linewidth broadening effects that are slow with respect to the radiative lifetime are usually termed spectral diffusion. They usually lead to a Gaussian lineshape and can be modeled by the so-called Ornstein–Uhlenbeck stochastic processes.
- Damping at high-excitation laser pulse areas is related to the increase of the excitation intensity and is termed excitation-induced dephasing (EID). The dominant EID arises from interactions with phonons.
- Photon dephasing and spectral diffusion processes deteriorate the two-photon interference (TPI) on a beamsplitter and therefore introduce errors in the quantum photonic protocols depending on TPI.

- The brightness of a quantum light source can be reduced by nonradiative recombination processes and fluorescence intermittency, i. e., an on-off intensity modulation, also called blinking.
- Blinking can be caused by nonresonant QD excitation processes, by charge carrier traps in close vicinity of the QD, or, e. g., by spin-flip processes to dark states.
- The presence of blinking leads to characteristic signatures in the population transients (bi- or multiexponential decays) and intensity autocorrelation transients (bunching).
- Carrier-refilling processes from nearby traps can lead to a secondary recombination inside the QD, which is delayed in time. This process affects the purity of single-photon emission and therefore represents a serious drawback for the application of a single-photon source.
- The application of an additional weak above-band laser may stabilize the charge environment around the QD. However, this may come at the price of increased dephasing processes.
- Charge controlled QD structures (e. g., p–i–n devices) possess superior quality. The charge inside the QD can be controlled by applying an external voltage, the emission line can be shifted by the Stark effect, and charge noise can be extremely low. Close to Fourier-limited emission lines have been emitted by these structures.
- Cavity mode emission of QD-microcavity systems can be observed even under large spectral detuning between the quantum dot and the mode. The effect is referred to as "nonresonant dot-cavity coupling (NRC)."
- Possible mechanism for NRC are phonon-assisted optical processes, nonresonant coupling mediated by multiexciton transitions, and Coulomb-assisted nonresonant coupling.
- Phonon-assisted optical processes are mainly relevant for small detunings (<3 meV) and may be present even for truly resonant pumping schemes.
- Multiexciton transitions have been observed for cavity feeding for detunings between +10 meV and −45 meV and may occur under nonresonant pumping conditions.
- For high carrier densities, Coulomb-assisted nonresonant coupling with delocalized carriers outside the QD prevails even at large detuning and leads to significant cavity feeding.
- With increasing temperature, thermally activated carrier loss processes from the QD into the surrounding wetting layer and/or barriers begin.
- For efficient high-temperature single-photon emission, the following properties are necessary: large electronic confinement and energy-level spacings between the quantized electron and hole levels, large biexciton binding energies, and low QD area densities.

9 Tuning the physical properties of quantum dots

9.1 Introduction

QDs are epitaxially grown in a solid-state matrix in a self-assembled process (see Chapter 12). Due to the statistical nature of the growth process, the QDs possess different sizes with a size distribution that can typically be described by a Gaussian distribution. Consequently, the electronic levels and optical transitions vary accordingly. Typical values for the PL linewidth of a quantum dot ensemble is in the range from several meV to tens of meV. Thus each QD possesses its individual electronic properties (optical transition energy, fine structure splitting, etc.). With regard to applications as quantum light emitters, this circumstance can constitute a serious drawback in case that a specific wavelength of a single quantum light source or several independent sources with identical wavelengths are required. Therefore the QD community has developed several tools to reversibly or permanently tune the optical transition energies of different excitonic complexes and their respective fine-structure splitting, which will be discussed below.

Several post-growth techniques for tuning the electronic and optical properties of QDs have been developed. The main purpose is to tune the absolute transition energy of excitonic complexes, preferentially over the full inhomogeneous linewidth of the QD ensemble, to change binding energies (e. g. biexciton binding energy), and to erase the exciton fine-structure splitting. We can distinguish between reversible and irreversible tuning techniques. Possible reversible tuning techniques are temperature tuning, applying DC lateral and vertical electric and magnetic fields, applying AC electric fields by shining a nonresonant laser onto the QD and strain tuning by piezoelectric actuators. Permanent tuning techniques relay on rapid thermal annealing (RTA) of the QDs, laser-assisted QD annealing, and laser-induced strain via local phase transitions, e. g., laser crystallization of a thin HfO_2 film on a surface close to the QD. Moreover, control of both emission energy and FSS is possible but requires the combination of several, independently controlled "tuning knobs" as will be discussed below.

9.2 Tuning by temperature

The easiest and most convenient way to tune the emission wavelength of a single quantum emitter is performed by changing the temperature of the crystal host matrix. The bandgap of a semiconductor typically decreases with increasing temperature (see Fig. 9.1). The reasons for this behavior are the change of electron–phonon interaction and the expansion of the lattice. Confinement effects appear to be rather small in this context, and models developed for the respective bulk materials provide reasonable results [143]. For many semiconductors, the temperature dependence can be described

https://doi.org/10.1515/9783110703412-009

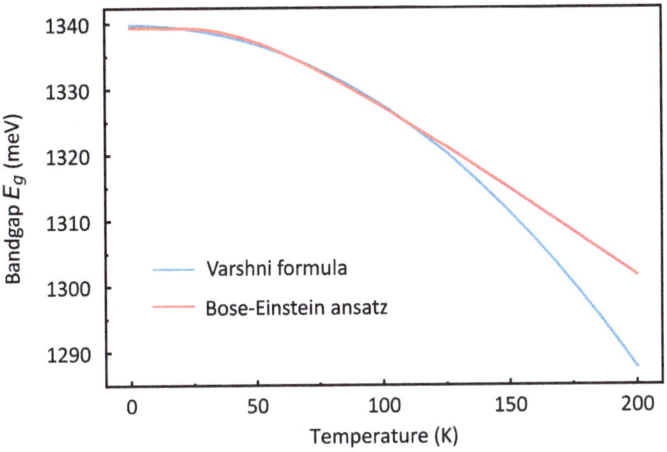

Figure 9.1: Bandgap vs. temperature.

by the empirical two-parameter Varshni formula [205]

$$E_g(T) = E_g(0) - \frac{\alpha T^2}{T + \beta}, \tag{9.1}$$

where $E_g(0)$ is the bandgap at zero temperature, and α and β are constants. A more precise formula is based on a Bose–Einstein-type ansatz and is given by [139]

$$E_g(T) = E_g(0) - \frac{1}{2}\alpha_B \Theta_B \left[\coth\left(\frac{\Theta_B}{2T}\right) - 1\right], \tag{9.2}$$

where α_B is a coupling constant, and $k_B\Theta_B$ is a typical phonon energy. This model reaches a better description of the fairly flat dependence at low temperatures. A more elaborate model [144] takes into account a more variable phonon dispersion, including optical phonons.

However, temperature tuning has some serious drawbacks. It leads to linewidth broadening, dephasing, and consequently to a reduction of photon coherence and indistinguishability (as discussed in Sections 5.2.8 and 8.3). Photon coherence and indistinguishability are the most important properties for most of the applications in the field of photonic quantum technology. Therefore alternative wavelength tuning processes have been developed and will be discussed in the following sections.

9.3 Applying electric fields – the quantum-confined Stark effect

A convenient way to adjust and manipulate the quantized states in a QD is the application of electric fields. The resulting effect is similar to the so-called Stark effect in atomic physics. In a bulk semiconductor, an electric field leads to a tilt of the band structure, the so called Franz–Keldysh effect, which gives rise to a slight redshift of the absorption spectra. A typical Stark shift for excitons is hardly observed since the exciton is weakly bound in 3D and quickly ionizes for increasing electric fields. The situation is different in QDs (quasi-zero-dimensional systems) since the external confinement potential given by the barrier layers prevents the ionization of, e. g., an exciton, and therefore stronger electric fields can be applied before exciton ionization (i. e., individual carriers tunnel out of the QD) is observed. In this case, large quadratic Stark shifts for the exciton can be observed in the absorption spectra due to a polarization of the exciton. Since this field effect is strongly dependent on the confinement, it is called the quantum confined Stark effect (QCSE). The energy shift ΔE in the framework of the QCSE in an electric field E is usually described as

$$\Delta E = p_{\text{perm}} E + \alpha E^2, \tag{9.3}$$

where p_{perm} is a permanent electric dipole, and α is the polarizability in the direction of the electric field E. The strain situation and the composition of the QD may lead to a slight separation of electron and hole giving rise to the permanent electric dipole p_{perm}. The first term leads to a linear Stark shift, whereas the polarizability of the exciton leads to an induced dipole moment $p_{\text{ind}} = \alpha E$ and thus to a quadratic Stark shift. The implications of the electric field and the confinement are depicted in Fig. 9.2 for both vertical and lateral electric fields.

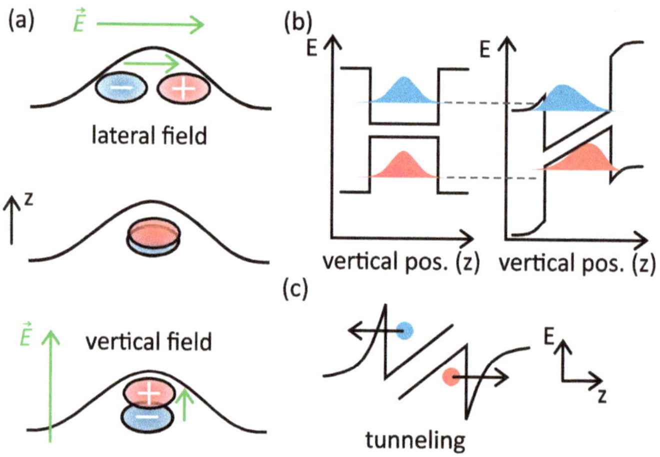

Figure 9.2: Schematic figure of the implications of an applied electric field in a QD. (a) The strain and composition of a QD may lead to a shift of the electron and hole with respect to each other along the z-direction at zero field ($E = 0$, center). With applied field ($\vec{E} \neq 0$), an additional polarization of the exciton is observed. The polarizability along the lateral direction (top) is usually larger due to the much larger lateral over vertical extension of the QD. (b) Sketch of the confinement potential and electron and hole wavefunctions in the z-direction without and with an applied lateral electric field. The change of the dipole moment gives rise to an induced dipole moment and a reduced spatial overlap of the electron and hole wavefunctions. (c) At high fields, carrier tunneling processes out of the QD into the barrier occur.

In self-assembled QDs the confinement along the growth direction (z-direction) is usually much more pronounced than in the growth plane, as discussed in Section 7.2. This leads to distinct differences in the electronic and optical properties of QDs for fields applied along the QD growth direction or perpendicular to it.

Let us start the discussion for in-plane electric field ($\perp$ to z). Since a QD is close to be radial symmetric in its growth plane, electron and hole wavefunctions exhibit a strong overlap concerning the radial coordinates, and consequently no permanent dipole moment exists. Thus only the quadratic Stark shift is commonly observed. It is important to note that the polarizability increases with larger dot diameters.

The situation is somewhat different for electric fields applied in growth direction ($\parallel$ to z). For example, assuming In(Ga)As QDs, the In profile and the strain in the QD give rise to a slight separation of the electron and hole wavefunctions (see Fig. 9.2(a)) inducing a permanent dipole moment. This gives rise to a linear Stark shift $p_{\text{perm}}E$. The polarizability, in contrast, is strongly reduced under vertical electric fields due to the stronger confinement potential. In summary, both linear and quadratic terms determine the wavelength shift of the exciton transition.

A further important consequence of the applied electric field results in a band tilting and a reduced spatial overlap of the electron and hole wavefunctions (see Fig. 9.2(b)). This leads to important implications. On the one hand, the radiative lifetime of the exciton is increased due to the reduced wavefunction overlap (see Eq. (7.10)). On the other hand, at large electric fields, carriers will tunnel out of the QD into the barriers due to the finite width of the confinement potential due to band tilting (see Fig. 9.2(c)). This represents an important nonradiative loss channel for the QD, which will reduce the quantum efficiency of a single-photon source and will also lead to a reduction of the exciton lifetime. This limits the range of useful energy shifts typically to several meV by the QCSE.

To avoid or at least reduce the tunnel processes, the effective barrier height can be increased by embedding the QD into a higher bandgap material. Figure 9.3(a) shows an example of a schematic band structure, where an InGaAs QD has been embedded into the center of a GaAs/AlGaAs quantum well [145]. Due to the increased carrier confinement, energy shifts of up to 25 meV can be observed (see Fig. 9.3(b)). The technique has been used to tune the QD emission line from Dot 1 into resonance with a reference QD (Dot A) (see Fig. 9.3(c)). This allowed the authors to demonstrate two-photon interference of their emission under coincidence gating.

It is important to mention that the actually observed electric field-dependent shift of an optical transition is not necessarily of the form given by Eq. (9.3). This is because the QCSE is also strongly dependent on the presence of charge carriers in the vicinity of a QD, since electric fields can be partially shielded by surrounding charges caused by charge traps, surface charges, or excess-free charge carriers.

Finally, it is worth noting that the tunnel process of the electron and the hole out of the QD into the barrier at strong electric fields can be used to implement an optically triggered single-electron turnstile device. The idea is to optically excite a QD, which is embedded into a photodiode structure, with a single exciton in a deterministic way with

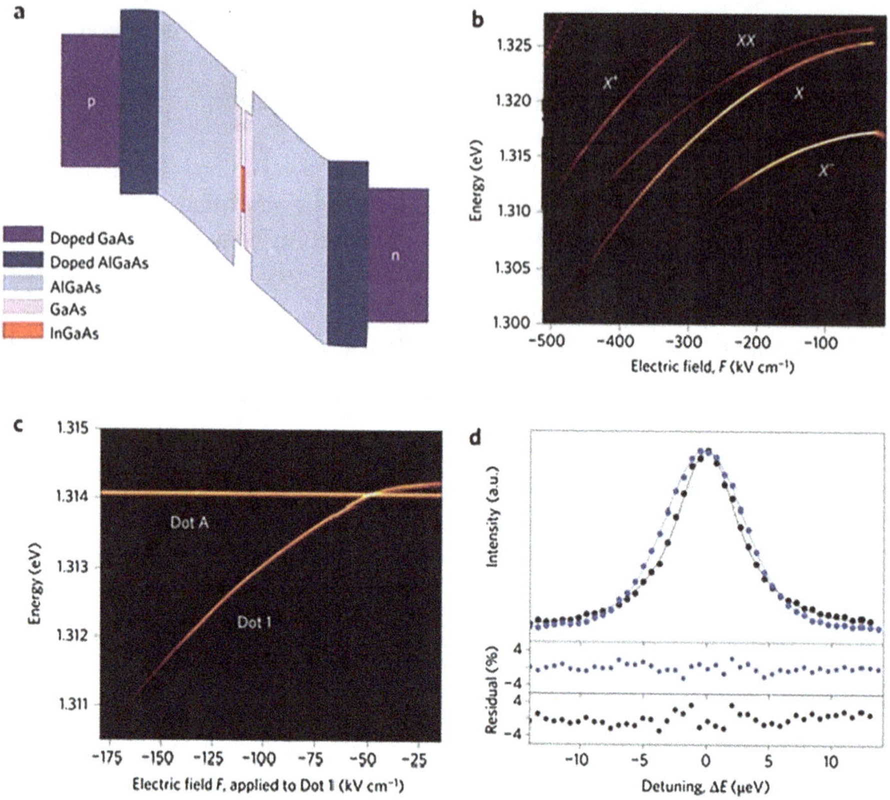

Figure 9.3: Design and spectral characteristics of an electric-field-tunable QD sample. (a) Schematic band structure of the tunable QD sample. The InGaAs QD layer is embedded at the center of a 10 nm GaAs quantum well clad with an $Al_{0.75}Ga_{0.25}As$ short-period superlattice. (b) Typical spectra of a single QD as the field is varied. Radiative transitions form different exciton complexes are visible. (c) Tuning a quantum dot emission line (Dot 1) to the same energy of a reference QD (line Dot A). (d) High-resolution spectra of Dot A (black lines) and Dot 1 (blue circles) at zero detuning. A Lorentzian spectrum (black curve) is observed for Dot A. Dot 1 shows a Gaussian component, and it is fitted with a Voigt profile (blue curve). Also shown are the residuals of the least-squares fit. Reprinted with permission of Springer Nature from [145].

a pulsed laser with repetition rate f. Due to the strong electric field, the electron and hole tunnel out of the QD. If the QD-photodiode is connected to an electric circuit, then the current is given by $I = f \times e$, where e is the elementary charge [236]. In this way, deterministic photocurrents can be generated.

9.4 Charge controllable samples

The actual control over the number of electron and holes in a quantum dot is essential for most of the applications in quantum information science. This control can be achieved with charge controllable samples. They allow a controlled QD charging via

Coulomb blockade, which then allows the controlled excitation of neutral or charged excitons under resonant excitation schemes (see Chapter 10). The Coulomb blockade[1] becomes possible since the thermal energy at 4 K (0.4 meV) is considerably smaller than the large on-site Coulomb energies (tens of meV). Moreover, the QD frequency can be tuned via the DC Stark effect. In addition, random charge fluctuations in the environment of the QD can be suppressed. This enhances the brightness of the source, suppresses unwanted blinking effects, and reduces dephasing. Consequently, the emitted photons possess long coherence times, and close to Fourier-limited photon emission has been reported. Depending on the layer design of the semiconductor device structure, we can distinguish three different cases. The first design exhibits a vertical tunnel structure, where the QDs are in tunnel contact with the Fermi sea of an n^+ layer. On top of the sample, a Schottky contact is present to apply the required voltage. The second and third designs are built on a p–i–n and n–i–n structures with the QDs situated inside the intrinsic layer (i), respectively. Figure 9.4 shows the basic idea and Coulomb blockade behavior of a vertical tunneling structure [219]. The operation principle is the following. By applying a voltage V_g to the gate we can tune the first confined electron level (s-shell) relative to the Fermi-energy (see Fig. 9.4(a), bottom). For low temperatures and large electric fields, the s-shell of the QD lies above the Fermi energy and is therefore unoccupied (Fig. 9.4(b)). When the s-shell lies slightly below the Fermi level, the s-shell is singly occupied (Fig. 9.4(c)), and even at more positive voltages V_g, it is doubly occupied. In this way a certain charge state can be selected in the QD, and after an optical excitation of an additional electron-hole pair neutral exciton X^0, single charged exciton X^{1-} or double charged exciton X^{2-} recombination can be observed (see Fig. 9.4(d)). The Coulomb-blockade operation becomes visible in the clear steps of the photoluminescence energy from the emission spectrum of the single QD. The DC-Stark tuning becomes obvious from the slight positive slope of the respective emission lines. It is important to note that within the Coulomb blockade regime, tunneling is suppressed in first order, but cotunneling[2] is still possible but strongly suppressed over several orders of magnitude in the center of the charging plateau [219].

Effective p–i–n diode structures have also been implemented for tuning the charge state of single QDs [234] and also used inside microcavities [186]. Here it is important to define the Fermi level around the s-shell of a QD while minimizing the free carrier losses in the mirrors. Otherwise, free carrier absorption will limit the cavity Q-factors. The authors were successful in fine tuning a QD exciton transition via the quantum confined Stark effect into resonance with the cavity mode. Moreover, it was shown that the

1 At low temperature, when the charging energy E_c of an island is larger than the thermal energy $k_bT \ll E_c$, the electron transport over this island is blocked by Coulomb repulsion, called he Coulomb blockade regime.

2 Cotunneling is an effect where within a very short time scale, an electron tunnels out of the QD, and another one tunnels in again, i. e., a quasi-simultaneous tunneling event of two electrons, whereas a first-order tunnel rate of a single electron is very low.

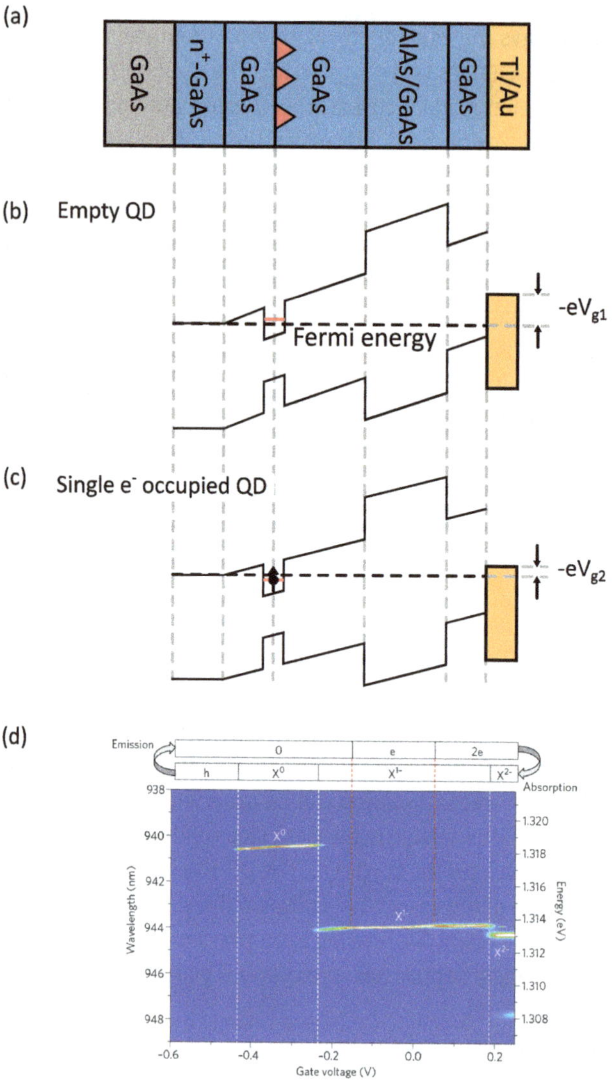

Figure 9.4: (a) Layer sequence of a charge loadable QD structure. The electron level of the QD is in tunnel contact with the Fermi sea in the n^+ layer. The blocking layer prevents electrons from moving to the sample surface. A metal contact (Ti/Au) forming a Schottky gate with the semiconductor allows control of the vertical electrical field and therefore enables the control of the relative position of the Fermi level to the first confined QD electron state. Typical data for the different layers are tunnel barrier (GaAs) 12–40 nm, capping layer 10–150 nm, blocking layer >100 nm, and the Schottky gate 5–10 nm. Schematic picture of the band diagram of the semiconductor structure: (b) At low temperature and at large built-in electric field, i. e., at a low voltage V_{g1}, the first confined QD electron state (red line) is above the Fermi level, and the QD is therefore empty. (c) At a larger voltage V_{g2}, the Fermi level is above the first confined electron state, and the QD is occupied by one electron. At more positive voltage, it is doubly occupied. (d) A photoluminescence map from a single QD in a vertical tunneling structure is shown as a function of V_g at a temperature of 4.2 K. Different charging events are clearly visible by steps in the PL energy. X^0, X^{-1}, and X^{-2} correspond to the neutral exciton, negatively charged exciton (trion), and the doubly negative charged exciton, respectively. Reprinted with permission of Springer Nature from [219].

charge noise was strongly reduced, and under resonant excitation conditions, indistinguishabilities of up to 0.995 has been reported.

Furthermore, n^+–i–n^{++} charge-controllable diode structures have been successfully demonstrated [191]. Despite that with p–i–n diode structures, excellent results have been obtained, some aspects could be advantageous using n^+–i–n^{++} structures. From the growth point of view, the presence of a p-layer in the structure of the diode embedding QDs requires doping with carbon atoms, which constitute an impurity during growth, an impurity which may impact the radiative efficiency of the device. Second, from the electrical device performance point of view, electrons have a 10-times higher mobility than holes, which would enhance the achievable speed in the application of an AC voltage to the device. The authors of the study have demonstrated single-photon emission, close to Fourier-limited linewidths, high two-photon indistinguishabilities together with spectral tuning of transitions, and a deterministic control of the observed charged states.

9.5 Applying magnetic fields – the Zeeman effect

Due to the spin of electron and holes, the QD excitons cannot only couple to electric fields, but also to magnetic ones. Therefore the eigenstates of QDs can be manipulated using the Zeeman effect, similar to what is known in atomic physics. Nevertheless, magnetic fields are typically not used in practice to tune the exciton transitions energies in quantum light source applications. The reasons are twofold: first, the interaction with magnetic fields leads to a splitting of the eigenstates, which results in shared oscillator strengths of the respective transitions. Second, bulky magnets complicate and prevent compact devices. However, the interaction between the carriers spins and magnetic fields is utilized to mediate spin-photon interactions, which can be used to establish spin-photon entanglement, and to generate photonic cluster states as discussed in the previous chapter. In this section, an introduction on the magnitude and splitting behavior for the most relevant excitonic excitations in two selected magnetic field directions will be given. A detailed discussion of the very rich spin physics in QDs is well beyond the framework of this book, and a comprehensive discussion is, for example, found in the books [124–126] and references therein. The general form of the interaction of electron S_e and hole spins J_h with an external magnetic field $B = (B_x, B_y, B_z)$ of arbitrary strength and orientation is given by

$$\hat{H}_{\text{Zeeman}} = \mu_B \sum_{i=x,y,z} (g_{e,i} S_{e,i} - g_{h,i} J_{h,i}) B_i, \tag{9.4}$$

where μ_B is the Bohr magneton, and $g_e(g_h)$ is the electron (hole) g-factor. Two principally different field configurations are typically discussed and applied. In the first one the field is oriented parallel to the QD growth direction ($B \parallel$ z, Faraday configuration), whereas in the second one, it is applied perpendicular to it (B $\perp$ z, Voigt configuration).

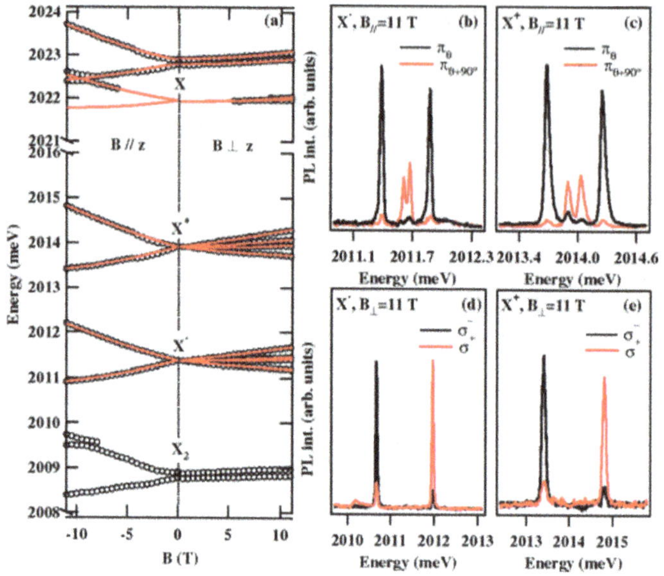

Figure 9.5: (a) Fine-structure of exciton, biexciton, and positively and negatively charged exciton emission of II–VI semiconductor (CdTe/ZnTe) QD in a magnetic field for both Faraday and Voigt geometries. (b) and (c) show the linearly polarized PL spectra of X^- and X^+ in a transverse magnetic field $B = 11$ T, respectively. (d) and (e) show the circularly polarized PL spectra of X^- and X^+ in Faraday configuration ($B = 11$ T). Reprinted with permission from [109]. Copyright (2023) by the American Physical Society.

The first configuration does not disturb the rotational invariance around the QD growth direction, whereas the second geometry destroys this symmetry leading to a totally different splitting and tuning behavior. For a comprehensive description, we have to take into account the Hamiltonian of the electron–hole exchange interaction, possible valence band mixing effects between heavy and light hole states, and the Zeeman Hamiltonian (see Eq. (9.4)) considering the respective field configuration [124]. The influence of magnetic fields on the different excitonic complexes in the two field configurations (Faraday (left) and Voigt (right)) is summarized in Fig. 9.5. In Faraday configuration, all excitonic lines ($X, X^+, X^-, X_2 = XX$ (biexciton line)) are split into doublets. For the exciton, an anticrossing of the bright and dark exciton states is observed around −9 T [109]. In Voigt configuration the broken symmetry over the angular quantum number J_z leads to a PL signal contribution from different exciton states. This becomes immediately obvious for the charged exciton case: here the PL signal splits into four lines. For neutral states, we can observe the dark states when applying relatively high magnetic fields. This is in contrast to observations in III–V semiconductors, where the dark excitons are already visible at low magnetic fields due to a smaller ratio of the exchange energy to the Zeeman energy. A detailed analysis of the splitting behavior allows one to obtain the g-factors parallel and perpendicular to the growth direction of the QD. As we can see from Fig. 9.5, the energy tuning range is ~1 meV for magnetic fields up to 10 T.

9.6 Applying strain fields

The tailoring of transition energies, biexciton binding energies, and excitonic fine-structure splittings of QDs by strain is a very powerful and widely used method. The possibility to introduce strain fields with controlled magnitude and in a reversible manner is very attractive for various QD light source applications. A typical approach consists of the integration of a semiconductor QD-containing membrane onto piezoelectric actuators, which convert an electric voltage into a mechanical deformation (strain). The mechanical deformation (strain) in materials leads to a change of the interatomic distances, which results in change of most of their physical properties, e. g., to a modification of the electron and hole energies, and therefore to a change of the relevant optical and quantum optical properties of the QD. The displacement of an element of the material can be described by the strain tensor, which contains six independent components (ϵ_{xx}, ϵ_{yy}, ϵ_{zz}, ϵ_{xy}, ϵ_{xz}, ϵ_{yz}). They are related to the stress components by the compliance tensor of the material. The connection between strain and stress in semiconductors is exhaustively treated in the book by Grundmann [61] and references therein.

Piezoelectric materials respond to electrical signals by mechanical deformations that can be accurately controlled by external voltages. The commonly used material for semiconductor membrane applications is $Pb(Mg_{1/3}Nb_{2/3})O_3 - PbTiO_3(PMN - PT)$ because of its large piezoelectric response. This material is also particularly well suited for miniaturized and machined cantilevers, which allow spatially localized applications, i. e., a high integration density. The following discussion in this chapter follows in extracts Ref. [120].

Figure 9.6 shows a sketch of a poled piezoelectric substrate: when an electric field is applied along the poling direction [001], both an out-of-plane deformation (strain $\epsilon_\perp$) and in-plane deformation (strain $\epsilon_\parallel$) is induced. Anisotropy values of $\epsilon_\parallel \approx -0.7 \times \epsilon_\perp$ can be achieved with compressive in-plane strain magnitudes up to $\epsilon_\parallel \sim -0.1\,\%$ for electric field strengths of $F_p = 10\,\text{kVcm}^{-1}$ [120]. In case of (001) oriented PMN-PT crystals, the similar values of the piezoelectric constants $e_{31} \approx e_{32}$ lead to similar in-plane

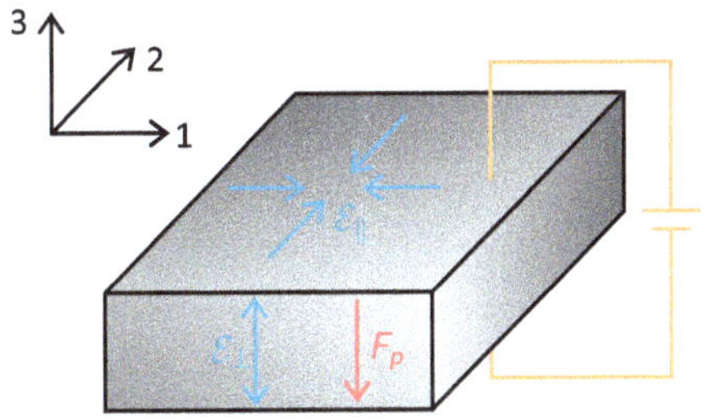

Figure 9.6: Schematic picture of a poled piezoelectric substrate where an electric field F_p is applied along the poling direction (“3”, i. e., [001]). Applying an electric field along [001] direction of the crystal produces out-of-plane and in-plane deformations, i. e., strain components $\epsilon_\perp$ and $\epsilon_\parallel$, respectively. Adapted from Martin Sanchez et al. [120].

deformation in both orthogonal directions ("1" and "2"; see Fig. 9.6). Also, other PMN-PT crystal orientations, e. g., (011), can be used to induce highly anisotropic strain fields due to an anisotropy in the piezoelectric constants $e_{32} > e_{31}$. Both compressive and tensile strains can be applied by simply controlling the sign (positive/negative) of the applied voltage. Details on device fabrication and working principles can be found in a topical review [120].

The basic idea is now the following. The strain from a poled piezoelectric substrate can be transferred to a semiconductor membrane (containing the QD) by physically attaching the membrane to the top of the piezoelectric substrate (see Figures 9.8, 9.9, and 9.11). The first step is to release a semiconductor membrane containing the optically active QDs from the substrate by selective wet chemical underetching.[3] In the second step the membrane is transferred on top of the piezoelectric substrate. Different techniques have been applied for this crucial step, including (i) gold thermo-compression bonding, (ii) polymer-based "soft" bonding, (iii) epitaxial growth, and (iv) direct transfer by van der Waals interaction. In short, the Au thermo-compression bonding is well established for semiconductor wafer bonding, and it is ideal for an efficient strain transfer due to the high Young modulus value of Au. It is important to note that sometimes a nonuniform bonding with air gaps is observed. In this respect, polymer-based bonding is a good choice since typically uniform bonding interlayers are observed, and the strain transfer can reach comparable values as for the Au thermo-compression bonding. Epitaxial growth is the best method for maximizing the strain transfer from the piezo to relevant layer. However, this method is only usable for a very limited number of materials due to the lattice mismatch between substrate and epilayer. The details and the pro and cons of the different techniques are fully explained in the review of Martin-Sanchez [120].

In the following, the discussion will focus on i) tuning the transition frequency of QD excitons, ii) tuning the exciton fine-structure splittings, and iii) establishing a resonance between a QD transition and a cavity mode. If we apply strain to the semiconductor lattice, then this causes changes in the bond lengths, which results in an altered band-structure and, in particular, in a modified energy bandgap of the semiconductor. For small strain values ($\epsilon \sim 0.01$), the shift of the band edges is linear with the strain, and for large strain values, it becomes nonlinear [61]. Figure 9.7 shows the tuning of QD emission lines as functions of the electric field applied to the piezoelectric actuator, i. e., vs the average in-plane strain $\epsilon_{||}$. For small voltages, the wavelength shift is nearly linear with the applied strain. As mentioned above, a compressive (tensile) biaxial strain results in an increase (decrease) of the QD emission energy. A total energy shift of ~10.5 meV is achieved.

3 The discussed approach allows for transferring of very thin membranes. Alternatively, mechanical thinning via polishing and subsequent transfer can be employed, even if it typically results in thicker membranes (and hence less achievable strain on the emitters).

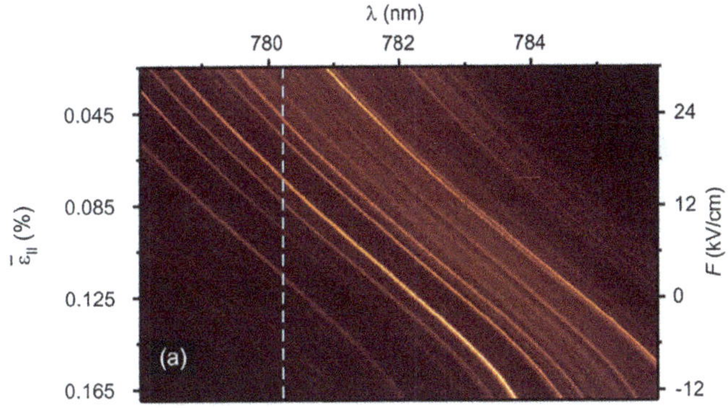

Figure 9.7: Color-coded PL intensity map of the exciton energy emission from several GaAs QDs, embedded in a nanomembrane and bonded on a PMN-PT actuator as a function of the electric field applied to the PMN-PT actuator. The dashed azure line represents a target wavelength (in the work, an atomic transition). Reprinted from [105] with the permission of AIP Publishing.

In the next example, we demonstrate the impact of strain tuning on a QD-cavity-waveguide device. The transfer of the cavity-waveguide device on a piezoelectric actuator enables a differential in situ wavelength tuning of the QD emission and cavity resonance via strain. Figure 9.8(a) shows an artistic illustration of the strain-tunable cavity-waveguide device. It consists of a ridge waveguide with a GaAs core layer with integrated self-assembled In(Ga)As QDs. The core layer has a thickness of 300 nm and a width of 560 nm, which enables single TE-mode operation of the device. The strain-dependent behavior of the sample is studied by applying a voltage on the piezoelectric PMN-PT substrate while recording the QD emission collected from the cleaved waveguide facet. Figure 9.8(b) displays a color-coded μ-PL intensity map of the strain-induced shift for a single QD emission line close to the fundamental cavity resonance [75]. With increasing voltage the QD emission line blueshifts across the cavity mode. In resonance a factor of 4 increase of the emission intensity is observed. A nearly linear tuning dependency over the given voltage range is observed with a calculated tuning efficiency of 2.3 μeV/V. In contrast, the spectral position of the cavity mode is less sensitive to the strain with a tuning efficiency of 0.58 μeV/V. This is a factor of ~4 smaller than for the QD and therefore enables differential tuning. The strain tuning techniques allows therefore a full compensation of the spectral mismatch between emitter and cavity resonance. This important property has also been demonstrated for other optical microcavities such as photonic crystal cavities and nanowire structures [120].

Maybe the most important application of strain fields is the possibility of erasing the excitonic fine-structure splitting. As outlined in Section 7.3.2, if the excitonic fine-structure splitting δ_1 is larger than the radiative broadening of the exciton line, then the radiative biexciton–exciton cascade has two distinguishable pathways to decay to the ground state, which leads to a two-photon polarization state with an additional phase factor between the two components (see Eq. (7.26)). In a typical time-integrated exper-

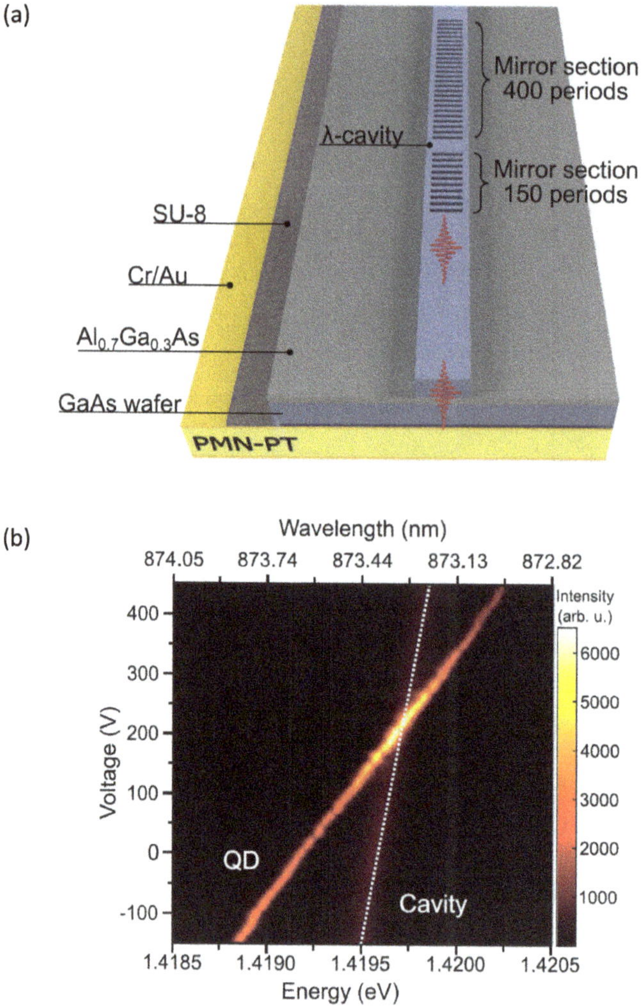

Figure 9.8: (a) Sketch of the strain-tunable cavity-waveguide device with embedded QDs. It consists of a Bragg grating cavity directly integrated in a ridge waveguide architecture. (b) Color-coded μ-PL intensity map performed at 4 K as a function of emission energy and applied voltage for a QD tuned into resonance with the cavity [75]. Reprinted from [75] with the permission of AIP Publishing.

iment, the integration over all events will randomize the phase between the two components in Eq. (7.26), effectively destroying the polarization entanglement between the two photons. Therefore the possibility of erasing the fine-structure splitting by external fields is of uttermost importance.

The fine-structure splitting can only be erased by an external field when the perturbation acts along specific directions aligned to the polarization axis of the exciton emission to modify the anisotropy of the QD confining potential. Otherwise, a lower bound in the fine-structure splitting is found due to level anticrossing. To modify the QD anisotropy, an anisotropic in-plane strain field has to be applied on the QD membrane. Here several possibilities exist: the QD membrane can be bonded on the side of a PMN-PT (001) substrate, which allows delivering highly anisotropic strain fields $\epsilon_{\parallel} \approx -0.7 \times \epsilon_{\perp}$ (see Fig. 9.6) or bonding it on a PMN-PT (011) oriented substrate, which also provides reasonable in-plane anisotropy of $\epsilon_{xx} \approx -0.37 \times \epsilon_{yy}$ [120]. However, in practice, anisotropic strain fields have also been found for QD membranes bonded on PMN-PT (001) substrates, where a nearly isotropic biaxial strain is expected due to equal in-plane piezoelectric constants. Several reasons have been discussed, among them, there are nonperfect PMN-PT substrates, which are polycrystalline with slight misorientation between different crystallites and bonding nonuniformities, which could affect the isotropy of the applied strain.

In principle, a single anisotropic external strain field applied perfectly along a specific exciton dipole direction could be sufficient to erase the exciton fine-structure splitting. In practice, the dipole direction is not always known in advance, and therefore it is convenient to implement another "tuning knob", e. g., an electric field. We will show below that this gives more flexibility in choosing a suitable QD but also can be used to either tune the emission energy via the QCSE or inject carriers in a LED structure. Figure 9.9(a) shows such a QD-device, where both strain and electric field tuning has been implemented simultaneously. It is demonstrated in Fig. 9.9(b) that the emission lines can be tuned across a spectral range larger than 30 meV when the two fields are varied on after each other. Furthermore, control over the binding energy of the biexciton (XX) is also demonstrated in Fig. 9.9(b).

In the following, we will show that the fine-structure splitting s (s and δ_1 are here used synonymously for the fine-structure splitting) can be reversibly controlled and erased. This can be achieved with a combination of vertical electrical fields along the [001] direction (F_d), together with in-plane anisotropic biaxial stress fields (F_p applied on PMN-PT). For this last strain, the principal directions have to be different from the [110] and [1-10] GaAs crystal directions. If S_1 and S_2 are the magnitudes of major and minor principal stresses, respectively, applied in the (001) plane, then the anisotropy of the in-plane anisotropic stress field is given by $\Delta S = S_1 - S_2$. Following the review article [120], the effective two-level Hamiltonian for the bright excitons takes the form

$$\hat{H} = [\eta + \alpha \Delta S + \beta F_d]\sigma_z + [k + \gamma \Delta S]\sigma_x, \tag{9.5}$$

where $\sigma_{x,z}$ are the Pauli matrices, and η and k account for the QD structural asymmetry. The parameters related to the external fields are α and γ (related to the elastic compliance constants renormalized by the valence band deformation potentials) and β (proportional to the difference of the exciton dipole moment). The diagonalization of the Hamiltonian results in the following values for the fine-structure splitting s and the polarization direction of the exciton emission ϕ:

$$s = [(\eta + \alpha \Delta S + \beta F_d)^2 + (k + \gamma \Delta S)^2]^{1/2} \tag{9.6}$$

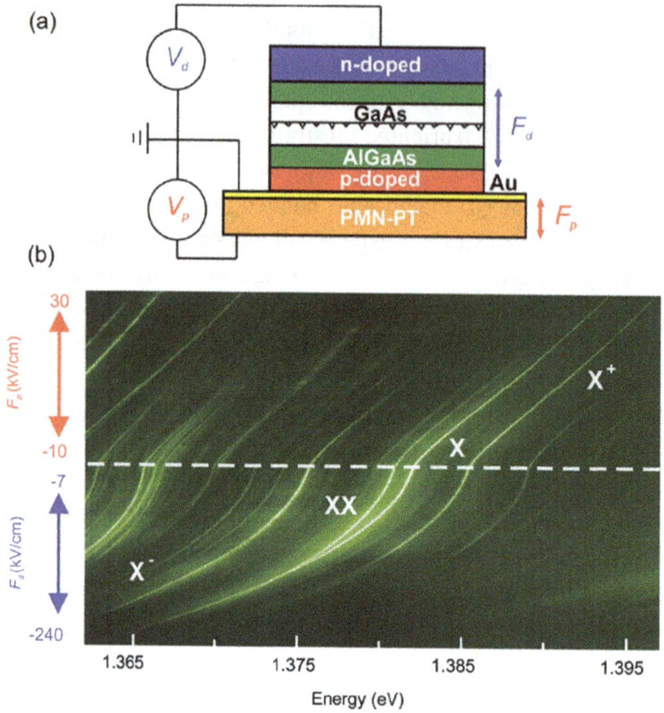

Figure 9.9: (a) Sketch of an n–i–p QD-nanomembrane device integrated on top of a piezoelectric actuator (PMN-PT). The piezoelectric actuator allows us to apply anisotropic biaxial strain in the (001) plane of the GaAs nanomembrane by tuning the voltage (electric field) V_p (F_p). In addition, a voltage (electric field) V_d (F_d) can be applied to the n–i–p membrane along the [001] direction. (b) Color-coded micro-PL map of a single QD as a function of F_p and F_d. Different charged exciton states can be observed. Reprinted with permission from [200]. Copyright (2023) by the American Physical Society.

and

$$\tan(\phi_\pm) = \frac{k + \gamma\Delta S}{\eta + \alpha\Delta S + \beta F_d \pm s}. \tag{9.7}$$

Equation (9.6) has a minimum at zero when F_d and ΔS take the values

$$\Delta S_{\text{critic}} = -\frac{k}{\gamma}, \quad F_{d,\text{critic}} = \frac{\alpha k}{\gamma\beta} - \frac{\eta}{\beta}. \tag{9.8}$$

Therefore it is always possible to adjust the fine-structure splitting s to zero for specific values for the in-plane stress F_p and out-of-plane electric fields F_d. This is exemplarily shown in Fig. 9.10, where experimental data for s and ϕ are depicted versus the applied electric field F_d for different in-plane strain fields F_p. With increasing out-of-plane electric fields F_d, a decrease of s to a minimum value followed by an subsequent increase is observed. With increasing F_p the minimum reachable value of s first decreases and

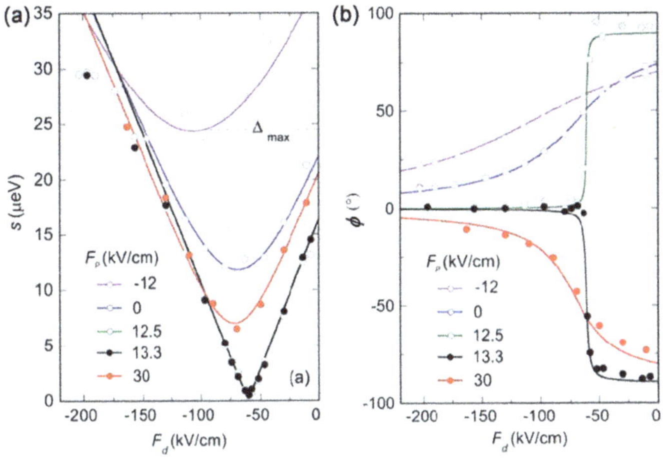

Figure 9.10: (a) Fine-structure splitting values s and (b) polarization angle ϕ as a function of applied electric field F_d for different applied strain fields, i. e., applied electric fields F_p on the PMN-PT piezo actuator (sample sketch; see Fig. 9.9). At certain critical fields $F_{d,\text{crit}}$ and $F_{p,\text{crit}}$, the fine-structure splitting s is tuned to zero. The solid lines are fit to the experimental data obtained using Eqs. (9.6) and (9.7). Reprinted with permission from [200] as displayed in [120]. Copyright (2023) by the American Physical Society.

then increases again. A minimum value for s close to zero is reached for $F_p = 13.3$ kV/cm. In this case the fine-structure splitting is erased. It is very interesting to study the corresponding phase angle ϕ. When F_p is increased, the ϕ curve shows sharper variations with F_d, and after the critical point where s reaches its minimum value close to zero, the direction of rotation is inverted, from anticlockwise to clockwise. This change in handedness of ϕ suggests that the critical point $s = 0$ has been crossed. The solid lines are fits to the experimental data using Eqs. (9.6) and (9.7). An excellent agreement is achieved, which supports the interpretation discussed above.

The following interpretation of the fine-structure tuning is instructive for the following discussion. One external perturbation (e. g., F_d) is used to align the polarization axis of the exciton emission along the axes of the second perturbation (e. g., F_p), which is then able to compensate completely for the difference of the asymmetry of the confining potentials of the two bright exciton eigenstates, i. e., it is able to tune the fine-structure splitting to zero. This insight points toward the fact that one external perturbation (e. g., strain field) can be sufficient to erase the fine-structure splitting as long as the polarization axis is aligned along the axis of the external perturbation. It is known that self-assembled InAs QDs on GaAs(001) substrates are often preferentially elongated in shape along the [1-10] GaAs crystal axes. This knowledge can be used for aligning the actuating direction of the piezoelectric crystal with respect to the GaAs crystal axes. However, the probability of a perfect alignment in practice is very low.

The previous discussion has shown that with two independent "tuning knobs", it is possible to reliably tune the exciton fine-structure splitting to zero, therefore enabling the generation of polarization-entangled photon pairs within time-integrated measurements. However, for some attractive applications, e. g., entanglement swapping, which is needed for a certain class of quantum repeaters or quantum teleportation applications, we would also like to have the additional freedom of freely choosing the emission wavelength of the photons. Therefore an additional "tuning knob" is necessary, which motivated the development of a novel class of micromachined piezoelectric actuators.

The basic idea is to achieve a full control of the in-plane strain tensor with especially designed piezoelectric actuators. Micromachined piezoelectric actuators allow us to control the magnitude, direction, and anisotropy of the exerted strain fields. Such an actuator essentially consists of the piezoelectric substrate cut into individually tunable strain segments, e. g., by a femtosecond laser technique. The fabrication details of the micromachined piezoelectric actuators can be found in the review of Martin Sanchez et al. [120]. Depending on the specific application, a different number of strain segments, so-called legs, are fabricated on the actuator. We typically distinguish between 2-, 3-, and 6-legged actuators. Figure 9.11 shows the principal sketch of a 2-legged, 3-legged, and a 6-legged actuator together with a sketch of the QD anisotropy. The QD membrane sample is bonded on the top side (Side A) of the actuators, e. g., by gold thermo-compression bonding. On the backside, only the legs are coated with Au so as to apply independent voltages on each of the legs. The top side is set to ground, and the individual voltages can be applied to each of the different legs, i. e., an electric field is locally induced across the piezoelectric material. This procedure induces in-plane strain fields on the bonded nanomembrane, and the QDs of interest are studied in the respective gaps between the legs.

In most reported experiments the bonded membranes exhibited prestrain, which was likely due to bonding and device processing. So far, a QD membrane bonded on the 2-leg device (see Fig. 9.11(a)) showed the highest PL-shift of a QD emission line of 41.5 meV for a range of applied electric fields up to 22 $kVcm^{-1}$. This large value was attributed to the small gap between the legs, which was ~80 μm. However, 2-leg devices fail in a full compensation of the prestrain when the induced stress in not aligned with the prestress axes. Therefore 3-leg devices were developed where the three legs are oriented along the directions of 60° away from each other (see Fig. 9.11(b)). With this design, we can independently control all three components of the in-plane stress tensor. This also means that arbitrary stress configurations can be generated. These include uniaxial and isotropic/anisotropic biaxial stress fields along any direction. In comparison with monolithic actuators, these devices allow us to apply up to one order of magnitude larger strain fields [120]. Even higher stress fields can be applied with 6-leg devices as demonstrated in [199]. For these devices (Fig. 9.11(c)), voltages are applied on pairs of aligned legs. This avoids lateral displacements at the center of the membrane, where the selected QD is used for quantum light generation. Maximum reported stress field anisotropies of 200 MPa in all stress directions (major stress angle) and hydrostatic stress values as large as 350 MPa have been reported. These properties make the 6-leg device a very versatile tool for strain engineering. In the following, we will show that the versatility of the 6-leg

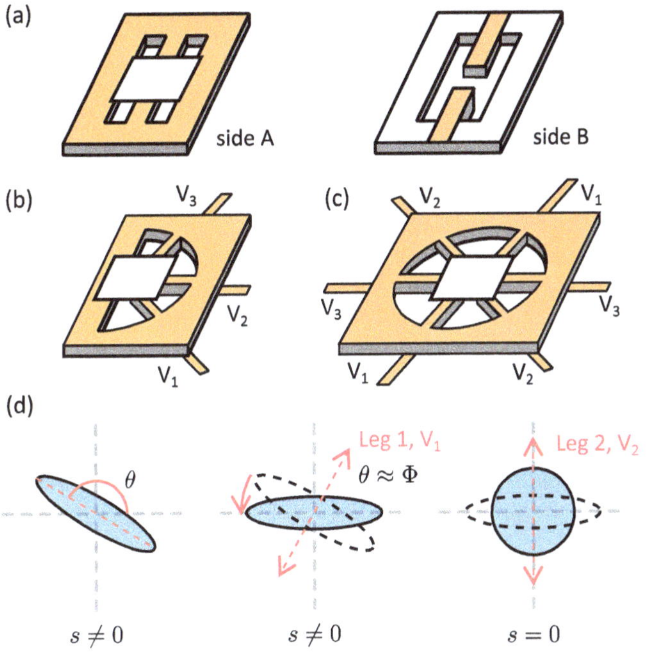

Figure 9.11: (a) Sketch of a 2-leg device seen from top (Side A) and bottom (Side B) with a bonded nanomembrane on the top side of the actuator. The contacts for applying the voltages on the two legs are defined at the bottom side. (b) Sketch of a 3-leg device and (c) a 6-leg device with a QD-containing nanomembrane on top of the PMN-PT based actuators. The contacts are also defined at the bottom (not shown) similar to the 2-leg device. (d) Schematic sketch of the QD anisotropy for a QD with $s \neq 0$. The deviation from a circle indicates the existence of an in-plane anisotropy in the QD confining potential. For a 3- or 6-leg device, in a first step the QD anisotropy is aligned along the actuating direction of leg-2 with leg-1 (V_1). In the second step the anisotropy is compensated by applying a certain voltage V_2 on leg-2. Image derived from [120] and [199].

device allows us to tune the exciton fine-structure splitting to zero and additionally tune the wavelength of the emitted photons. The tuning procedure is based on a three-step approach. In the first step a certain voltage is applied on one of the legs to align the polarization angle of the exciton emission along the actuating direction of the second leg (see Fig. 9.11(d)). In the second step the latter is used to tune the fine-structure splitting to zero. Finally, the third leg is used to change the initial QD anisotropy: this means that repeating once more the first two steps enables the adjustment $s = 0$ at a different energy of the exciton transition. This iterative method has been applied on an InAs QD with relative exciton polarization angle with respect to actuating leg-2 direction of 19° (see Fig. 9.12). It is demonstrated that the exciton emission energy can be continuously tuned over a range of 7 meV with $s < 1\,\mu$eV. The maximum achievable tuning range depends on the specific device, i. e., on the strain, which can be applied before the sample cracks. This flexibility can be used to generate energy-tunable polarization-entangled

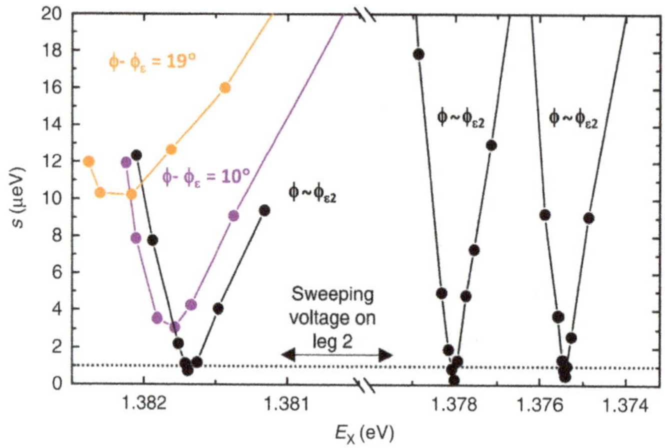

Figure 9.12: Experimental fine-structure splitting vs the exciton emission energy for an InAs QD with relative exciton polarization angle with respect to actuating leg-2 (major strain angle $\phi_{\epsilon 2}$) $\phi - \phi_{\epsilon 2} = 19°$. The exciton polarization angle can be tuned along the actuation direction of leg-2 by properly tuning the voltage on leg-1 (V_1), so that $s = 0$ can be reached by sweeping the voltage on leg-2 (V_2). Reprinted from [120] and [199].

photons from a QD embedded in a nanomembrane [199]. Moreover, since the strain tuning alone gives now this flexibility, a vertical electrical field can be used to inject carriers into the semiconductor membrane. This would allow us to fabricate electrically driven energy-tunable polarization-entangled photon devices.

9.7 Permanent tuning methods

So far, several reversible tuning techniques have been discussed. Now permanent tuning methods will be shortly mentioned, i. e., laser crystallization of a thin HfO_2 film on a surface close to a QD, will be discussed in more detail. With rapid thermal annealing, the composition of all QDs within the QD ensemble can be changed all together, which provides long-range wavelength tuning. However, this does not allow the controlled tuning of individual QDs. In contrast, laser processing allows for local thermal annealing, but the required high temperatures (>650°) can damage the semiconductor structure, leading to a degradation of the QD emission.

In the following, we will discuss a permanent strain tuning method, which is based on local laser-assisted HfO_2 film crystallization [60]. This method provides a breakthrough in many aspects with respect to scaling up the number of QDs in an integrated quantum photonic platform. The fundamental idea is to perform a local laser-induced modification of the crystal structure of a thin deposited layer of HfO_2. This introduces a controlled mechanical strain in specific areas of photonic structures with submicrometric spatial resolution, which provides a local strain field around a preselected QD and

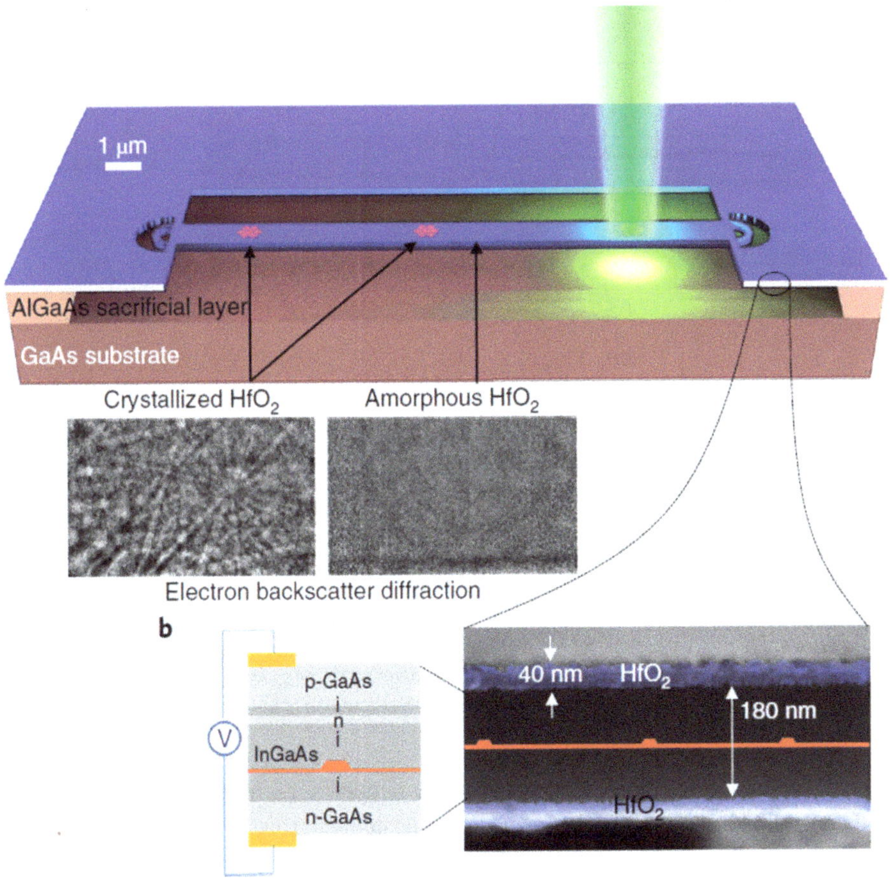

Figure 9.13: (a) The sketch displays a green laser beam, which is used to locally heat a GaAs bridge waveguide to crystallize a small region of a HfO_2 film. The idea is to shift the emission energy of a QD located in the center of the waveguide underneath the laser focus. Electron backscatter diffraction images on (left, crystallized) and away (right, amorphous) from positions that have been laser heated. (b) Right: cross-sectional SEM image showing a conformal ~40 nm HfO_2 film. Left: GaAs membrane heterostructure with n–i–n–i–p regions to deterministically charge the QDs with electrons. Reprinted with permission of Springer Nature from [60].

can bring its emission energy to a preselected value. Figure 9.13(a) shows a schematic sketch of a QD containing bridge waveguide together with the energy tuning approach. The strain-based tuning technique has been used for a GaAs waveguide containing a low density of InAs QDs (~1 μm^{-2}). A green laser beam (532 nm) is applied to locally heat the GaAs bridge waveguide to crystallize a small region of a HfO_2 film. Since only a small volume of the structure is heated, laser crystallization can be performed with the sample being kept at cryogenic temperatures (~6 K). This has the big advantage that the achieved QD energy tuning can be quickly checked by measuring the QD μ-PL directly

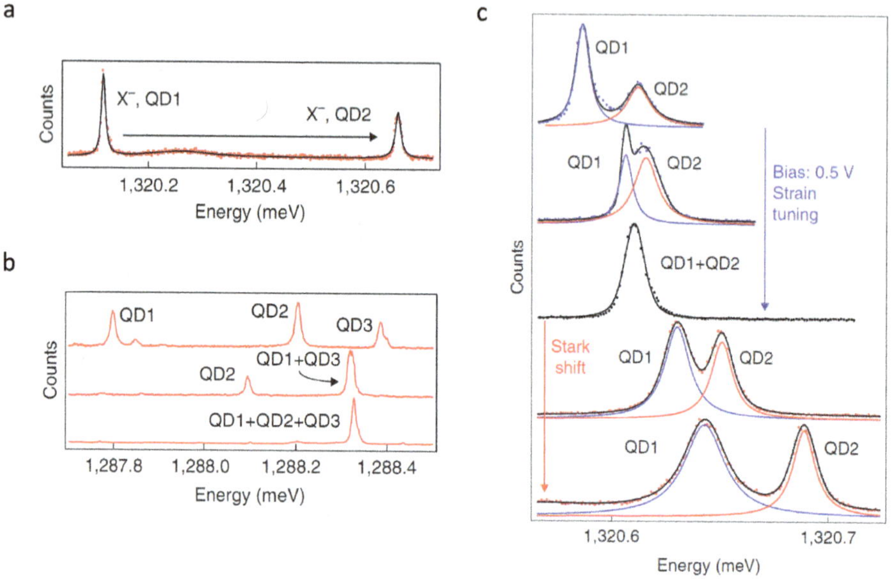

Figure 9.14: (a) High-resolution spectra of two charged exciton transitions from two QDs before the tuning. (b) Three QDs negative charged excitons in high-resolution spectroscopy and respective tunings. (c) Spectra of the two transitions as in (a) in the last stage of strain tuning (three spectra on top). The two bottom spectra are tuned with applied electric field. Reprinted with permission of Springer Nature from [60].

after the applied annealing cycle, again at low temperature. In this way a precise and well-controlled tuning can be performed by repeatedly performing this procedure. Figure 9.14 shows an example where two and three QDs were tuned into resonance with the technique described above. In this case a photonic crystal waveguide was used for the tuning experiment. The sample was additionally contacted to apply a certain voltage to spectrally tune the two QDs via the Stark effect. The trion emission energies (X^-) of the two QDs (QD1 and QD2) in Fig. 9.14(a) were initially 0.54 meV out of resonance. In a first step the two QDs were strain-tuned into resonance at a fixed bias of 0.5 V. The top three spectra in Fig. 9.14(c) show the three tuning steps with both QDs in resonance with the third spectrum. In the second step the two QDs were tuned out of resonance via the Stark effect, which allows a reversible tuning discussed in one of the previous sections. In a different photonic crystal waveguide, even three QD emission lines were tuned into resonance (see Fig. 9.14(b)), and the authors demonstrated a quantum interaction via superradiant emission from the three quantum dots [60]. Most importantly, this approach allows tuning InAs quantum dot emission energies over the full inhomogeneous distribution (~65 meV) with a step size down to the homogeneous linewidth (~ 1 µeV) and a spatial resolution better than ~1 µm.

9.8 Summary

- Temperature tuning of QD transitions is an easy possibility to tune the emission wavelength of a QD light source but leads to linewidth broadening, dephasing, and, consequently, to a reduction of photon indistinguishability.
- A convenient way to adjust and manipulate the energy of quantized states in QDs is via the quantum confined Stark effect by application of electric fields. The energy shift is described by $\Delta E = p_{\mathrm{perm}} E + \alpha E^2$, where p_{perm} is a permanent electric dipole, and α is the polarizability of the exciton state.
- In QDs with large carrier confinement potentials, Stark shifts of up to 25 meV can be observed. At higher fields, carriers tunnel out of the QDs into the barrier material, which represents a nonradiative loss channel.
- Charge controllable samples allow a controlled QD charging via Coulomb blockade, and the QD frequency can be tuned via the quantum confined DC Stark effect. Moreover, random charge fluctuations in the environment of the QD can be suppressed. This enhances the brightness of the source, suppresses unwanted blinking effects, and reduces dephasing.
- In a magnetic field in Faraday configuration ($B \parallel z$), all excitonic lines are split into doublets. In Voigt configuration ($B \perp z$) the dark excitons can contribute to the PL signal, because the symmetry with respect to the angular quantum momentum number J_z is broken. A typical tuning range of ~1 meV is achieved for magnetic fields up to 10 T.
- Strain tuning via piezoelectric actuators is a very powerful method for tailoring transition energies, biexciton binding energies, and exciton fine-structure splittings.
- Monolithic piezoelectric actuators allow tuning ranges of ~10 meV and enable a differential in situ wavelength tuning of a QD emission and microcavity modes to finally tune QD emission lines into resonance with a cavity mode.
- QD devices with simultaneous strain and electric field tuning allow for control over binding energies, spectral tuning larger than 30 meV, and elimination of the fine-structure splitting.
- Micromachined 6-leg piezoelectric actuators allow us to tune the exciton fine-structure splitting to zero and additionally tune the emission wavelength of the emitted photons. This flexibility can be used to generate energy-tunable polarization-entangled photons from a QD in a nanomembrane.
- The permanent tuning method, which is based on local laser-assisted HfO_2 film crystallization, allows tuning InAs QD emission over the full inhomogeneous distribution (~65 meV) with a step size down to the homogeneous linewidth (~1 μeV) and a spatial resolution better than ~1 μm.

10 Optical and electrical excitation methods of quantum dots

10.1 Introduction

In most excitation schemes a single quantum dot is carried to an excited state before a subsequent radiative decay process occurs and a photon is emitted. In other words, electron and holes are generated inside QDs or captured from the surrounding layers into the QD, and in a second step, they recombine radiatively. Exceptions are, e. g., Raman schemes, where a single photon is scattered by a single quantum emitter. The Raman schemes transcend the temporal separation of the excitation and decay process and promise a high degree of photon indistinguishability. In general, we can distinguish between optical and electrical excitation schemes.

Optical excitation schemes are very versatile, and many different schemes have been developed and studied for quantum dots. Typically, a pulsed laser source is used for the triggered generation of the excited states, which under particular conditions results in the triggered emission of single photons at certain selected wavelengths (photons on demand). Continuous-wave excitation schemes[1] and the resulting emitted light fields are, for example, discussed in the book [126]. It is meaningful to differentiate between resonant and nonresonant excitation schemes.

Electrical excitation schemes allow small, robust, and compact designs and therefore promise scalability and on-chip integration. The electrical layout and sample structure of an electrically driven single-photon source is accomplished with standard semiconductor technology based on p–n diode structures. Here a QD is embedded in the intrinsic layer (nearly undoped region) of p–i–n structures, integrated in planar optical cavities, micropillar or photonic crystals, etc. and excited by a continuous-wave or pulsed electrical current.

10.2 Optical excitation schemes

In practice the properties of the emitted photons from single-photon sources, like coherence time, indistinguishability, polarization, and time jitter, critically depend on the chosen excitation process. A large number of different optical excitation schemes have been studied and developed. Here we will discuss the most prominent ones, which have also been widely used in the past. We distinguish between nonresonant and resonant excitation schemes. Figure 10.1 shows an overview of the most applied schemes, which will be discussed in detail below.

1 Here we mainly focus on pulsed excitation techniques, which allow for on-demand emission of quantum light. However, continuous-wave (cw) excitation can also be useful when synchronization of signals is not required.

https://doi.org/10.1515/9783110703412-010

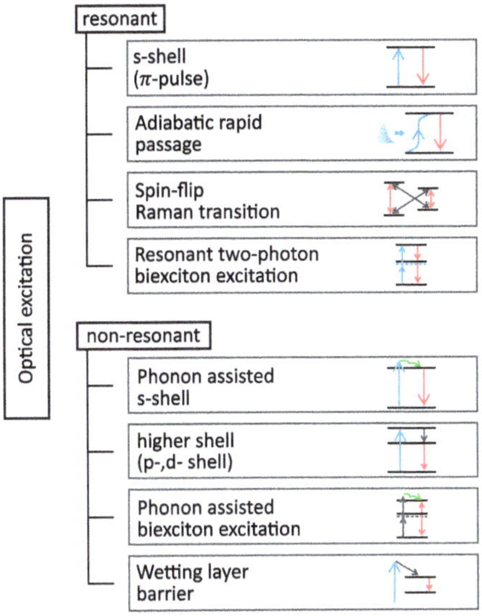

Figure 10.1: Optical schemes of resonant and nonresonant exciton and biexciton excitation.

10.2.1 Nonresonant optical excitation

Nonresonant excitation schemes can be understood considering the band structure of a self-assembled QD (see also introduction of Chapter 8 and Fig. 8.1). Figure 10.2 shows a sketch of the band structure of a QD together with different excitation and radiative recombination channels. Using a short laser pulse (of the order of several ps to several tens of ps) with an excitation energy above the bandgap of the barrier or wetting layer will mainly create charge carriers (electron and holes) into these layers (Fig. 10.2(left)). Some charge carriers will subsequently be captured by nearby QDs and relax to the lowest energy levels within a short time scale (~ few to hundreds of picoseconds). Due to the statistical nature of the carrier capture process, even or odd carrier numbers in the QD are possible. The detailed occupation depends on the pump power and on the specific environment of the QD. As an example, the case of an even number of carriers in the dot is discussed. The radiative recombination of this multiexcitonic state occurs in a cascade process of successive radiation recombination processes of the multiexciton states NX, $(N-1)X, \ldots,$ the biexciton $2X$ (equivalent notation XX), and the exciton $1X$. In the case of odd number of carriers, the corresponding charged recombination processes occur. The energy of the photons emitted during recombination depends significantly on the number of charge carriers existing in the QD due to Coulomb interactions, enhanced by strong carrier confinement (see also Chapter 7). All transitions possess typical radiative lifetimes in the range of several 100 ps up to ~1 ns. Since the recombination times of the free carriers in the barrier and wetting layer are typically shorter than inside the QD, the reexcitation probability of the QD is low, and each excitation pulse can lead almost only to one photon emission event at the corresponding NX transition (see also Section 7.3.4).

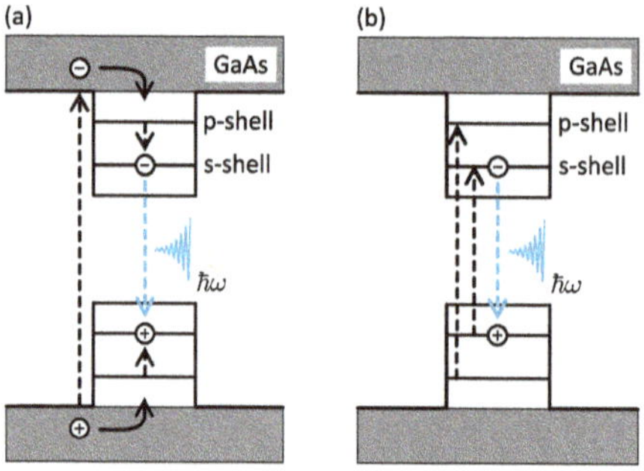

Figure 10.2: Sketch of the bandstructure of a QD together with different excitation schemes. (left) Nonresonant excitation schemes are depicted, i. e., above barrier bandgap or wetting layer (not shown) excitation. The carries are trapped by the QD and relax to the lowest energy level. (right) Quasi-resonant (p-shell) and resonant (s-shell) excitation schemes. The electron–hole pair that will form is directly created in the respective shell.

Thus specific transitions from the cascade can be selected by spectral filters, e. g., $2X$ and $1X$, and used to generate a regulated single-photon stream at the corresponding photon energy.

Due to the random nature of the carrier capture process, the number of captured carriers in the quantum dot is given by a Poisson distribution. To achieve a high quantum efficiency close to unity for a certain transition, the pump power has to be adjusted well above the average occupation number of the respective excited state. This has serious implications for the time jitter of the emission process and the resulting linewidth of the emission line. The time, jitter is determined by the carrier capture times and the relaxation rates between the different shells within the dot. A typical jitter is in the range between several tens of picoseconds up to several nanoseconds. At high pump powers the emission linewidth is influenced by carrier–carrier scattering processes (see Sections 5.2.6 and 8.3). Both effects lead to a reduction of the photon indistinguishability.[2]

Another drawback of a nonresonant excitation process is that charge carriers can be captured and released by adjacent traps or defect centers in the vicinity of the QD. This might lead to fluctuating electric fields nearby the QD causing fluctuations in the emission wavelength between different excitation pulses: this is known as spectral diffusion (see Sections 5.2.7 and 8.3), a major line broadening process for quantum dot transitions. This will further reduce the indistinguishability of the photons. Moreover, such released carriers might be captured by the QD after a first photon emission process has already occurred. Such an unwanted reexcitation process would lead to a secondary emission process at the respective transition and, consequently, decrease the single-photon purity, detectable by an increase of a measured $g^{(2)}(0)$-value in a photon correlation measure-

2 An additional time jitter on the emission process provides an uncertainty on the photon arrival time on the BS used of the interference, hence reducing the achievable TPI visibility.

ment. Likewise, a delayed carrier capture process from the barrier can also lead to a decrease in single-photon purity (see Section 8.4).

Furthermore, the time-averaged biexciton/exciton emission is unpolarized since both exciton fine-structure components contribute equally to the emitted light. Thus for applications where well-defined polarized photons are required, for example, in quantum cryptography, a polarization filter is necessary to select, e. g., linearly polarized light. This reduces the efficiency of the source by a factor of 2 (see Section 5.2.4).

The advantage of the nonresonant excitation scheme is its easy implementation. The excitation laser does not need a specific photon energy (except larger than the barrier bandgap) and not to be energy tunable. Furthermore, from a spectroscopic point of view, the excitation laser is spectrally well separated from the emitted photons, being then easy to filter (see Section 5.2.1).

10.2.2 Quasi-resonant optical excitation

A quasi-resonant excitation process into a higher shell, e. g., p-shell, has some advantages over the nonresonant pumping scheme. This opens the possibility of a controlled occupation of charge carriers in the p-shell (see Fig. 10.2 (right)). After relaxation into the s-shell, single-photon emission can occur. As a consequence, pure dephasing processes by carrier–carrier scattering are drastically reduced since the charge carriers are exclusively generated within the desired dot. Furthermore, the influence of charge fluctuations in nearby traps is also reduced. The time jitter is only determined by the p- to s-shell relaxation process, which is typically much faster than the radiative lifetime of the transition. Moreover, e. g., the carrier relaxation process from an excited trion state to the trion ground level occurs by a spin-preserving optical phonon process. This allows, for example, applications where spin-photon entanglement is required, such as photon cluster state generation (see Section 7.3.6). Furthermore, the excitation laser light can be easily spectrally separated from the emitted single photons.

10.2.3 Resonant optical excitation

It has been proven that resonant optical excitation schemes deliver photons with the highest quality. These excitation schemes are coherent and avoid any time jitter in the photon emission process. Therefore high efficiencies and high degrees of photon indistinguishability can be reached, which are essential for many applications in photonic quantum technology. Different types of resonant excitation schemes for quantum dots have been studied. Among them, there are resonant exciton and trion excitation, adiabatic rapid passage schemes, spin-flip Raman transitions, and resonant two-photon biexciton excitation (see Fig. 10.1). All of them have their pros and cons, which will be discussed below. The discussion will focus on triggered pulsed excitation schemes that allow the on-demand emission of single photons and photon pairs, which is the most relevant scheme for quantum photonic applications. Just recently developed more complex two laser excitation schemes, using a phase-locked dichromatic electromagnetic field [70] or the so-called swing-up of quantum emitter population (SUPER) scheme [93], are potential attractive coherent excitation schemes. These methods posses no spectral overlap with the desired optical transition, and the scattered laser light can therefore be easily spectrally filtered. However, these schemes are so far almost unexplored and are therefore not considered in this introductory textbook.

10.2.3.1 π-pulse excitation

The resonant excitation process allows a direct excitation of a specific state of the quantum dot and therefore also gives the highest control of the subsequent emission process. At the same time, a sophisticated suppression of the laser excitation light is necessary, which typically complicates the practical realization of light collection. Moreover, the different basic excited states of a quantum dot, e. g., exciton, trion, and biexciton, require different resonant excitation and stray-light suppression techniques. In the following, we will shortly discuss polarization filtering, geometrical filtering, and spatial filtering, which are widely in use (see also Chapter 5).

One of the most common methods is based on a cross-polarized excitation and detection scheme (see Fig. 10.3(a)). Such a scheme can be applied if the signal has an orthogonal polarization component with respect to the excitation laser. In this case, one polarizer can be placed in the excitation arm while a cross-polarized analyzer is used in the detection arm. This can in principle also be achieved by a single polarizing beam splitter. Actual setups typically incorporate additional waveplates and polarizers to compensate for any polarization change due to sample- or setup-related polarization distortions. A drawback of the cross-polarized setup is that the brightness is limited to at most 50 % due to the rejection of photons orthogonally polarized to the collection mode. This limit can be overcome by using polarized cavities and utilizing an asymmetric Purcell effect, as we will discuss below. The loss of 50 % of the photons is also avoided by a geometrical

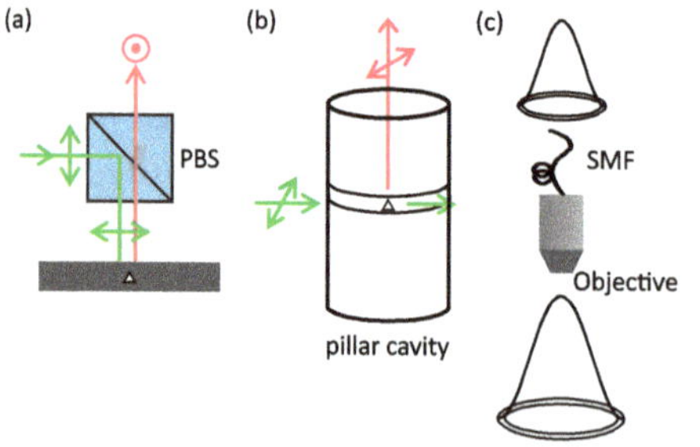

Figure 10.3: Laser suppression techniques. (a) Polarization filtering. (b) Orthogonal excitation and detection geometry. (c) Spatial filtering using a single-mode fiber.

filtering technique (see Fig. 10.3(b)). If both the edge and the top of the photon source are optically accessible, then an orthogonal excitation and detection geometry can be implemented. This is, e. g., typically possible for waveguide structures and micropillar cavities. A very efficient technique relies on spatial filtering using a single-mode fiber in the detection path (see Fig. 10.3(c)). The working principle is similar to a confocal filter, i. e., it restricts the collection of randomly scattered laser light into the collection mode (see Section 5.2.1). Of course, a combination of these methods could be very beneficial.

It is important to distinguish the resonant excitation of an eigenstate of the system from the excitation of a superposition of eigenstates. In the following, we will discuss the relevant optical dipoles, selection rules, and transient decays for the most relevant schemes. We start the discussion with the resonant optical excitation of an exciton with a certain fine-structure splitting ($\delta_1 \neq 0$) (see Fig. 10.4(a)). The exciton fine-structure splitting results in two nondegenerate exciton states with orthogonal linear dipoles (H, V) (see Fig. 10.4(a,b) and Chapter 7). The excitation of a single eigenstate of the system is firstly discussed. For example, a V-polarized laser interacts only with the V-dipole, and therefore only V-polarized light is emitted from the excitonic state as long as the spin orientation remains essentially stable during the radiative lifetime. This is usually the case, and thus a monoexponential decay and a maximum detection rate, i. e., equal to the excitation rate, can be in principle detected under V-polarization. This scheme has been successfully applied in combination with geometrical filtering techniques.

In contrast, with a laser polarization orientation of $\varphi = 45°$ with respect to the main QD dipoles (H, V) (see Fig. 10.4(b)) and using a pulsed laser that is resonant with both fine-structure components, i. e., with a laser linewidth larger than δ_1, a superposition of both fine-structure components is excited. This will result in a precession of the exciton spin. In this case, quantum beats, i. e., intensity oscillations in the photon emission can be observed under linear polarization detection (see Fig. 10.4(f)). It is important to note that in a cross-polarized detection scheme, the emission is delayed from the excitation with a time scale inversely proportional to the fine-structure splitting. Moreover, in an idealized case, the maximum detection rate is expected to approach 50 % of the exciton preparation rate for large precession frequencies ($\gg 1/\tau_{rad}$) and equal contribution of both fine-structure components. This represents a serious limitation in brightness.

The situation for the trion state is different. The trion possesses two degenerate circular-polarized dipole transitions (see Fig. 10.4(c) and Chapter 7). The energy scheme can be also expressed with the quantization axis parallel to H (see Fig. 10.4(d)). Each exited state is then connected to both ground states through two optical transitions (H, V) with linear orthogonal polarization. From this scheme, under the assumption of a mixture of spin up and spin down states for the electron in the ground state (no spin initialization), a linear-polarized optical excitation, e. g., a V-polarized optical pulse populates both trion states. Each of them radiates with 50 % probability in any linear polarization direction (e. g., H). In this case the rise of the emission is governed by the excitation laser pulse width, and the monoexponential decay is determined by the radiative lifetime of the transition (Fig. 10.4(e)). However, in the cross-polarization setup the maximum count

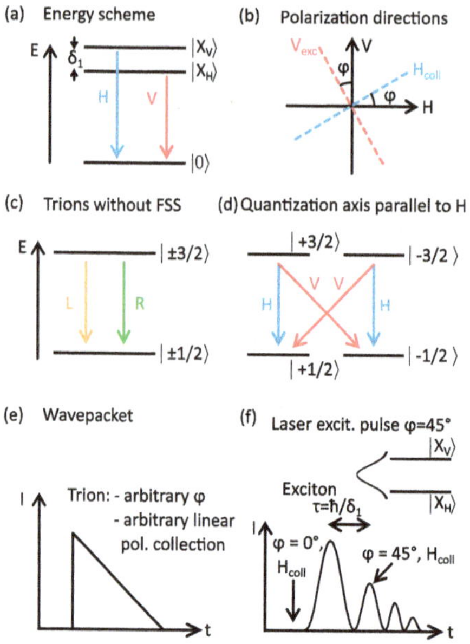

Figure 10.4: Sketch of energy schemes, optical dipole transitions (H, V, R, L), polarization directions of laser excitation (V_{exc}) and collection (H_{coll}), and temporal shapes of photon wavepackets after certain exciton and trion excitation schemes. (a) Energy scheme of an exciton with a fine-structure splitting δ_1. (b) Polarization directions: exciton dipole axes V and H, and polarization of laser excitation and collection, V_{exc} and H_{coll}, respectively. (c) Energy scheme of the trion with the respective circular polarized transitions R and L. (d) Trion energy scheme shown in the basis with quantization axis parallel to H. Each state is connected to two ground states through two optical transitions (H, V). (e) Temporal shape of the trion wavepacket. (f) Temporal shape of exciton wavepackets for different polarization angles of the excitation laser ϕ and detection directions (V_{coll}, H_{coll}).

rate is also limited to at most 50 % due to the rejection of photons orthogonally polarized to the collection polarization. For an exciton with zero fine-structure splitting, the situation is similar to the above-discussed trion state.

For achieving both high brightness and high photon indistinguishability the QDs are typically embedded in photonic microcavities (see Chapters 6, 13, and 14). Due to fabrication imperfections or on purpose, the cavities can possess a nonperfect circular symmetry, and therefore the cavity introduces an anisotropy leading to two linearly polarized fundamental cavity modes, labeled V_{cav} and H_{cav} in the following. In the former case the energy difference between the modes (Δ_{HV}) is typically smaller than the spectral linewidth of the modes ($\Delta_{HV} \ll \Delta_{mode}$); nevertheless, this causes an effective birefringence in the system. Despite now a more elaborate description is necessary for a cross-polarized detection scheme due to three possible different polarization directions, namely the orthogonal modes V_{cav} and H_{cav}, the quantum dot dipole directions V and H, and the polarizer directions V_{exc} and H_{coll}, the main result for an exciton and trion emitter is similar to that discussed above, i. e., a delayed emission process for the

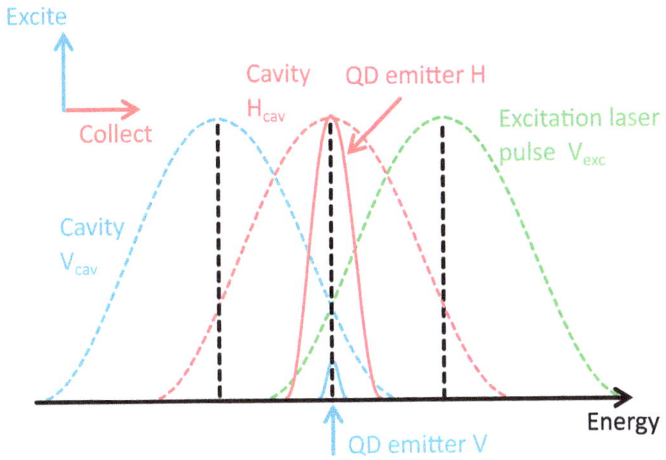

Figure 10.5: Sketch of the excitation scheme of a polarized single-photon source by resonantly exciting a quantum dot emitter in a birefringent microcavity. The asymmetric cavity exhibits two non-degenerate cavity modes V_{cav}, H_{cav} and the QD transition is resonant with the H-mode of the cavity (H_{cav}). The excitation laser is blue-detuned and V-polarized (V_{exc}). The V-polarized laser couples off-resonantly to the emitter' transition via the V-polarized cavity mode (V_{cav}). Due to the coupling, the QD can be optically excited even though a considerable higher power than for resonant π-pulse excitation is required. The Purcell effect enhances the emission into the resonant H-mode of the cavity. The single photons are finally collected with a polarizer in H-direction (H_{coll}).

exciton but not for the trion, and an overall reduction of the maximum brightness of 50 % for both transitions. The relevant physics has been described, e. g., by Ollivier [141].

In the latter case ($\Delta_{HV} \gtrsim \Delta_{mode}$), the modes are energetically well separated, which can be used in a very advantageous way for achieving ultra-bright and pure single-photon sources utilizing a clever cross-polarized stray-light suppression technique. Figure 10.5 shows the theoretical scheme of a polarized single-photon source by resonantly exciting a quantum dot emitter in a birefringent microcavity. The asymmetric cavity supports two nondegenerate cavity modes V_{cav} and H_{cav}. The basic principle of the scheme is exemplarily explained with a QD trion state, where the transition is resonant with the H-mode of the cavity (H_{cav}). An excitation laser is blue-detuned with respect to both cavity modes and exhibits a V-polarization. The spectral tails of the pulsed laser and the V-polarized cavity mode (V_{cav}) overlap at the trion transition energy. Thus the trion can be optically excited even though a considerable higher power than for resonant π-pulse excitation is required. Due to the Purcell effect, the cavity redistributes the spontaneous emission of the QD in the H- and V-polarizations with ratio $F_H/F_V = 1 + 4(\Delta_{HV}/\Delta_{mode})^2$ [217]. The single photons are finally collected with a polarizer in the H-direction (H_{coll}). In this way the excitation laser is effectively filtered out from the collection path. Figure 10.6 shows the polarized single-photon efficiency for a series of realistic Purcell factors and (cavity mode) splitting-to-linewidth ratios. For example, a photon extraction efficiency of ~90 % can be reached for a Purcell factor

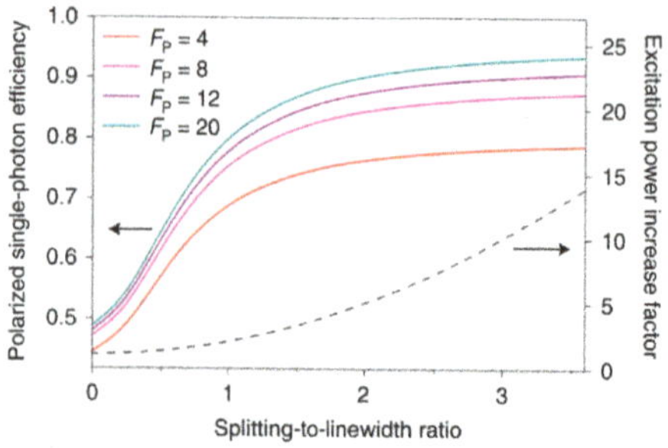

Figure 10.6: The graph displays the efficiency of preparing polarized single photons as a function of the ratio of the cavity splitting to the cavity linewidth for four selected Purcell factors F_p. The dashed line displays the increase factor of the excitation laser power required for a deterministic π-pulse, compared to the case of a micropillar with circular cross-section. Reprinted with permission of Springer Nature from [217].

of 20 and a splitting-to-linewidth ratio of 2.5. At the same time a seven times higher excitation power is necessary. It is important to note that this excitation and emission scheme is applicable in all photonic structures where two nondegenerate linearly polarized fundamental cavity modes can be induced, such as micropillars, circular Bragg gratings (bull's eye), microdisks, and photonic crystals (see also Chapter 14). Using this technique, excellent overall single-photon source properties have been already published, including record high overall single-photon efficiencies (>60 %), single-photon purities (>97 %), and photon indistinguishabilities (>97 %) [196, 217].

10.2.3.2 Adiabatic rapid passage

An alternative way for resonant excitation is using the adiabatic rapid passage (ARP) technique with frequency-chirped pulses (Fig. 10.7(a)). The main advantage relates to the robustness of the generation with respect to the standard π-pulse resonant excitation process. The ARP technique is largely unaffected by variations in the excitation power and emitter dipole moments. The drawback is an additional complexity in optical pulse generation. The basic idea of the scheme can be understood in a dressed level picture, where the two eigenstates of a two-level system interacting with a resonant laser field are no longer the emitter ground state $|g\rangle$ and the excited state $|e\rangle$, but rather their coherent superposition $|\pm\rangle$, i. e., dressed states (see Fig. 10.7(b) [223]). When the frequency of the excitation laser pulse is tuned across the excited state, with a rate much slower than the peak Rabi frequency, the two-level system starts from the ground state and evolves adiabatically to the excited state along a dressed state. For example, applying a positively chirped pulse, the laser sweeps from low to high frequency, and the two-level system evolves along the lower dressed state $|-\rangle$ to its excited state. For a negatively chirped pulse, the excited state is reached via the $|+\rangle$ state. Figure 10.7(c) shows the recorded single-photon count rate obtained from a trion state (X^-) recombination inside a planar microcavity as a function of excitation pulse area

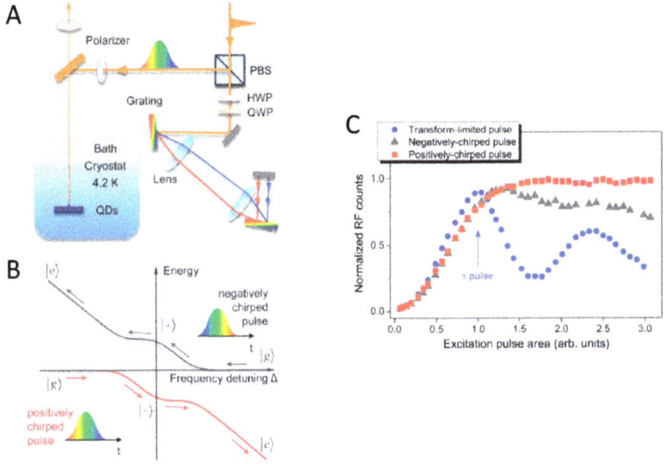

Figure 10.7: (a) Schematic scheme for chirped pulse generation and optical excitation of a QD inside a cryostat. (b) Eigenenergies of a two-level system in interaction with a positively (left half) and a negatively (right half) chirped laser pulse. The strong interaction of a resonant laser pulse with the two-level system leads to new dressed states $|\pm\rangle$ formed by a coherent superposition of ground $|g\rangle$ and excited states $|e\rangle$ of the two-level system. For far red (blue) frequency detuning of the excitation laser, the states $|+\rangle$ and $|-\rangle$ reduce again to $|e\rangle$ ($|g\rangle$) and $|g\rangle$ ($|e\rangle$) states. (c) Comparison of the single-photon count rate behavior versus pump power for three different types of excitation pulses, i. e., Fourier limit pulses and negatively and positively chirped pulses. Reprinted (and adapted) with permission from [223]. Copyright (2023) American Chemical Society.

for excitation with transform limited pulses (3-ps duration), negatively and positively chirped pulses (each 30 ps). In the case of transform-limited pulses, the expected oscillatory behavior (blue circles) is observed where the maximum is reached at the π-pulse. The observed damping for higher pulse areas could be caused by excitation-induced dephasing. For negatively chirped pulses, a steep increase is observed until a maximum is reached at 1.5π, followed by a gradual decrease at higher pulse areas. The decrease at higher pulse areas is attributed to a phonon-assisted relaxation (emission) process from the $|+\rangle$ to $|-\rangle$ state, which breaks the ARP. As a result, the efficiency of single-photon generation is reduced. In contrast, for positively chirped excitation pulses, the gradual decrease at high pump powers is not observed since the relaxation from the $|-\rangle$ to $|+\rangle$ state would require a phonon absorption process, which is very unlikely at low temperatures (~4.2 K). The plateau-like area of the single-photon intensity from 1.5 to 3.0π represents an excitation area robust against fluctuations of driving pulse area, with highest possible single-photon generation rate. Moreover, the authors of this study have shown that the emitted photons posses a vanishing two-photon emission probability of 0.3 % and a raw (corrected) two-photon Hong–Ou–Mandel interference visibility of 97.9 % (99.5 %) [223]. These properties makes the ARP technique highly attractive, especially for creating wavelength multiplexed multiple single-photon sources due to the robustness against small excitation wavelength, power, and dipole moment variations.

10.2.3.3 Spontaneous spin-flip Raman transitions

Another very versatile way of single-photon generation is based on a trion state in a QD utilizing spontaneous spin-flip Raman transitions. This process allows a tailored single-photon waveform, i. e., controlled temporal profiles with duration exceeding the intrinsic QD lifetime by up to three orders of magnitude together with reductions in spectral linewidth [10]. However, the price to be paid for this flexibility is a more complex experimental setup including two lasers with corresponding electro-optic modulators and applied magnetic field, as we will see below.

The generation scheme is based on a three-level system, i. e., a Λ-system built-up by a trion and two single-charged QD ground state states (two different spin configurations), and by applying a magnetic field in Faraday geometry along the growth direction (z-direction). The scheme is exemplarily discussed on the basis of a QD charged with a single-hole state. The two possible hole spin states are $|\Uparrow\rangle_z$ and $|\Downarrow\rangle_z$, and the corresponding two trion states are $|\Uparrow\Downarrow\uparrow\rangle_z$ and $|\Downarrow\Uparrow\downarrow\rangle_z$, consisting of two spin-paired holes and an electron spin (see Fig. 10.8(a)). The degeneracies between the ground and excited states are lifted in a magnetic field due to the corresponding out-of-plane electron and hole g-factors (see also Section 9.5). The optical selection rules allow only vertical spin-preserving transitions (① and ④) with orthogonal circular polarization, respectively. This is in contrast to the situation where a magnetic field is applied perpendicular to the growth direction (Voigt profile), as discussed for the generation of spin-photon entanglement in Section 7.3.5. However, the nonspin-preserving transitions (② and ③) are also weakly allowed due to heavy-hole-light-hole mixing or to hyperfine interaction. This allows an effective optical spin pumping process, which can be used to prepare a specific spin state followed by driving a spin-flipping transition within the Λ-system to generate a single-photon state.

The idea is the following: there are two possible Λ-systems in a charged QD, which can be utilized to generate a single-photon state via spin-flip Raman transitions. A Λ-system consists of a trion state ($|\Uparrow\Downarrow\uparrow\rangle_z$ or $|\Downarrow\Uparrow\downarrow\rangle_z$), where each possesses an allowed fast (Γ) and a weakly allowed slow (γ) spontaneous transition rate to two different ground states ($|\Uparrow\rangle_z$ and $|\Downarrow\rangle_z$). Exemplarily, the generation scheme with the state $|\Uparrow\Downarrow\uparrow\rangle_z$ is discussed. Optical spin pumping is achieved by resonantly driving the allowed transition ① until the state spontaneously decays via the weak spin-flipping transition into state $|\Downarrow\rangle_z$. This represents the spin initialization process. In the next step a single photon can be generated by driving the weakly allowed diagonal spin-flipping transition ③ with a second laser pulse. A single photon is generated on driving the hole spin from the spin down to the spin up state via a Raman transition (red photon in upper scheme; see Fig. 10.8). A detailed theoretical characterization of the wave packet of a photon emitted by such a single Λ-system is given by Müller et al. [134]. It turns out that the asymmetric branching ratio for the transitions discussed above $\gamma/(\Gamma+\gamma) \ll 1$ ensures that the purity of the Raman photon is hardly affected.

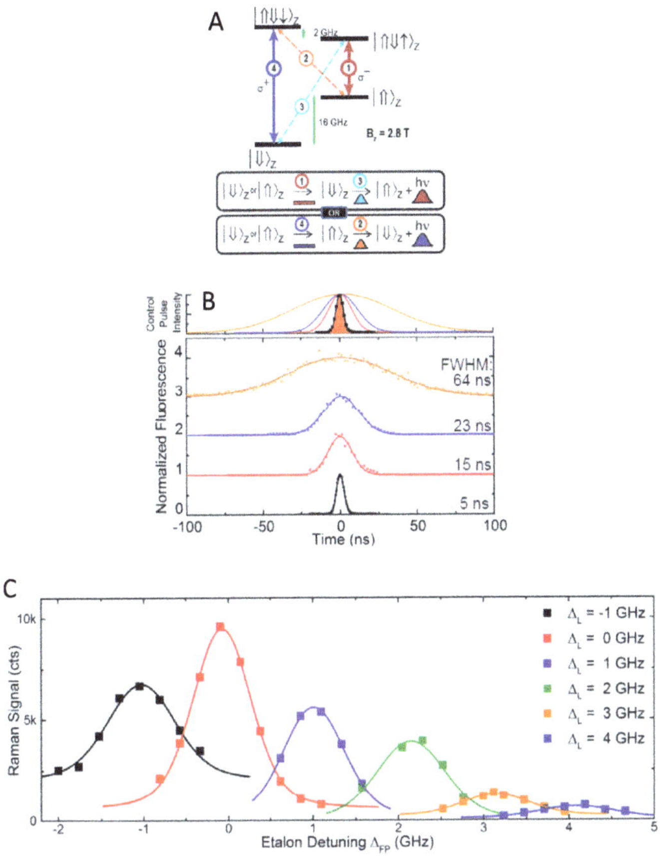

Figure 10.8: (a) Sketch of the energy level scheme of quantum dot charged with a single hole in a magnetic field in the Faraday geometry (field along the growth direction) and the corresponding optical transitions. Transitions ① and ④ are optically allowed and posses orthogonal circular polarizations. Transitions ② and ③ are weakly allowed. The inset illustrates the temporal sequence used for generating a red or a blue single Raman photon with controlled wave form. (b) Gaussian photon wave forms from Gaussian control pulses with FWHM of 5, 15, 23, and 64 ns. (c) Spectra of 50-ns Gaussian single photons measured for different control-laser detunings Δ_L. Reprinted with permission from [10]. Copyright (2023) by the American Physical Society.

For example, using Gaussian control pulses, Gaussian single photons of full width at half-maximum duration ranging from 5 to 64 ns have been generated (see Fig. 10.8(b) and for experimental details, [10]). Furthermore, the wavelength tunability has been demonstrated by introducing different excitation frequency detunings Δ_L from the $|\Downarrow\rangle_z \rightarrow |\Downarrow\Uparrow\uparrow\rangle_z$ resonance (see Fig. 10.8(c)). Here the Raman signals shift linearly with laser detuning Δ_L, and the peak amplitude follows a Lorentzian profile in Δ_L. However, the expected decrease in linewidth with Δ_L was not observed, most probably due to charge noise in the environment of the QD.

10.2.3.4 Two-photon biexciton excitation

A very advantageous possibility for the optical excitation of biexcitons in QDs is based on a resonant coherent two-photon excitation process via a virtual intermediate state. The optical selection rules prevent the direct excitation of the biexciton via a one-photon resonant absorption process since both the ground and biexciton states are spin zero states. Instead, a linearly polarized laser energetically tuned in between the exciton and biexciton resonances fulfill both the energy and polarization selection rules via a resonant two-photon absorption process. The energy scheme, together with the excitation and recombination processes, is sketched in Fig. 10.9(a) (see also Section 7.3.2). The result of a corresponding μ-PL experiment on a single QD is displayed in Fig. 10.9(c), where a linearly polarized laser (FWHM = 95 μeV) in resonance with the TPE virtual state of the biexciton has been applied. The intensities of the biexciton and exciton lines are similar, because only one biexciton can be prepared after each pulse, and the exciton is populated through the radiative cascade after recombination of the biexciton. The large peak between the exciton and biexciton lines is due to some residual scattered laser light. Since the stray light does not spectrally overlap with both exciton and biexciton lines, a laser background-free signal of both lines can be collected without any polarization suppression technique, as often used for resonant experiments on excitons. This allows polarization-dependent cross-correlation measurements on the biexciton–exciton cascade necessary for all applications that rely on polarization-entangled photon pairs. For comparison, Fig. 10.9(b) shows a μ-PL spectrum of the same QD under above-bandgap excitation. Besides exciton and biexciton lines with different intensities, an additional trion line is visible. The appearance of different emission lines (X, XX, T) with different

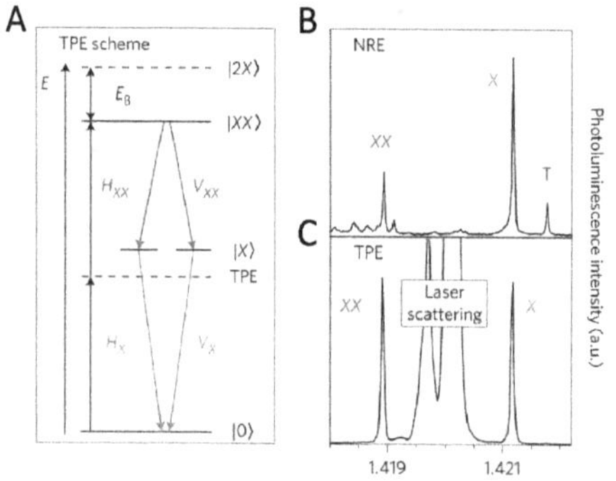

Figure 10.9: (a) Two-photon excitation (TPE) scheme for a quantum dot exhibiting no excitonic fine-structure splitting. For the two-photon excitation, a laser is brought into resonance with the virtual TPE state, which lies between the biexciton and exciton emission lines. After the generation of the biexciton state, two radiative recombination paths are possible to the ground state |0⟩ via one of the two bright exciton $|X\rangle$ states. (b) Photoluminescence spectra of the QD for nonresonant excitation (NRE), i. e., above-GaAs-bandgap excitation, and (c) for two-photon excitation. Reprinted with permission of Springer Nature from [133].

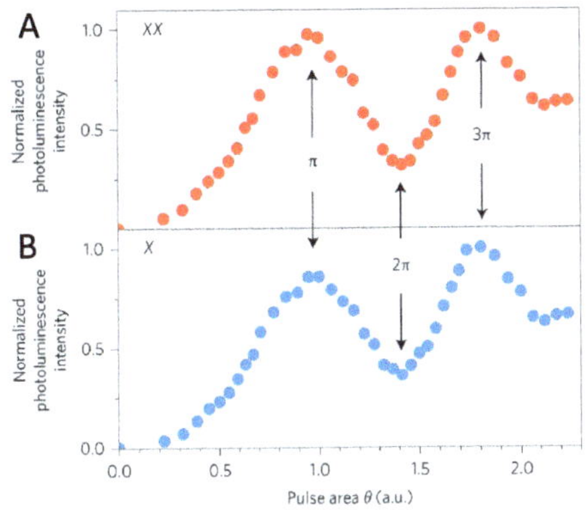

Figure 10.10: (a) Integrated intensities for the biexciton (a) and exciton (b) emission lines under resonant TPE excitation versus the square root of the pump power (proportional to the excitation pulse area). The abscissa is normalized in π-units so that the first biexciton intensity maximum is reached for a pulse area equal to π. Reprinted with permission of Springer Nature from [133].

intensities is caused by the statistical nature of the carrier capture process inside the QD. Thus on-demand emission of a specific photon is not possible under these nonresonant excitation conditions. In contrast, the coherent and deterministic (π-pulse) TPE generation process is reflected in the observation of well-pronounced Rabi oscillations for both the biexciton and exciton emission intensities (see Fig. 10.10). High biexciton and exciton occupation probabilities of 98 % and 86 % have been estimated from the detailed analysis of the data [133]. The observed damping of the Rabi oscillations is mainly due to phonon-induced decoherence processes, which increase when the pulse area is increased. The slight monotonous increase superimposed to the oscillations has been attributed to the slight chirp of the excitation pulse or to some incoherent contribution in the excitation process.

The main advantages of the TPE process: i) no phonon relaxation processes are needed for the biexciton preparation, ii) no or less charge carriers are created in the vicinity of the QD avoiding carrier–carrier scattering processes, iii) no spectral overlap of the scattered laser light with exciton and biexciton transitions, and iv) a fully coherent and deterministic generation process is realized. In summary, this can result in the deterministic generation of high-purity and indistinguishable entangled photon pairs, as already demonstrated by different research groups [84, 116] (see also Section 7.3.3).

10.2.4 Phonon-assisted excitation methods

The following sections will describe phonon-assisted excitation processes of exciton and biexciton states in a quantum dot, which can reach close to unity QD inversion probability. Both processes are incoherent excitation schemes combining the advantage that they are insensitive to small excitation power fluctuations, and laser stray-light suppression is possible by spectral filtering. At the same time, it has been shown that the emitted photons can possess similar high single-photon purity, coherence times [18], and indistinguishabilities [195] as their resonant counterparts discussed in the previous chapters.

10.2.4.1 Phonon-assisted exciton excitation

The physics of the excitation process is best described in a dressed state picture, where the optical pulse dresses the ground and excited states of the respective optical transition (see Fig. 10.11, [155]). A circular polarized laser pulse with a certain positive detuning (~ 1 meV) is tuned to the phonon sideband of the exciton transition. With the presence of a strong laser pulse, the two bare states are optically dressed (see Fig. 10.11(a) (ii) and (iii)). In the rotating frame, $|0\rangle$ comprises the crystal ground state and the incoming laser photons. $|X\rangle_R$ implies the exciton state and the laser field with one photon less. Figure 10.11(b) depicts the dynamics of the phonon-assisted excitation process in the rotating frame. The state $|0\rangle_R$ initially coincides with the higher-energy dressed state α. During the pulse duration, the system relaxes between the dressed states α and β through the emission of a longitudinal-acoustic phonon. In this way a continuous transfer of population from α to β is achieved. Finally, a strong occupation of the excited state is achieved after the switch-off of the excitation pulse. In the lab frame, this process is described by a photon absorption process together with generation of an exciton and the

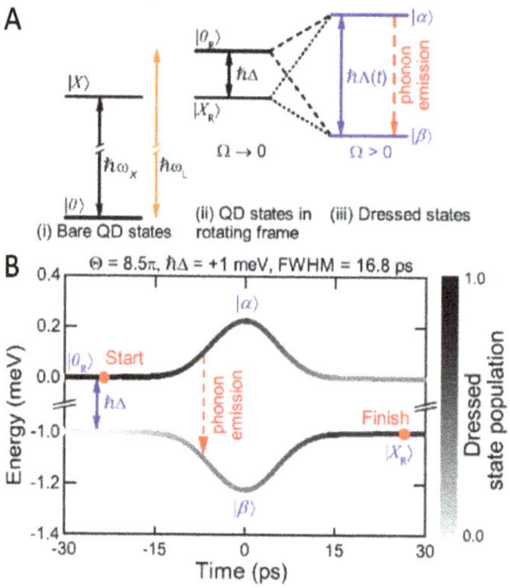

Figure 10.11: (a) Sketch of the energy diagram for QD bare states in (i) lab frame and (ii) rotating frame. $|0\rangle$ ($|0\rangle_R$) and $|X\rangle$ ($|X\rangle_R$) refer to ground and exciton state in lab (rotating) frame. $\hbar\omega$ is the exciton transition energy, $\hbar\omega_L$ is the excitation laser energy, $\hbar\Delta$ is the positive detuning from the exciton transition, and Ω is the Rabi frequency. Optically dressed states are indicated as α and β. (b) Time evolution of the QD dressed states during the absorption of a laser pulse (8.5π and $\hbar\Delta = +1$ meV). The instantaneous population of the states is indicated by the gray scale [126]. After Quilter et al. [155].

emission of a phonon. Using this scheme, Thomas et al. [195] have achieved the emission of fully polarized single photons, with a measured degree of linear polarization of up to 99.4 % and a high population inversion of 85, %. Furthermore, they demonstrated a single-photon source with polarized first lens brightness of 50 %, a single-photon purity of 95 %, and single-photon indistinguishability of 91 %.

10.2.4.2 Phonon-assisted biexciton excitation

In the same manner, the biexciton state can also be excited using a phonon-assisted two-photon excitation scheme. From the physical point of view, the process is equivalent to the phonon-assisted exciton excitation, and the state preparation can therefore also be attributed to carrier–phonon coupling, i. e., a phonon-induced relaxation as the bare electronic states become dressed by the laser field. This relaxation can lead to a thermal occupation of the photon dressed states, which for positive laser detunings yields a high biexciton population. A careful choice of detuning and laser pulse duration can optimize the state preparation fidelity, as it will be shown in the following.

First, it is instructive to compare typical μ-PL spectra and biexciton occupation probabilities for different pumping schemes. Figure 10.12 shows microphotoluminescence spectra of a single QD after (a) above barrier pumping, (b) resonant two-photon excitation, and (c) phonon-assisted two-photon excitation. Pronounced exciton and biexciton lines can be seen in all spectra with additional laser scattering for the resonant two-photon and phonon-assisted biexciton excitation. For the latter two schemes, biexciton and exciton lines posses comparable intensities as expected due to the nature of the radiative biexciton–exciton cascade, as explained earlier. It is now interesting to compare the power dependence of the biexciton occupation probability as a function of the pump pulse area and in dependence of laser detuning (see Fig. 10.13). Pulsed excitation (laser pulse width of 13 ps) on the two-photon resonance, i. e., with zero detuning, led to the

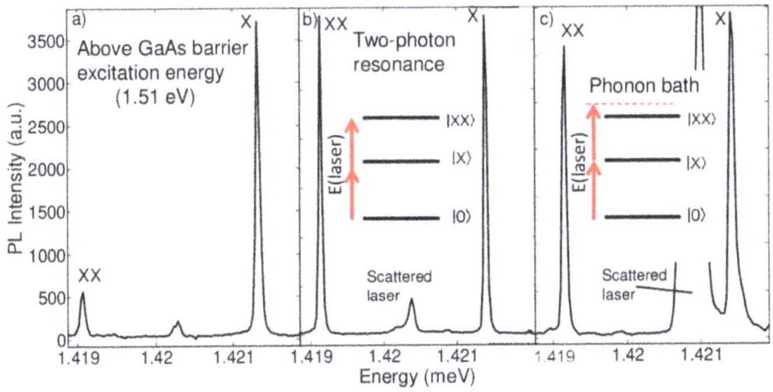

Figure 10.12: Single QD emission spectrum (a) recorded under nonresonant above bandgap excitation, (b) under resonant two-photon biexciton state excitation, and (c) under phonon-assisted 13 ps pulsed, 0.65 meV positively detuned excitation [18]. Reprinted with permission from [18]. Copyright (2023) by the American Physical Society.

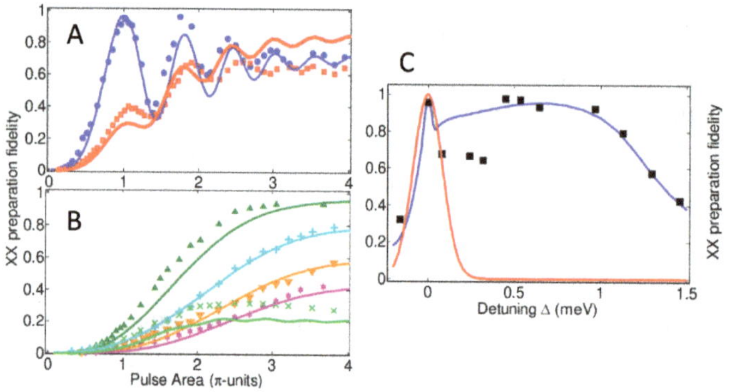

Figure 10.13: (a) Biexciton occupation C_{XX} as obtained from comparison of simulations vs renormalized pulse area with excitation pulse length of 13 ps superposed to path-integral simulation results for different laser detunings from the two-photon biexciton resonance: in resonance represented in blue, 0.08 meV detuning in red, (b) 0.65 meV detuning in green, 1.1 meV detuning in light blue, 1.3 meV detuning in dark yellow, 1.5 meV detuning in magenta, and −0.1 meV detuning in light green. (c) Maximum biexciton occupation reached for different laser detunings from the two-photon biexciton resonance. The points represent occupations measured in the experiment, and the full lines are results of the simulations (in blue with phonon coupling included in the calculations, in red without any phonon coupling) [18]. Reprinted with permission from [18]. Copyright (2023) by the American Physical Society.

well-known Rabi oscillations (Fig. 10.13(a), blue) as a function of excitation power. All solid lines are solutions from a numerically exact real-time path-integral simulations (see [18] and references therein). By comparison with the measured data the biexciton occupation probability C_{XX} is found to be 0.94 for the π-pulse. The observed decrease of the Rabi frequency with increasing pulse area is due to the two-photon excitation process, whereas the damping of the amplitude is due to coupling of phonons. It is important to note that the average value of the oscillations is above 0.5 due to a small chirp of the excitation pulse. According to simulations, this chirp had very small effect on off-resonant laser pulses discussed below. The measurements under a certain laser detuning are displayed in Fig. 10.13(a–c). With increasing positive detuning, the Rabi oscillations are damped, and a smooth increase in the occupation probability is observed until a detuning dependent saturation level is finally reached. Figure 10.13(c) shows the maximum biexciton occupation probability as a function of laser detuning. The state initialization is most efficient between 0.45 and 1 meV detuning, and a maximum value of 0.95 is achieved for a 0.65 meV detuning, similar to the resonant π-pulse excitation. The maximum in the biexciton occupation is obtained when the most pronounced phonons are in resonance with the relevant dressed-state transition (see Section 8.2). In this regime the occupation is not only stable against small changes of the excitation frequency, but also shows a pronounced region where the biexciton population stays unaffected by fluctuations of the laser power. In this respect, this regime is advantageous compared to resonant two-photon π-pulse excitation. For larger detunings, the preparation becomes less efficient. For negative detunings, the energetic order of the dressed states is reversed, and thus the biexciton state is no longer the final state of relaxation at low temperatures. Thus, for negative detunings, a transition between the dressed states requires a phonon absorption process, which is less probable than phonon emission at low temperatures.

It is important to note that the excitation pulse duration has to be long enough to allow the relaxation step between the dressed states. Otherwise, for a too short pulse, efficient relaxation is not possible, which results in a less efficient biexciton preparation. For example, in the above-presented study, it has been reported that the biexciton occupation probability stays below 0.62 for a 7-ps-long excitation pulse. Furthermore, it is also shown that the coherence time of the photons with the phonon-assisted scheme was comparable with the coherence time of the resonantly generated photons. In summary, the phonon-assisted two-photon excitation scheme combines the advantage of the robustness against power fluctuations, small changes in excitation frequency, and the possibility to spectrally filter out the pump laser.

10.3 Electrical excitation schemes

To excite QDs electrically, they have to be embedded into a structure that enables a current flow through the device to excite the QDs. So far, most of the electrically driven single-photon sources have the design of single-photon light emitting diodes (SPLED).

The SPLED is typically based on the integration of a layer of low-density quantum dots within the intrinsic region of a p–i–n junction to provide the necessary electron and hole current to populate the QDs with electrons and holes, which then radiatively recombine spontaneously with the emission of single photons (see Fig. 10.14). The physics of a p–n- and p–i–n-junctions is described, for example, in the textbook of Sze [192]. In short, without applying a voltage (Fig. 10.14(a)), charge carriers cannot be injected into the QD. By applying a voltage in the forward direction (+ to p and – to n) the potential difference, and thus the band bending, is reduced. As a result, a current can flow through the junction, charge carriers can be captured by the QDs, and radiative recombination will take place (Fig. 10.14(c)). Here the carriers are injected via diffusion over the p–n junction in the barrier material, and therefore the injection mechanism is similar to nonresonant optical pumping. This means that the carrier capture process is of random nature and the number of captured carriers in the quantum dot is given by a Poisson distribution.

A resonant electrical excitation scheme can be applied by providing voltage biases below the level that would allow charge carriers to flow through the whole intrinsic region, as has been demonstrated in charge-tunable devices for electron and/or hole injection devices [97, 220]. There the applied bias is tuned such that the carriers in the n- and p-doped regions are resonant with energy of electron and hole levels inside the QD or Quantum Well [97], allowing resonant tunneling injection (Fig. 10.14(b)). So far this regime is almost unexplored but may lead to a high level of control of charge injection

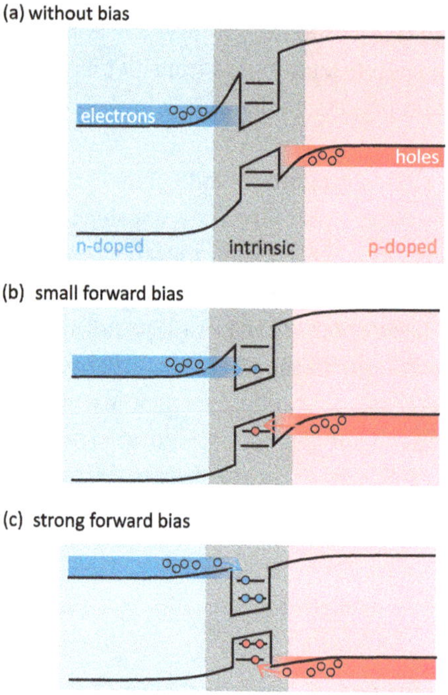

Figure 10.14: (a) Schematic of the band structure of a p–i–n diode with a QD embedded in the intrinsic region of a p–i–n junction (a) without and (b,c) with applied bias in forward direction. In (b) the applied bias is tuned such that the carriers in the n- and p-doped regions are resonant with electron and hole energy levels inside the QD, allowing resonant tunneling injection. In (c) the band bending is further reduced, and electrons and holes are injected into the intrinsic region and captured by the QD. They subsequently relax down the lowest shell (s-shell) and radiatively recombine, which leads to photon emission.

and therefore also to higher photon indistinguishabilities. The following discussion focuses on the nonresonant pumping scheme (Fig. 10.14(c)).

The ideal SPLED would require to deliver a single electron–hole pair to a single QD and collect the subsequently emitted single photon. In other words, we have to restrict and confine the electrical current to excite only one target QD and collect only all emitted single photons from this target QD. In practice, this holy grail has not been reached so far.

Different methods, i. e., spatial and spectral filtering techniques, have been developed so that preferentially only one or a few QDs are excited and the emission from only one QD transition is collected, thereby allowing the collection of single photons. One of the first and easiest solutions to the problem is the implementation of an opaque metallic film patterned with micron-sized apertures on the surface of the device, which allows the dominant emission to be collected from only one QD (see Fig. 10.15) [232]. This design has the obvious disadvantage that the electrical current is not confined to a single QD and the area of the current flow is given by the opening (or eventual mesa) size. One of the first SPLEDs measured several tens of microns, and the resulting current for exciton

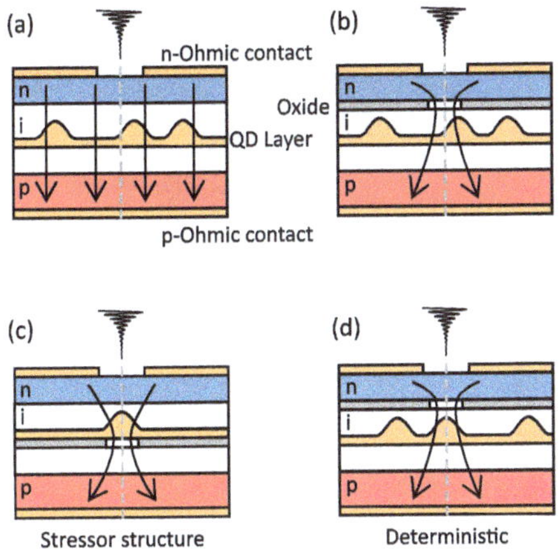

Figure 10.15: Different types of single-photon emitting diodes with schematic cross sections of QD layer, aperture sizes, and current flow (red lines). These structures can also be implemented in smaller, laterally confined, mesa devices (a) A SPLED with an opaque metallic film patterned with micron-sized apertures. (b) An oxide-aperture SPLED. (c) An oxide aperture acting as a buried stressor is forcing site-controlled QD growth above the aperture. (d) An oxide-aperture defined above a preselected QD.

saturation was in the range of up to 6 μA [232]. In this case, more than 10^4 electrons have to flow through the junction per photon emission event.

A step forward was realized by the introduction of a current blocking layer with a small aperture opened in the center of the device, ideally above a single QD. This was implemented by the introduction of an oxide aperture for current confinement, a strategy originally developed for ultra-low threshold vertical cavity surface emitting lasers (VCSELs) (see Fig. 10.15(b)). This can be done by creating an insulating ring of aluminium oxide within a larger mesa by oxidation of an aluminium arsenide layer in a high-temperature humid atmosphere, leaving a small unoxidized aperture in the center. Apertures down to a submicron diameter can be realized by this technique. In this way the current for exciton saturation was decreased to ~10 nA, which corresponds to only tens of carriers injected per emitted photon [42].

However, for an optimal device, i. e., lowest pump currents, low background emission from nearby QDs and highest mode coupling efficiencies only a single QD should be in the center of the aperture. This is not realizable with the first two approaches due to the random spatial distribution of QDs inside the QD layer. This results from the probabilistic nature of the self-assembled growth in the Stranski–Krastanov growth mode and also applies for conventional droplet epitaxy approaches. To solve this issue, a deterministic fabrication process is necessary (see Chapter 12). Two different approaches are possible, a site-controlled QD growth or a fabrication process based on deterministic lithography around a preselected QD. Figure 10.15(c) shows a schematic sketch of a buried stressor approach. Here an oxide aperture acts as a buried stressor structure and forces a site-controlled QD growth. Figure 10.16(a) shows electroluminescence (EL)

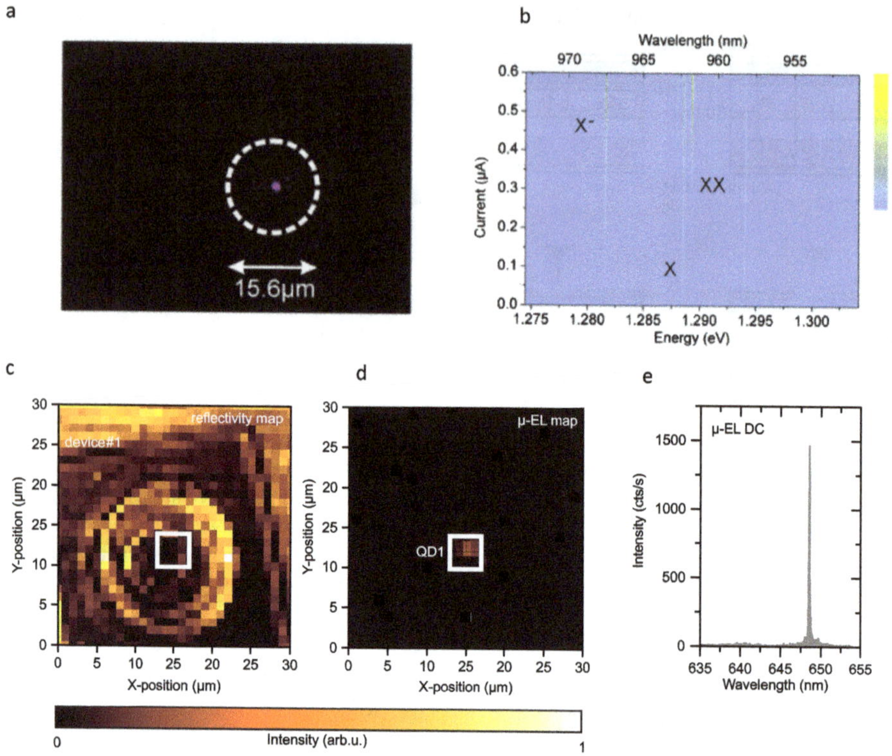

Figure 10.16: (a) Electroluminescence (EL) recorded by a digital camera using an IR filter and (b) EL spectra recorded from the aperture region as a function of the injection current. Reprinted from [201] with the permission of AIP Publishing. (c) Reflectivity scan of the aperture area of the deterministically fabricated SPLED. (d) μ-EL map of the device and (e) corresponding μ-EL spectrum. Reprinted from [171] with the permission of AIP Publishing.

emission from a single QD in such a device as recorded by a digital camera [201]. One single emission spot is observed. Figure 10.16(b) displays the corresponding μ-EL spectrum with emissions from exciton-, biexciton-, and charged exciton transitions as a function of the injection current. The onset of exciton emission is achieved for currents of 0.01 µA.

Figure 10.15(d) displays a sketch from a SPLED with an oxide-aperture defined directly above a preselected QD. Such a device has been fabricated with a deterministic lithography technique and operated under pulsed electrical carrier injection (see Fig. 10.16). Figure 10.16(c) shows a reflectivity map of the device allowing us to observe the topography of the device. At the same time the electrical emission from a single QD is recorded close to the center of the device (see μ-EL map in Fig. 10.16(d)), and the corresponding spectrum is shown in Fig. 10.16(e). The spectrum is dominated by one single recombination line from the preselected QD. Typical excitation parameters are an excitation repetition rate of 200 MHz, an excitation pulse width of 200 ps, a DC voltage of $V_{DC} = 2.2$ V, and a pulse amplitude of $V_{pulse} = 3.9$ V. These deterministic approaches will probably support the wide scale use of SPLEDs in the future.

It is important to note that electrical-based excitation schemes are extremely flexible, and versatile pumping schemes have been explored. For example, if the DC bias voltage V_{DC} is chosen to be far below the turn-on voltage V_T of the diode, then the charge carriers can quickly tunnel out of the QD potential between the driving pulses. This scheme quenches the optical emission but reduces the time jitter of the photon emission process. This allows us to increase the possible excitation repetition rate above the inverse radiative lifetime of the relevant QD transition. In this way, SPLED operation at 2 GHz [67] and 3 GHz [131] excitation rates have been demonstrated.

Moreover, actively resetting the biexciton exciton cascade by excitation repetition rates larger than the inverse sum of the biexciton–exciton lifetime enables the generation of entangled photon pairs with higher fidelity and emission rate than the optimum continuously driven state. Electrically driven entangled photon pairs with an average fidelity of 79.5 % at a clock rate of 1.15 GHz have been demonstrated [131].

Furthermore, it is important to note that electrically driven single-photon sources have not yet reached the high level of photon indistinguishability due to their nonresonant pumping schemes. So far, resonant electrical tunnel-injection of carriers into a single QD is almost unexplored but may lead to higher photon indistinguishabilities.

10.4 Summary

- The advantage of nonresonant excitation is its easy implementation, and the excitation laser stray-light is spectrally well separated from the emitted photons, being then easy to filter. The drawback is a certain time jitter of the emitted light determined by carrier capture times and relaxation rates, a broadened spectral linewidth of the emitted photons due to carrier-carrier scattering and spectral diffusion. Both effects lead to a reduction of photon indistinguishability. Furthermore, the time-averaged emission is unpolarized.
- A p-shell excitation process avoids carrier generation in the vicinity of the QD, thus reducing linewidth broadening effects. The time jitter is only determined by the p- to s-shell relaxation process, and it is often nearly spin conserving. The excitation laser can be easily spectrally filtered.
- Resonant optical laser excitation schemes are coherent and deliver photons with highest quality. High excitation efficiencies and high degree of photon indistinguishabilities can be reached (both close to one).
- π-pulse excitation allows a direct excitation of a specific quantum dot state and gives a high control of the subsequent emission process. At the same time a sophisticated suppression of the laser excitation light is necessary, which typically complicates the practical realization of light collection.
- Utilizing a nonperfect circular symmetry for a cavity introduces an anisotropy leading to two linearly polarized fundamental cavity modes. Such a cavity with an em-

bedded quantum dot can be used in a very advantageous way for achieving ultra-bright and pure single-photon emission under π-pulse excitation.
- The adiabatic rapid passage technique with frequency-chirped pulses is largely unaffected by variations in the excitation power and emitter dipole moments. The drawback is an additional complexity in optical pulse generation. High Hong–Ou–Mandel interference visibilities have been achieved.
- Spontaneous spin-flip Raman transitions allow a tailored single-photon waveform, i. e., controlled temporal profiles with duration exceeding the intrinsic QD lifetime together with reductions in spectral linewidth. The drawback is a more complex setup including two lasers with electro-optic modulators and applying a magnetic field.
- The biexciton can be directly created by a resonant coherent two-photon absorption process via a virtual intermediate state. The laser background is spectrally detuned from the exciton and biexciton transition lines and can therefore be filtered without relying on a polarization filtering technique (i. e., suitable for the generation of polarization entangled photon pairs).
- Incoherent phonon-assisted exciton and biexciton excitation processes can reach close to unity QD inversion probability. Both processes combine the advantage that they are insensitive to small excitation power fluctuations and laser stray-light suppression is possible by spectral filtering. Long coherence times and high indistinguishabilities can be reached.
- A single-photon emitting diode (SPLED) contains a layer of low-density quantum dots within the intrinsic region of a p–i–n junction to provide the necessary electron and hole current to populate the QD with electron and holes.
- A SPLED is small, robust, and compact and therefore promises scalability and on-chip integration. Furthermore, flexible and ultra-high repetition rates (few-GHz regime) are possible. However, electrically driven single-photon sources have not yet reached the high level of photon indistinguishability as optically pumped sources.

11 Photon indistinguishability and two-photon interference

11.1 Introduction

An ideal single-photon source emits a regular stream of identical photons (as discussed in Chapter 5). This means that the photons are identical in all their physical properties, e. g., spectral bandwidth, pulse width, polarization, carrier frequency, and mode profile, i. e., in the language of quantum optics, they are indistinguishable. The model of an isolated two-level system that is repeatedly excited to the upper level followed by a photon emission process, which returns it to the ground level, represents such a source of indistinguishable photons. Photon indistinguishability is an important precondition for most of the discussed applications in quantum information science, e. g., for photonic quantum computing, photonic simulations, and quantum communication. In practice the central element of a quantum light source, i. e., the QD, is not a simple two-level system, which may limit the photons indistinguishability (see Section 7.3.3), and it is furthermore embedded in a solid state matrix and therefore experiences interaction with its environment (see Chapter 8). This coupling to the environment leads to dephasing and spectral diffusion, which also limits the photon indistinguishability (see Chapter 5).

The photon indistinguishability I can be directly derived from the measured visibility V_{HOM} of a Hong–Ou–Mandel (HOM) experiment, i. e., a two-photon interference experiment, where two individual photon wavepackets are sent to two different input channels of a beamsplitter, and a quantum interference process takes place. The obtained visibility V_{HOM} of this quantum interference is identical to the indistinguishability I under ideal conditions, i. e., symmetric beamsplitter, perfect spatial mode overlap, and perfect single-photon purity. The experimental method and the determination of I under nonideal conditions is discussed in Section 5.2.8.

Here we focus on the different physical effects that influence I. In our discussion, we will distinguish HOM experiments with photons originating from one and the same source and from two remote sources.

This distinction is necessary since the environmental influences that disturb the indistinguishability of the individual photons stemming from a single quantum emitter occur on very different times scales. Therefore the resulting visibility of a two-photon interference measurement from photons of the same stream strongly depends on the chosen time separation between these photons used for the experiment. Thus the visibility is in general a time-dependent property, and the full linewidth of the emitter, which expresses all interactions with the environment, is not an adequate measure to predict the indistinguishability in this case.

In contrast for two remote sources, we have to consider that the interaction of the respective emitters with their environments is totally uncorrelated. In a time integrating two-photon interference experiment, all environmental influences acting on both emitters sum up, and the resulting visibility depends on the full linewidths of both emitters.

https://doi.org/10.1515/9783110703412-011

11.2 Wave-packet description for quantum dot-emitted single photons

Let us start the discussion assuming the radiative recombination of an undisturbed single emitter, i. e., a Fourier-limited single-photon source. The photons emitted by such a source have an exponentially temporal shape. The single-photon wave-packet can be written as

$$\xi(t) = \frac{1}{\sqrt{\tau_{\text{rad}}}} H(t) e^{-i\omega t} e^{-t/2\tau_{\text{rad}}} \quad \text{with} \int dt |\xi(t)|^2 = 1, \tag{11.1}$$

where τ_{rad} is the radiative decay time, $H(t)$ is the Heaviside step function, and $\omega = 2\pi\nu$ is the angular carrier frequency of the photon. Figure 11.1 (a) illustrates this wavepacket. For this Fourier-limited source, both frequency and phase are constant over time, which results in a lifetime limited coherence time $\tau_c = 2\tau_{\text{rad}}$ (otherwise defined $T_2 = 2T_1$ in Chapter 5) and a purely Lorentzian spectral profile. In this case the overlap integral $|J|^2 = 1$ (see Eq. (4.47)), and thus the visibility of a two-photon interference process is $V_{\text{HOM}} = 1$, i. e., we have photons with perfect photon indistinguishability $I = 1$.

The electronic excitation of QDs is coupled to vibrational modes, i. e., phonons of the host material, and can also experience interactions with electric and magnetic field fluctuations in their solid-state environment, which in turn broaden the emission line (see Chapters 1, 5, and 8). Due to their random nature, an adequate description is often based on a model assuming a random walk in emission phase and frequency (see also [212, 213]). These effects give rise to pure dephasing (PD) and spectral diffusion (SD). All perturbations causing frequency fluctuations can be classified by their characteristic interaction time τ_p. In the following, very fast interactions ($\tau_p \ll \tau_{\text{rad}}$) will be considered, which are characteristic for pure dephasing processes caused by carrier–phonon or carrier–carrier scattering. Such processes lead to a brief jump from the momentary frequency ν_0 to ν_1 and nearly instantaneously back to ν_0 (see Fig. 11.1(c)). Therefore the phase of the photon is changed in the order of $\sim(\nu_0 - \nu_1)\tau_p$ (see Fig. 11.1(d)), but the average frequency of the wavepacket is hardly affected (see Fig. 11.1(b)). On the other side, long interaction times ($\tau_p \gg \tau_{\text{rad}}$) are well known for spectral diffusion processes, and the spectral jumps rarely occur during the lifetime of a photon but very likely between the individual photon emission processes (see Fig. 11.1(b,c)). The wavepacket description of a single photon has to be adapted to consider pure dephasing:

$$\xi(t) = \frac{1}{\sqrt{\tau_{\text{rad}}}} H(t) e^{-i(\omega t + \phi(t))} e^{-t/2\tau_{\text{rad}}} \quad \text{with} \int dt |\xi(t)|^2 = 1, \tag{11.2}$$

where a time-dependent phase factor $\phi(t)$ for PD is considered, and it is assumed that the angular carrier frequency ω remains constant for the duration of a single photon pulse, which is valid as discussed above. Thus the frequency occasionally jumps in between the photon emission events (SD), causing inhomogeneous line broadening, which can

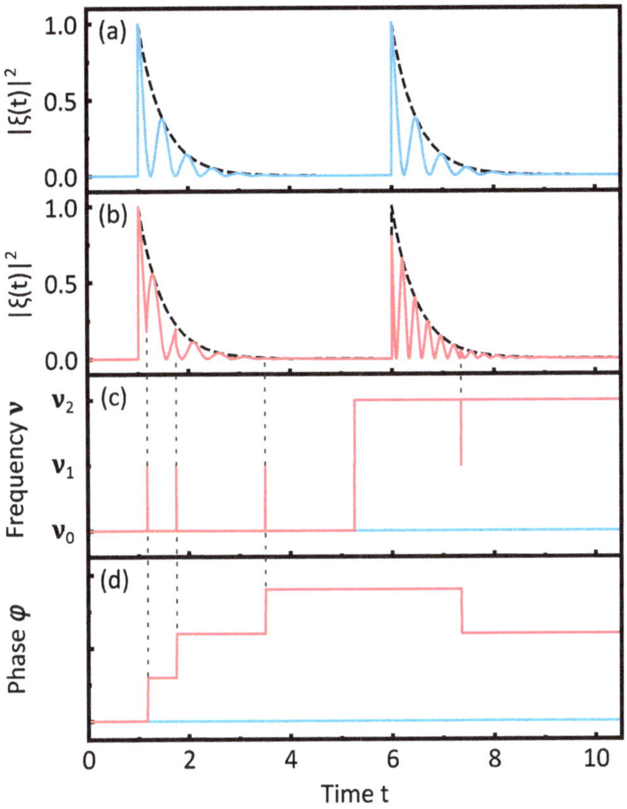

Figure 11.1: Illustration of wave-packets without (blue) and with (red) the influence of perturbations. (a) Wave-packet envelope $|\xi(t)|^2$ (black dashed line) and oscillatory part $|\mathrm{Re}(\xi(t))|^2$ (blue solid line) of Fourier-limited wave-packets. (b) Wave-packet envelope $|\xi(t)|^2$ (black dashed line) and oscillatory part $|\mathrm{Re}(\xi(t))|^2$ (red solid line) under the influence of perturbations. (c) Frequency jumps with short interactions (delta like peaks) cause a random walk in the phase (d, red), whereas frequency jumps with long interaction time rarely occur during the photon lifetime, but lead to a frequency random walk between emitted photons. For simplicity all quantities are given in arbitrary units. After B. Kambs and C. Becher [91].

be described by a Gaussian function and can then be used for statistical averaging, as we will discuss below.

11.3 Two-photon interference of photons from remote sources

Two-photon interference (TPI) with high visibility from remote solid-state single-photon sources is a real technological challenge due to the manifold of possible interactions of the emitter with the environment. Here we start the discussion by assuming pure radiative recombination of a two-level emitter, which is resonantly pumped into the excited state (e. g., exciton or trion in a QD (see Chapter 10)). This means that the emitter is directly excited and no relaxation process from a higher level has to be considered. It has been demonstrated that this is the most favorable excitation scheme for achieving highly indistinguishable photons. In the first step, pure dephasing will also be neglected, i. e., Fourier-limited photons are assumed, but spectral diffusion will be considered. In the second step, pure dephasing will be included.

HOM with Fourier-limited photons and spectral diffusion

Accordingly, the photon wavefunctions $\xi_i(t)$ from the two emitters ($i = 1, 2$) can be written as

$$\xi_i(t) = \frac{1}{\sqrt{\tau_i}} \cdot H(t) \cdot e^{-\frac{t}{2\tau_i} - i2\pi\nu_i t}, \tag{11.3}$$

where τ_i denote the radiative lifetimes, ν_i are the instantaneous emission frequencies, and $H(t)$ is the Heaviside function. Moreover, the QDs are subject to spectral diffusion appearing as a jitter of their emission frequencies. Typically, this leads to an inhomogeneous broadening of the spectral line shape following a normal distribution $p_i(\nu_i)$ around the mean frequency $\nu_{c,i}$ of the QD with standard deviation σ_i according to

$$p_i(\nu_i) = \frac{1}{\sigma_i \sqrt{2\pi}} \cdot \exp\left\{\left[-\frac{1}{2}\left(\frac{\nu_i - \nu_{c,i}}{\sigma_i}\right)^2\right]\right\}. \tag{11.4}$$

An excellent and extensive discussion of the analysis of HOM experiments with remote QDs is given in [91]. The key steps of the derivation of the HOM visibility are discussed in the following.

The derivation starts from a well-established formalism describing the HOM experimental situation with the photon fields $\xi_{1,2}(t)$ at the two inputs of a symmetric BS (50:50) [110] (see also Eq. (4.43)). Therein, the probability $P(t_0, \tau)$ with which both input photons leave the BS through distinct output ports and become detected at times t_0 and $t_0 + \tau$ is given by

$$P(t_0, \tau) = \frac{1}{4}\left|\xi_1(t_0 + \tau)\xi_2(t_0) - \xi_2(t_0 + \tau)\xi_1(t_0)\right|^2. \tag{11.5}$$

Using the wavefunctions (11.3), the second-order cross-correlation $g^{(2)}(\tau)$ can be evaluated as

$$\begin{aligned} g^{(2)}(\tau) &= \int_{-\infty}^{+\infty} P(t_0, \tau)\, \mathrm{d}t_0 \\ &= \frac{1}{4(\tau_1 + \tau_2)} \\ &\quad \times \left(e^{-\frac{|\tau|}{\tau_1}} + e^{-\frac{|\tau|}{\tau_2}} - 2 \cdot e^{-\frac{|\tau|}{2T}} \cos(2\pi\Delta\nu\,\tau)\right), \end{aligned} \tag{11.6}$$

where the instantaneous emission frequency displacement is described by $\Delta\nu = \nu_1 - \nu_2$ and $1/T = 1/\tau_1 + 1/\tau_2$. During a long-time measurement, $\Delta\nu$ does not stay constant, but it is subject to jitter due to the independent spectral diffusion processes of both QDs. For a measurement, which takes much longer than the time both emitters need to explore their frequency ranges (Eq. (11.4)), the probability ρ to find a given splitting $\Delta\nu$ is simply given by the cross-correlation of p_1 and p_2:

$$\rho(\Delta\nu) = \int_{-\infty}^{+\infty} p_1(\nu)p_2(\nu + \Delta\nu)\,\mathrm{d}\nu = \frac{1}{\Sigma\sqrt{2\pi}} \cdot \exp\left\{\left[-\frac{1}{2}\left(\frac{\Delta\nu + \delta\nu}{\Sigma}\right)^2\right]\right\}, \tag{11.7}$$

where $\delta\nu = \nu_{c,1} - \nu_{c,2}$ is the relative displacement of both mean emission frequencies, and $\Sigma^2 = \sigma_1^2 + \sigma_2^2$ defines the width of ρ. Accordingly, the measured cross-correlation $\mathcal{G}^{(2)}(\tau)$ in the long-time limit is obtained by a weighted average of Eq. (11.6) using $\rho(\Delta\nu)$ leading to

$$\mathcal{G}^{(2)}(\tau) = \int_{-\infty}^{+\infty} \rho(\Delta\nu) g^{(2)}(\tau, \Delta\nu)\,\mathrm{d}\Delta\nu = \frac{1}{4(\tau_1 + \tau_2)} \times \left(e^{-\frac{|\tau|}{\tau_1}} + e^{-\frac{|\tau|}{\tau_2}} - 2 \cdot e^{-\frac{|\tau|}{2T}} e^{-2\pi^2\Sigma^2\tau^2} \cos(2\pi\delta\nu\,\tau)\right). \tag{11.8}$$

This is the final result for describing the central peak around $\tau = 0$ for a remote TPI correlation measurement.

The visibility of HOM-interference is defined by $V_{\mathrm{HOM}} = 1 - 2 \cdot P$, where P is the overall probability of both photons going separate ways after meeting at the BS. Accordingly, P can be calculated by integrating Eq. (11.8) with respect to the timelag τ. Using the variable $z = (2\pi\delta\nu + i/2T)/(2\pi\sqrt{2}\Sigma)$, the visibility evaluates to

$$V_{\mathrm{HOM}} = \frac{\mathrm{Re}[w(z)]}{\sqrt{2\pi}\Sigma(\tau_1 + \tau_2)}, \tag{11.9}$$

where the Faddeeva function $w(z)$[1] ([1] and Chapter 7) is used to express the result. Equation (11.9) is the definition of a Voigt line shape as a function of the detuning $\delta\nu$, whose width is given by the homogeneous contributions $\tau_{1,2}$ and inhomogeneous contributions $\sigma_{1,2}$. Thus the visibility of a remote HOM experiment is determined by the joint spectral properties of both emitters.

1 The Faddeeva function is a scaled complex complementary error function: $w(z) := e^{-z^2}\,\mathrm{erfc}(-iz)$.

HOM considering pure dephasing and spectral diffusion

In the following, we extend the discussion by including both spectral diffusion and pure dephasing, where SD and PD are assumed to act independently on the emitters. The discussion follows again parts of [91]. After a collision process (e. g., exciton–phonon scattering), the photon picks up a certain random phase, which gives rise to a Brownian-motion-like diffusion of the phase known as PD. Assuming a photon at time t_0 with phase ϕ_0, the probability to find it at a later time $t_0 + \tau$ at a phase $\phi_0 + \Delta\phi$ is given by

$$p_{\mathrm{PD}}(\Delta\phi) = \frac{1}{\sqrt{4\pi D|\tau|}} \cdot \exp(-\Delta\phi^2/4\pi D|\tau|), \tag{11.10}$$

where D is the diffusion constant. Equation (11.10) is adopted from the classical theory of a diffusing particle in one dimension, where D relates to the emitter dephasing rate $\Gamma_{\mathrm{deph}} = 2D$ ($\Gamma_{\mathrm{deph}} = 1/T_2^*$). The derivation of the HOM visibility follows the same route as described above, except that we have to perform an additional statistical averaging step by using Eq. (11.10) for phase averaging. The HOM visibility is then given by [91]

$$V_{\mathrm{HOM}} = \frac{\mathrm{Re}[w(z)]}{\sqrt{2\pi}\Sigma(\tau_1 + \tau_2)}, \tag{11.11}$$

where $w(z)$ is again the Faddeeva function [1]. The expression for the visibility has the same form as Eq. (11.9), but now it includes also the pure dephasing rates of both emitters in the argument z, i. e., $z = (2\pi\delta\nu + i\gamma)/(2\pi\sqrt{2}\Sigma)$ with $\gamma = \gamma_1 + \gamma_2$ and $\gamma_{1,2} = 1/(2\tau_{1,2}) + \Gamma_{\mathrm{deph},1,2}$. Figure 11.2(a) shows an example of the calculated visibility as a function of frequency detuning of two different single-quantum emitters (for the emitter parameters, see figure caption). It is obvious that even for zero frequency detuning, only a limited visibility of $V = 28\,\%$ can be reached for moderate dephasing and spectral diffusion. Figures 11.2(b) and (c) show the corresponding cross-correlation functions around the zero delay peak for relative detunings of $\delta\nu = 0$ GHz and $\delta\nu = 3$ GHz, respectively. For zero frequency detuning, the area of the zero delay peak under the red curve (which shows the classical limit) is reduced with respect to the classical case ($V = 28\,\%$) and further shows a characteristic dip at delay time zero. At $\tau = 0$ the correlation function is always zero regardless of how different both input photons are. A value larger than zero indicates a time jitter in the experiment or a slow detector response time. For the detuning of $\delta\nu = 3$ GHz, an oscillatory behavior is observed for the correlation function, which is due to quantum beating between the photons, and the overall HOM visibility is reduced to 1 %.

HOM considering only pure dephasing

In the following, we consider only pure dephasing, which leads to a more simplified expression where the HOM visibility only depends on the individual radiative lifetimes,

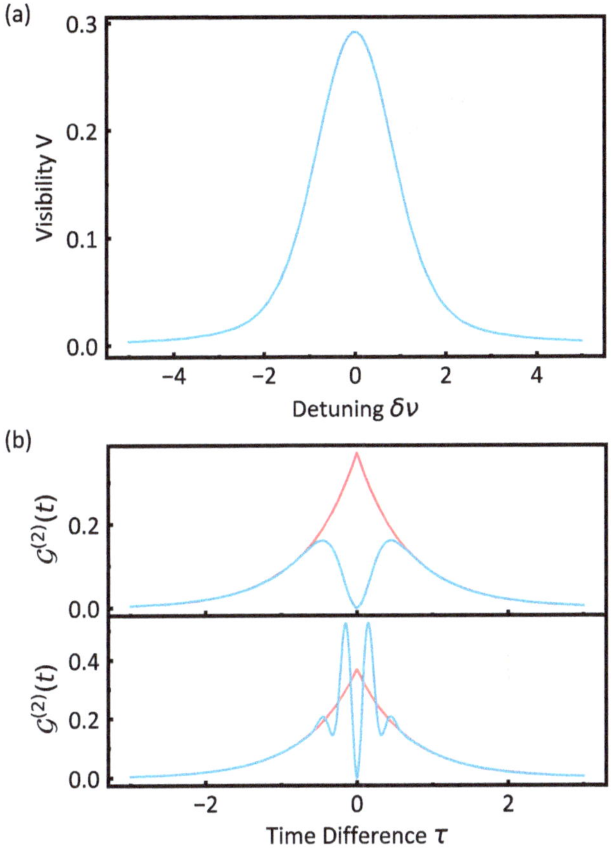

Figure 11.2: (a) Calculated visibility (after Eq. (11.11)) as a function of frequency detuning for two emitters with $\tau_1 = 700$ ps, $\tau_2 = 650$ ps, $\sigma_1' = 1.4$ GHz, $\sigma_2' = 0.8$ GHz (with $\sigma' = 2\sqrt{2\ln 2}\cdot\sigma$), $\Gamma_{\text{deph},1} = 600$ MHz, and $\Gamma_{\text{deph},2} = 300$ MHz. The corresponding cross-correlation functions $\mathcal{G}^{(2)}(\tau)$ at relative detunings of $\delta\nu = 0$ GHz and $\delta\nu = 3$ GHz are shown in (b) and (c), respectively (solid blue). The red curves show the classical limit for completely distinguishable photons. Adapted from Kambs and Becher [91].

dephasing rates, and the frequency detuning. The visibility is then given by [91]

$$V_{\text{HOM}} = \frac{4}{\tau_1 + \tau_2} \cdot \frac{1/\tau_1 + 1/\tau_2 + 2\Gamma_{\text{deph},1} + 2\Gamma_{\text{deph},2}}{(1/\tau_1 + 1/\tau_2 + 2\Gamma_{\text{deph},1} + 2\Gamma_{\text{deph},2})^2 + 16\pi^2\delta\nu^2}. \tag{11.12}$$

It is interesting to note that for two independent emitters exhibiting the same radiative lifetimes $\tau_{\text{rad}} = \tau_1 = \tau_2$, the dephasing rates $\Gamma_{\text{deph}} = \Gamma_{\text{deph},1} = \Gamma_{\text{deph},2}$, and emitting at the same frequency ($\delta\nu = 0$), Eq. (11.12) can be further simplified and given in a compact form:

$$V_{\text{HOM}} = \frac{\tau_c}{2\tau_{\text{rad}}} \tag{11.13}$$

with $1/\tau_c = 1/2\tau_{rad} + \Gamma_{deph}$. It is obvious that for independent Fourier-limited sources ($\Gamma_{deph} = 0$), the visibility becomes $V_{HOM} = 1$. Coming close to this regime is essential for most multisource quantum photonic implementations.

It is important to note that it has been recently shown that quantum interference of single photons from remote QDs can reach very high two-photon visibilities of $V_{HOM} =$ 93 % [235]. This result has been realized by using gated GaAs QDs. This system exhibits an ultra-low level of noise, i. e., small exciton dephasing and weak spectral diffusion. This development opens the way to create coherent single-photon sources in a scalable way.

11.4 Two-photon interference of photons from one and the same source

To characterize the indistinguishability I of a single-photon source, a TPI measurement is performed with photons generated from the same source within a certain short time separation δt_{exc} (typically, between one up to hundreds of nanoseconds). This time separation is often given by the repetition rate of the excitation laser (see discussion in Section 5.2.8). Since spectral diffusion and pure dephasing typically possess a certain time evolution (see Section 8.3), the resulting measured visibility strongly depends on the chosen time separation δt_{exc}, and it is therefore generally a time-dependent property of the source ($V_{HOM} = V_{HOM}(\delta t_{exc})$).

HOM considering pure dephasing

In high-quality samples, spectral diffusion is not present at least for short time periods ($\tau_{period} < 100\,\text{ns} - 1\,\mu\text{s}$). Thus the wavelength and the spectral linewidth is constant over this time period, and spectral diffusion related line broadening effects occur at longer times (see Fig. 11.3). Then Eq. (11.13) can be used to estimate the indistinguishability of a resonantly excited source for photon separation times between δt_{exc} and τ_{period}:

$$V_{HOM} = \frac{\tau_c}{2\tau_{rad}} = \frac{\gamma_{rad}}{\gamma_{rad} + 2\Gamma_{deph}} \tag{11.14}$$

with $\gamma_{rad} = 1/\tau_{rad}$ and Γ_{deph} the dephasing rate, here assumed to be higher than $1/\delta t_{exc}$. A possible dephasing mechanism and its reduction in QDs are discussed in Chapter 8.3. If we can enhance γ_{rad}, e. g., by using a microcavity and the Purcell effect (see Chapter 6) we can decrease the detrimental effect of pure dephasing and finally reach visibilities close to 1. By embedding QDs into micropillar cavities, values for the indistinguishability of up to 99.56 % have been achieved for photons emitted shortly after each other (few ns) [186]. Furthermore, values of $V_{HOM} \sim 97.5\,\%$ have been achieved for time periods of $\tau_{period} \sim 1.5\,\mu\text{s}$ [196]. This time period can then be used to provide N photons

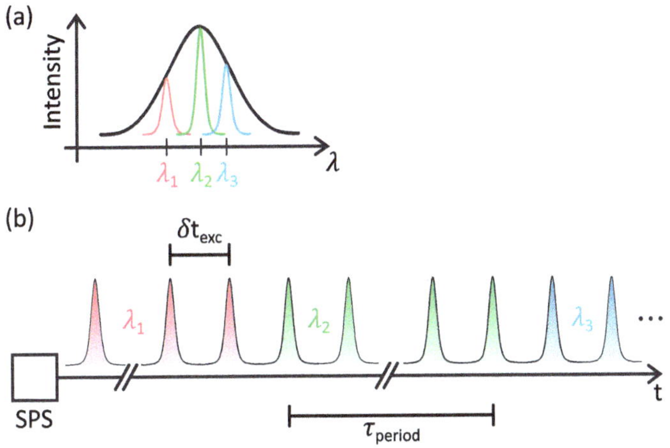

Figure 11.3: Illustration of the generation of N photons strings that possess the same wavelength and linewidth. (a) Sketch of an inhomogeneously broadened emission line (black) and individual spectral components (red, green, blue). The color codes of the individual components match between the figure parts (a) and (b). (b) Sketch of a single-photon source (SPS) emitting a stream of photons with repetition rate $1/\delta t_{exc}$ having the same wavelength and linewidth (same color) over a time period of τ_{period}. Fluctuating electric and/or magnetic fields lead to discrete linewidth jumps ($\lambda_1 \rightarrow \lambda_2 \rightarrow \lambda_3$) on a timescale of τ_{period}.

($N \sim \tau_{period}/\delta t_{exc}$) with a certain constant indistinguishability ($I = V_{HOM}$) useful for applications of a demultiplexed single-photon source (see discussion in Section 8.3). This would allow us to generate thousands of photons with single-photon creation at a repetition rate of 1 GHz.

HOM considering pure dephasing and spectral diffusion

However, in general, the time dependence of dephasing and spectral diffusion processes cannot be neglected. In this case a more general treatment is necessary. Heindel et al. (see Chapter 6 in [126]) introduced a time-dependent dephasing rate $\Gamma_{deph}(\delta t_{exc}) = \Gamma'(\delta t_{exc}) + \gamma_{ph}$, which describes dephasing due to spectral diffusion $\Gamma'(\delta t_{exc})$ and phonon coupling γ_{ph}. The used model approximates the QD as a two-level system, and dephasing is introduced by a phenomenological dephasing description, i. e., by including a stochastic force that shifts the transition energy of the QD. This stochastic force includes the dephasing due to spectral diffusion and phonon interaction. The spectral diffusion reveals a strong dependence on the time separation δt_{exc} and includes a finite memory effect with specific correlation time τ_{SD}, which can be regarded as τ_{period} discussed above. Thus, if $\delta t_{exc} \ll \tau_{SD}$, then the effect of spectral diffusion can be neglected. Based on these considerations, Heindel et al. calculate the time-dependent visibility as a function of photon time separation δt_{exc}, spectral diffusion, and phonon dephasing, which is

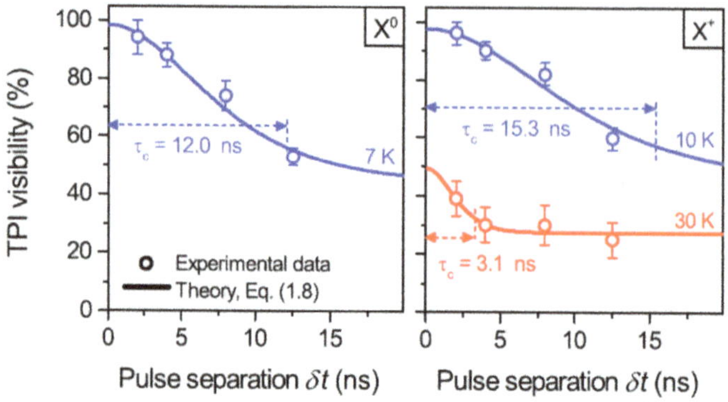

Figure 11.4: Two-photon interference visibilities of consecutively emitted single photons versus time δt elapsed between the emission processes. Experimental data for the exciton X^0 (left) and the positive trion X^+ (right) state are quantitatively described by Eq. (11.15). A characteristic temperature-dependent correlation time (here $\tau_c = \tau_{SD}$) is observed. Reprinted with permission of Springer Nature from [126], Chapter 6.

considered as temperature-dependent $\gamma_{ph}(T)$ [126]:

$$V_{HOM}(\delta t_{exc}, \tau_{SD}, T) = \frac{\gamma_{rad}}{\gamma_{rad} + \Gamma_0'\{1 - \exp[-(\delta t_{exc}/\tau_{SD})^2]\} + \gamma_{ph}(T)}, \tag{11.15}$$

where Γ_0' describes the maximal amount of dephasing induced by spectral diffusion. Figure 11.4 displays measured TPI visibilities of consecutively emitted single photons versus the time delay δt_{exc} between the emission processes of a QD. At low temperatures (7 K and 10 K) and small δt_{exc} relatively high TPI visibilities ($V_{HOM} \sim 94\,\%$) are observed. With increasing δt_{exc}, the TPI visibilities decrease fast on a time scale of 15 ns, which reflects relatively fast spectral diffusion processes in these QD samples. At higher temperature (30 K), a lower value of V_{HOM} at small time separation δt_{exc} together with a smaller τ_{SD} is observed. Here strong temperature-dependent phonon-induced dephasing occurs.

HOM considering relaxation and pure dephasing

Now we will consider the situation shown in Fig. 11.5, where the system is initially excited to a higher excited state $|p\rangle$ followed by a relaxation process with rate r into the emitting state $|e\rangle$. Such a relaxation process will introduce a time jitter in the photon emission process ($|e\rangle \rightarrow |g\rangle$), which will influence the visibility of the TPI process at a beamsplitter. The general wavepacket of the photon considering both dephasing with a pure dephasing rate $\Gamma_{deph} = 1/T_2^*$ and relaxation with a relaxation rate r can be written as [165]

$$|\Psi\rangle \sim \int_{t_0}^{\infty} dt \cdot \exp[-(\gamma_{rad}/2)(t - t_0) + i\Phi(t)]\hat{a}^\dagger(t)|0\rangle \tag{11.16}$$

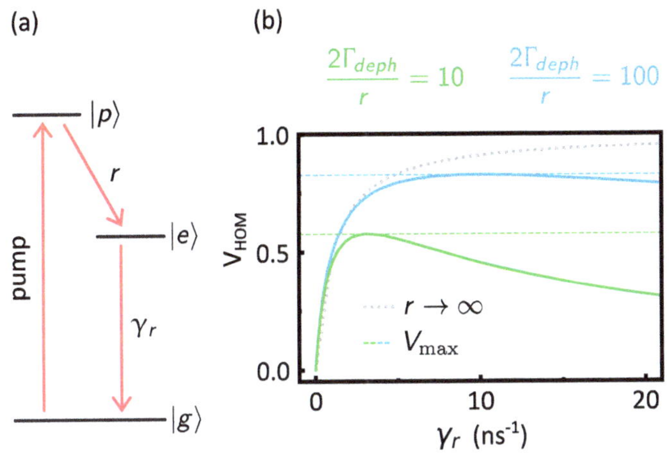

Figure 11.5: (a) Energy level scheme with relaxation between two excited states ($|p\rangle$ and $|e\rangle$) and photon emission to the ground state $|g\rangle$. (b) Simulated visibility (Eq. (11.17)) as a function of the radiative decay rate γ_{rad} for different ratios of the relaxation rate r and the dephasing rate Γ_{deph}.

with the radiative decay rate $\gamma_{rad} = 1/\tau_{rad}$ and a random variable t_0 defined by the probability density $r\,e^{-rt_0}$ for $t > t_0$; $\Phi(t)$ describes a random function taking into account stationary dephasing processes, and it is related to the pure dephasing rate by $\langle e^{i\Phi(t)}e^{-i\Phi(t+\tau)}\rangle = e^{-\Gamma_{deph}\tau}$. For the photons in Eq. (11.16), the mean magnitude-square overlap between the wavepackets of two photons is equal to the visibility and is given by [165]:

$$V_{HOM} = \frac{\gamma_{rad}}{\gamma_{rad} + 2\Gamma_{deph}} \cdot \frac{r}{r + \gamma_{rad}}. \tag{11.17}$$

From this result it becomes clear that a truly resonant excitation scheme of the emitter is advantageous since it avoids the time jitter from the relaxation process. As $r \to \infty$, the visibility is again equal to $V_{HOM} = \gamma_{rad}/(\gamma_{rad} + 2\Gamma_{deph})$ (see Eq. (11.14)). When relaxation and pure dephasing are both present, there is an interesting compromise with respect to γ_{rad} if we can adjust this parameter, e. g., with a tunable Purcell effect inside a microcavity (see Fig. 11.5). For $\gamma_{rad} = \sqrt{2r\Gamma_{deph}}$, we achieve from Eq. (11.17) a maximum visibility of

$$V_{HOM}^{max} = \left(1 + \sqrt{\frac{2\Gamma_{deph}}{r}}\right)^{-2}. \tag{11.18}$$

If we assume reasonable values for the relaxation and dephasing rates for a quantum dot, e. g., $r = (10\,\text{ps})^{-1}$ and $2\Gamma_{deph} = (1\,\text{ns})^{-1}$, then the highest achievable visibility is $V_{HOM}^{max} \approx 83\,\%$ for a radiative recombination rate of $\gamma_{rad} = (100\,\text{ps})^{-1}$.

11.5 Summary

- An ideal single-photon source emits a regular stream of identical photons. This means that the photons are identical in all their physical properties, e. g., the spectral bandwidth, pulse width, polarization, carrier frequency, and mode profile, i. e., in the language of quantum optics they are indistinguishable.
- A QD is embedded in a solid-state matrix and therefore experiences interaction with its environment. This coupling to the environment leads to dephasing and spectral diffusion, which limits the photon indistinguishability.
- Resonant optical excitation is the most favorable excitation scheme for achieving highly indistinguishable single photons.
- By embedding a quantum dot into a cavity and utilizing the Purcell effect values for the indistinguishability of ~97.5 % have been achieved for photons emitted in a time period of ~1.5 μs.
- Quantum interference of single photons from remote QDs has reached very high two-photon visibilities of $V_{HOM} = 93\,\%$.

Part III: **Technologies and devices**

12 Fabrication of QDs and nanofabrication methods

12.1 Introduction

One of the practical aspects that makes quantum dots highly appealing for large-scale implementations is their semiconductor nature. Indeed, implementing quantum light sources and devices based on QDs can benefit from the highly developed semiconductor technology, which encompasses material growth and fabrication. In this chapter, we will provide an overview on the most utilized techniques for quantum dot deposition and clean room nanofabrication. Particular attention will be given to deterministic techniques, which allowed higher fabrication yield and highly controlled properties of the realized light sources.

12.2 Epitaxial growth of semiconductors

Semiconductor quantum dots are usually grown on solid-state substrates employing molecular beam epitaxy (MBE) or metal-organic vapor-phase epitaxy (MOVPE), as schematized in Fig. 12.1.

As a brief overview, MBE deposits materials employing solid-state targets, which are heated up to the sublimation point and then sent to condensate over the target substrate material. The deposition chamber is kept in high vacuum (typically $<10^8$ mbar), and, together with high purity targets, high material quality and low defects due to unwanted impurities in the grown crystal are ensured. Thanks to the slow deposition rate, highly controlled growth conditions can be achieved (also because of several available methods for in situ crystal analysis). Nonetheless, this growth method is relatively slow and requires a careful maintenance to preserve the high achievable crystal purity with respect to the vapor counterpart. Indeed, metal-organic vapor-phase epitaxy (also named metal-organic chemical vapor deposition, MOCVD) employs a reactor, which is kept at lower vacuum conditions (often >10 mbar) than MBE. The materials in MOVPE are ultra-pure metal-organic precursors (liquid or solid), which are carried in the gas phase with the help of a carrier gas into the reactor chamber. The gases are decomposed thermally via a pyrolysis process, which frees the materials needed to be deposited on the substrates. In MOVPE the deposition process is influenced by gas pressure, temperature (typically higher than in MBE), flux, and ratio of the elements (in the exemplary case of GaAs-based system, ratio of group III and V elements). This epitaxy method is largely employed in semiconductor industry thanks to its fast deposition rate. Furthermore, operating at high temperatures ensures a high crystal quality in the deposited semiconductors, making this technique suitable, for example, for the realization of high reflectivity (low optical loss) distributed Bragg reflectors. Still, operating at lower vacuum levels together with eventual unwanted products coming from the pyrolysis (as exemplary carbon) can impact the optical properties of single-quantum emitters: indeed, this can result in the

https://doi.org/10.1515/9783110703412-012

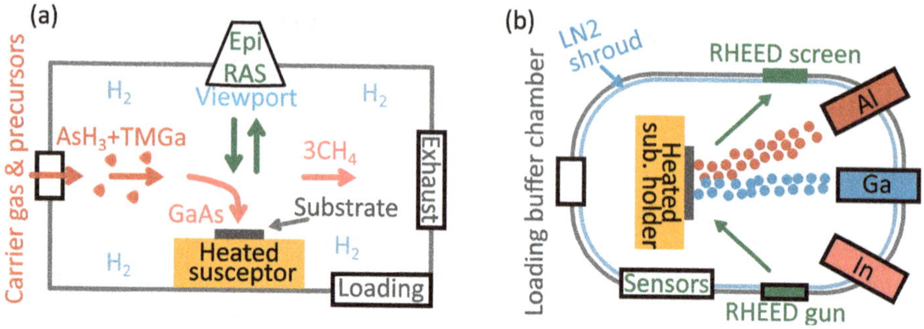

Figure 12.1: (a) MOVPE schematics. Metal-organic compounds are transported by a carrier gas (e. g., hydrogen) in the reactor chamber together with the hydrides. The gaseous species pyrolysis happens at the sample surface. A viewport allows for visualization of the sample growth, where, for example, reflectance anisotropy spectroscopy (Epi RAS) can be performed. Residuals from the thermal decomposition are removed through the exhaust port. Substrates and samples can be loaded/unloaded via the loading chamber. (b) MBE schematics. The sample is loaded via a loading buffer chamber allowing for maintaining the high vacuum in the reactor while inserting and removing samples. The heated substrate holder can be controlled via an external manipulator. Reflection high-energy electron diffraction (RHEED) can be conducted (using gun and detection screen) to obtain information from the surface during growth. Further detectors (as residual gas analysis or diffuse reflectance spectroscopy) are shown as "sensors." Three effusion cells (aluminum, gallium opened, and indium closed) are shown as examples. A liquid nitrogen shroud maintains the surface surrounding the area of deposition cold to ensure optimal control of the environment.

formation of local defects or unwanted presence of dopants that can induce dephasing or spectral diffusion in the QD emission. On the contrary, controlled high-quality MBE growth allowed for producing single-quantum dots prone to very limited dephasing and spectral diffusion, therefore suitable for photonic quantum technologies. Recent efforts demonstrated that these two techniques can be even combined, utilizing MOVPE for the realization of high-quality DBRs, whereas the MBE was employed for the growth of QDs surrounded by defect-free material and embedded into diode structures [191].

Apart from the growth methods themselves, producing QDs requires a careful control of the growth conditions, in particular, regarding strain. When epitaxial growth is performed, the lattice constants between the substrate and the deposited material may differ. This lattice mismatch plays a central role in the formation of QDs.

When the lattice constant of the layer a_l formed during the growth is smaller than that of the substrate bulk material a_s, the strain that creates at the interface is tensile; conversely, when $a_l > a_s$, we observe the accumulation compressive strain (Fig. 12.2). This distortion of the atomic bonds generates elastic energy, which depends on the volume of the structure under stress. This accumulated energy is then released when the deposited material surpasses a certain critical height, resulting in the formation of a crystal defect. Below this critical thickness, the grown layers can have the same lattice constant as the substrate, named pseudomorphic structure as in Fig. 12.2(b). If the lattice-mismatch of the heterostructure is large, then the deposited layer starts to re-

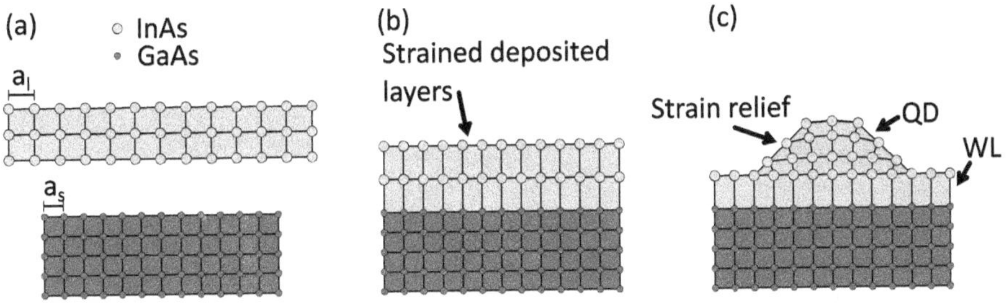

Figure 12.2: (a) Exemplary materials InAs and GaAs with different lattice constants (GaAs substrate marked as a_s, InAs as a_l). (b) Pseudomorphic growth of InAs on the GaAs substrate resulting in compressive strain on the deposited two-dimensional layers. (c) Under the conditions as described in the text, strain relief results in the formation of a 3D island. The quantum dot (QD) is marked as the so-called wetting layer (WL) underneath it.

lax the strain energy by forming misfit and threading dislocations, resulting in a partially plastically relaxed layer. The lattice parameter approaches the unstrained value of the deposited layer. This metamorphic structures have to be formed in such a way that the dislocations do not penetrate, or penetrate as less as possible, to the surface disturbing the formation of the QDs. Looking at growth of III–V materials as a case study, gallium arsenide (GaAs) based systems are very often employed. With these materials, high-reflectivity DBRs can be realized by depositing alternating layers of AlAs and GaAs. These two materials have different crystal lattices, although quite similar, and this heteroepitaxy requires the precise adjustment of mismatch and accumulated strain. Considering MOVPE growth, gas pressure, flux, and process temperature can be utilized for this adjustment. In particular, the process temperature impacts the growth rate and surface kinetics, which further influences the growth characteristics as briefly discussed in the following. For the heteroepitaxy of AlAs/GaAs, we can employ hydrides and metal-organic precursors as arsine (AsH_3), trimethylgallium (TMGa), triethylgallium (TEGa), and trimethylaluminium (TMAl).[1] The growth temperature is precisely controlled at the sample stage, whereas the temperature of the sample surface serves as a key region for the growth control and engineering. Indeed, the temperature of the solid/gas interface controls the pyrolysis of the precursors and therefore the amount of material depositing on the solid-state surface. Once again, the temperature impacts the kinetics of the atoms and molecules and their behavior during growth. Indeed, adatoms can be adsorbed between the lattice atoms, incorporated at the edges of the grown material, diffuse along the surface, or experience desorption from the crystal. The interplay between substrate, interface, and layer energies will result in different growth modes, i. e., Frank–van der Merwe, Vollmer–Weber, or Stranski–Krastanov [49, 189, 211]. The latter is

1 Other examples of precursors for atoms being part of the group V is phosphine (PH_3), whereas for group III, trimethylindium (TMIn) can also be employed. To provide doping atoms, monosilane (SiH_4, n-doping) and dimethylzinc (DMZn, p-doping) can be employed.

of key importance since the lattice mismatch allows for transitioning from a 2D layer-by-layer growth (as in Frank–van der Merwe) to a 3D growth (similar to Vollmer–Weber, but with less surface variation and defects) having the formation of strain energy-induced three-dimensional islands. This is a key for the growth of high-quality quantum dots. As an example, depositing indium and arsenide on the GaAs surface results in compressive strain that enables the formation of In(Ga)As QDs (Fig. 12.2(c)). A typical approach during deposition relies on keeping one element concentration very high (for example, As), and the growth rate is precisely controlled adjusting the concentration of the second atom forming the compound (i. e., gallium for GaAs or indium for InAs). Once the QDs are formed, the sample is capped with additional layers of material (in this case study, GaAs) to avoid the presence of dangling bonds next to the QD.

Together with the very exciting possibilities provided by Stranski–Krastanov grown quantum dots (typically, In(Ga)As), QD growth via droplet etching has proven to be a very valuable tool for the realization of strain-free highly symmetric dots, therefore resulting in small fine-structure values making them appealing as sources of entangled photon pairs. One interesting approach for this scope is based on using Al droplets deposited on top of AlGaAs (also demonstrated with Ga and In droplets on different substrates). Instead of the necessary strain discussed above, Al droplets can react with the sample surface creating shallow depressions in it. This is explained by diffusion of As from the AlGaAs surface to the Al droplet (due to gradient in As concentration between the droplet and the material) resulting in the etching of the AlGaAs. Refilling these depressions (holes) with gallium arsenide creates an unstrained GaAs island. Overgrowth with AlGaAs finalizes the QD structure. A more detailed description of the chemistry of the process is beyond the scope of this book and can be found, for example, in [30].

12.3 Lithography: optical, electron-beam, and deterministic techniques

This section is devoted to the description of the most commonly utilized lithography approaches, from optical to electron-beam and related techniques for the sample preparation and processing. An important part is reserved for the description of deterministic lithography setups, which constitute important techniques for the realization of high-performance solid-state sources of quantum light.

12.3.1 Resists, spin-coating, exposure, and development

The first step of transferring a lithographically defined pattern in the solid-state material is based on the sample coating with specific kinds of polymeric materials named resists. These polymers react upon irradiation of light (typically visible or UV) or electrons [71]. Local energy absorption by specific chromophoric sites can lead to a break of the polymer chains, which results in a change of the material solubility upon wet chemical attack (the so-called development). In other words, a specifically designed developer can selectively attack different areas of the resist with different rates. In particular, these polymers can be divided in two categories: positive and negative resists (see the following discussion and Fig. 12.3). Upon interaction with light, the former increases its solubility, whereas the latter decreases it when immersed in the developer. This means, in the case of a positive resist, that only the areas of the polymer that have been irradiated will get selectively removed by the developer, whereas the other areas are less prone to be wet-chemically removed.

Before exposure, a homogeneous film of resist has to be deposited on the sample. This is typically done employing the so-called spin coaters. These machines are constituted by a chamber equipped with a rotating stage. The sample is first placed on the stage (referred to as chuck), where it is usually kept in place by vacuum pulled via holes present in the holder. Afterward, some drops of resist are deposited on the sample,[2] followed by the activation of the spinning. Thanks to centripetal force, the resist spreads on the sample. For a controlled and reproducible processing as described in the following, the thickness of the spin-coated resist has to be defined and controlled. The film thickness depends on various properties of the resist itself as viscosity, drying rate, surface tension, and, clearly, on the spinning parameters (acceleration, rotation speed, and duration of the spinning), which are precisely controlled by the spin-coater itself. Often, after spin-coating, the resist is "baked" on a hotplate (in air or nitrogen atmosphere) to remove solvent traces present in the resist. Typically, the manufacturer provides several information on the resist handling, as the achievable thickness for specific spin-coating speed, acceleration, and duration, as well as the most suitable developer for the removal. The choice of the resist strongly depends on the following processing, i. e., if optical or electron beam lithography will be used, if a hard mask will be deposited, and which kind of etching will be employed.

After the spin coating, the sample is ready for transferring the chosen mask design into the resist. To realize the desired pattern on the polymer, optical or electron beam lithography can be utilized, with their respective resist, depending on the required resolution (from several microns to several nanometers, respectively, for the two techniques). In case of electron beam lithography, the energies of the electrons suffice to directly break the bonds of the resist starting possible reactions. These exposure processes allow for selectively etch areas of the resist with respect to others when the sample is immersed in the developer, allowing then for transferring the pattern in the polymer itself.

In case of UV lithography a hard metal mask is typically utilized: these masks are typically constituted by a metal pattern defined on the surface of a glass substrate. To

2 The amount of resist to be placed depends strongly on the sample size, the resist viscosity, and the film thickness that we want to achieve.

reach the highest possible resolution (ultimately limited by the wavelength of the exposing light), the side of the mask substrate where the metal is deposited is brought in close contact with the resist to limit diffraction at the edges that can impact the sharpness of the pattern transferred in the resit. Machines called mask aligners are employed for this purpose. They can control the mask position with respect to the sample, their respective distance, and the duration of the UV exposure (often provided by opening a shutter placed in front of a UV lamp). This process is schematically depicted in Fig. 12.3, where both positive and negative resists are shown in more detail.

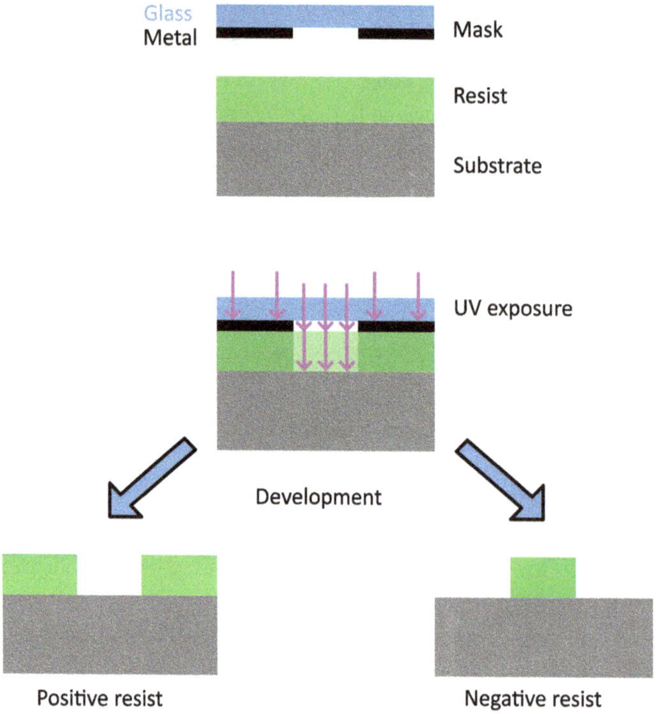

Figure 12.3: Exemplary UV lithography process. The hard mask is aligned with respect to the sample (which has been spin coated first with resist defining a homogeneous film) and brought into contact. UV light is utilized for the exposure: the metal mask screens locally the resist from light, whereas only defined areas are irradiated. The light absorption changes the local solubility upon development. The mask is now defined on top of the sample for positive or negative processes.

To reach a higher resolution, electron beam lithography can be utilized (pushing the resolution from several microns for optical lithography to several nanometers). Here a high voltage is utilized for supplying current to a filament, which emits electrons via the thermionic effect (see Fig. 12.4). Electromagnetic lenses are employed to focus and deflect the electrons through an aperture, with typical size of a few microns, that is used for the beam definition. Differently from scanning electron microscopes (SEM), an

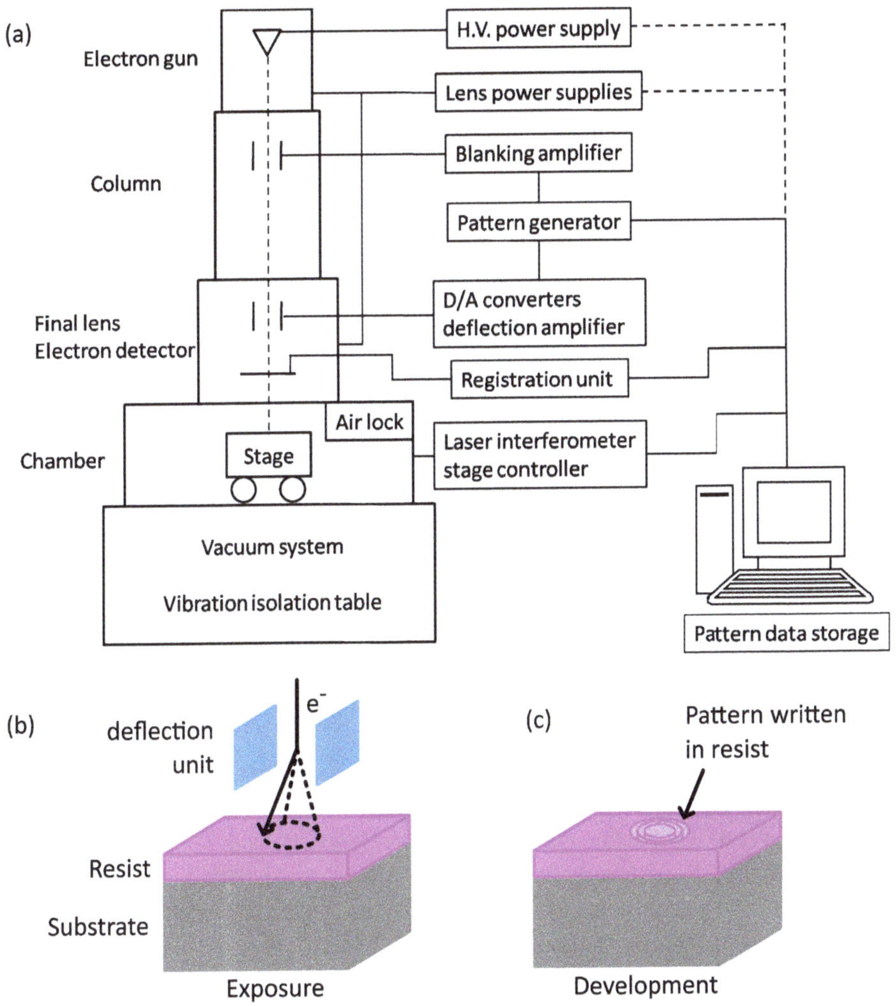

Figure 12.4: (a) Sketch of the main elements of an electron beam lithography machine [150]. In the column a simplified electromagnetic lens system with respective power supplies is sketched. See text for further details. (b) Schematic of the exposure process: the electron beam is deflected on the resist according to the chosen pattern (alternatively, the sample is moved, whereas a minimal deflection of the electron beam is utilized only to compensate for nonideal sample movement). Typically, the pattern is transferred by writing line by line similarly to Fig. 12.13, even if depending on the machine, polygons may be handled differently. (c) Pattern transferred in the resist, visible after development (as for UV resists, it can be a positive or negative process).

e-beam lithographer is equipped in the electron column of a blanking amplifier, which can switch off (blank) the beam via deflection while maintaining the electron gun always on to reach higher electron flux stability. This capability of selectively switching on and off the beam is a key for defining the pattern in the resist. The so-called pattern generator is connected to the blanking amplifier and to the digital-to-analog (D/A) converter and deflection amplifier. In some models, electron detectors are also present similarly to an SEM for eventual sample imaging (often required during e-beam alignment stage).[3] Accurate sample positioning is achieved by a motorized stage, regulated via a precision laser interferometer control system. The entire exposure process is performed under vacuum to ensure optimal performance and minimize contamination. Typically, vibration isolation tables are utilized for ensuring mechanically stable operation of the lithographer. The choice of exposure parameters is a function of various factors as the resist employed, its thickness, the e-beam aperture, the features size to be written, and the required resolution. Here the pattern is defined via software, and the electron beam is scanned on the sample accordingly; modern machines move the sample with respect to the beam with minimal electron beam deflection (to avoid distortion) for final adjustment over the sample position. Blanking the beam becomes a key to move across the sample without exposing the resist, having the electron beam reaching the surface only when needed. It is worth noting that when writing patterns close to each other, the electron beam can diffuse laterally resulting in unwanted overexposure of the already processed areas. This is typically avoided during the pattern file preparation stage, since calculations can be performed for compensating this cross-talk between the areas, depending on the beam parameters and pattern geometry.

12.3.2 Etching: RIE and ICP-RIE

Once the mask has been defined in the resist, the sample can undergo subsequent processing. Generally, the pattern defined in the resist needs to be transferred in the solid-state sample. For this purpose, etching techniques can be applied. In most cases the resist is utilized to protect the sample areas that do not have to be etched. A key here is to ensure that the resist-protected areas are not etched until the pattern is fully transferred in the sample. Therefore the etching rate has to be higher on the sample than on the resist (or the polymeric film has to be thick enough). This is schematically shown in Fig. 12.5(a,b). In the figure, ions from ionized gases are depicted as etchants.

These elements can be generated in a *Reactive Ion Etching* (RIE) setup (Fig. 12.5(c)), typically categorized as a *dry-etching* technique. A radio frequency field (typical around 13.56 MHz) is used to ionize the gases and create a plasma. Electrons and ions are accelerated back and forth in the electric field; electrons are moving faster than ions, and they hit one electrode charging it negatively, typically the bottom electrode, whereas

3 For completeness, e-beam utilizing high acceleration voltages may use different approaches for the sample visualization (i. e., additional optical microscopy outside the chamber for prealignment, followed by e-beam scanning on predetermined sample positions), since it is difficult to realize a sample image with highly accelerated electrons.

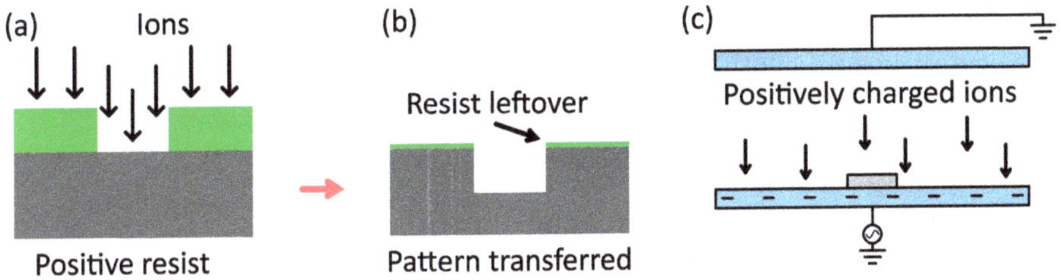

Figure 12.5: (a) Ions impinging on the sample protected by a resist (in a positive process as in Fig. 12.3). The resist acts as a mask against the mechanical action stemming from the accelerated ions. (b) After etching, the pattern formed in the resist is transferred to the sample. Some resist leftover may be still present on the surface and can be removed by chemical or plasma cleaning. (c) Simplified sketch of a RIE reactor. Electrodes with respective electrical connection are shown. A radio frequency field is applied to the bottom electrode, having the top one connected to ground [150].

the top one is grounded. A DC bias forms, and it superimposes to the radio frequency field attracting the positively ionized ions toward the bottom electrode. Together with a mechanical etching process arising from the impact of the ions, other species can be employed as radical reactants to provide a chemical attack of the sample surface. The interplay between mechanical action and chemical reactivity changes the isotropy of the etching: whereas the mechanical impact of the ions perpendicularly to the surface results in vertical etching (anisotropic etching, as for example provided by accelerating argon ions in GaAs etching), the presence of highly reactive radicals results in etching also in the direction parallel to the sample surface (isotropic etching as, for example, decomposing the silicon tetrachloride ($SiCl_4$) in the plasma into $SiCl_x$ and Cl_x radicals). Whereas in chemical etching, typical high plasma densities and pressures are employed to allow for high supply of radicals, low pressures are used in mechanical etching to avoid collisions between ions. Higher plasma densities can be achieved utilizing *inductively coupled plasma reactive-ion etching* (ICP-RIE) setups (see Fig. 12.6). Differently from RIE, these machines are equipped with an additional ICP coil for the generation of the ion plasma. ICP-RIE can control separately the current of ions and their energy, allowing for more flexibility in the process control. In conclusion, gas pressure, flux, chemical species, mechanical/chemical etching ratio, and field power emerge as important parameters for controlling the etching process.

An alternative approach for semiconductor and solid-state etching is based on *wet chemical* approaches (i. e., in the liquid phase instead of gas as before). This approach does not require expensive machines, since it is only necessary to immerse the sample into etching solutions. Nevertheless, reactive ion etching approaches allow for different forms of anisotropic etching. Additionally, wet chemical approaches represent a valuable fabrication tool and allowed reaching, for example, extremely low roughness in semiconductor processing, having the surface quality after etching comparable with wafers ready for epitaxy (roughness smaller than one nanometer) [43, 169]. This kind of processes applied to semiconductors can be understood as a first oxidation of the sample

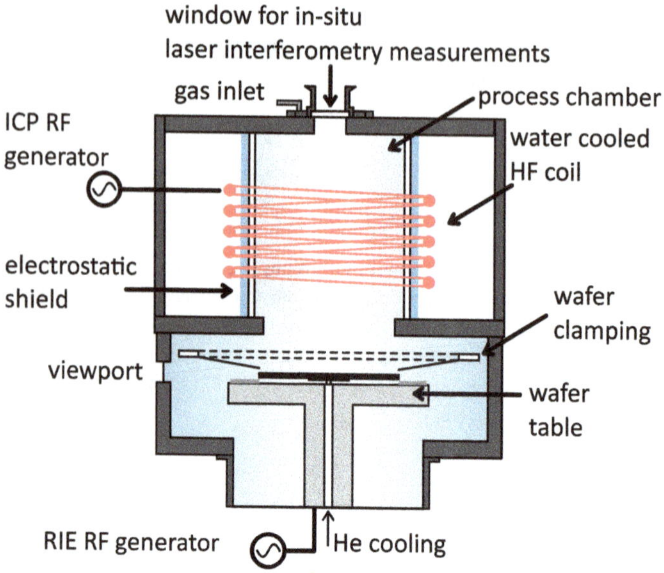

Figure 12.6: Sketch of an ICP-RIE chamber and the key constituting elements [160, 168]. Gases are inserted in the chamber via the inlet. The radio frequency (RF) generator ionizes them creating a plasma. 13.56 MHz is a typical frequency. This field further controls the plasma density, which is accelerated toward the sample by the ICP coil.

surface, which results in an increased solubility of the crystal structure against the attacking etching solution. In the exemplary case of GaAs, hydrogen peroxide (H_2O_2 as oxidant), together with aqueous solution of sulfuric acid (H_2SO_4), has been employed [43]. For more detailed information on the etching techniques, we refer, for example, to [160].

12.3.3 Dielectric material and metal deposition

In some aggressive etching processes, the resists themselves may not be resistant enough to protect the sample entirely. For example, a resist can be attacked too fast and be removed by the etchant before the semiconductor material is fully processed. In these cases, a hard mask can be defined to further protect the selected sample areas during dry or wet chemical etching. This type of mask can be realized with dielectric or metallic layers deposited on the sample. To ensure that the hard mask follows the same pattern as the lithographically defined one, a process called *lift-off* is often employed. This method is also utilized for defining metallic contacts on the sample surface.

First, the sample is covered with resist (Fig. 12.7(a); the process includes a double resist layer as explained below), and lithography can be used to define the pattern. The use of a proper developer removes the unwanted resist (exposed or not depending on positive or negative process; see Fig. 12.7(b) for positive resist). Afterward, a dielectric

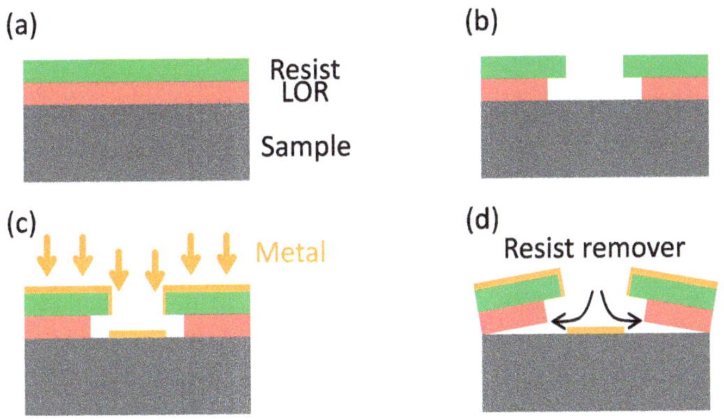

Figure 12.7: (a) Double resist layer spin coated on the sample. (b) Exposure and development (exemplary positive process): whereas the resist has the pattern as defined by the mask, an additional underetching appears in the second layer. (c) Metal (or dielectric) deposition. In addition to material on the horizontal surface, thin layers can also stick to the vertical surfaces of the resist. (d) Lift-off utilizing a resist remover. The wet-chemical attack allows penetrating under the top resist layer, which would be challenging if one single layer is used, in particular, in the presence of metal on the sidewalls [150, 168].

or metal layer is deposited via evaporation, plasma deposition, chemical vapor deposition, or sputtering (Fig. 12.7(c)). For nonprotected areas, the newly deposited layer is in direct contact with the sample, whereas for the others, it will be on top of the resist. It is important that after development and before metal deposition, all unwanted resist is removed from the surface (otherwise, it might result in lift-off of all deposited material). Then wet chemical processes can be employed to remove the resist, which will together carry away (lifting-off) the deposited metallic or dielectric layer (Fig. 12.7(d)).

As exemplary shown in Fig. 12.7(c), during metal deposition, a thin layer of material can also stick on the vertical surface, making the resist penetration for the lift-off challenging. Therefore the lift-off resist has to be thicker than the deposited material thickness. One approach is based on the successive spin coating of two layers of different resists. Typical for silicon and GaAs-related materials, a suggested combination from the resist manufacturers is S1818 and LOR resist types (for optical lithography processes). After development, the LOR surface is curved (called underetched), enabling an even better penetration of the resist remover (for example, NMP PG remover). Whereas the S1818 or S1805G2 are photosensitive resists, the LOR family is not photosensitive and acts only as a soluble spacing layer. Among the LOR family, various kinds of resists enable different thickness (around 500 nm for the LOR5A and around 700 nm for the LOR7B) and different dissolution rates (LOR7B desolves faster than the LOR5A, allowing also deeper undercuts) upon exposure of metal ion-free developers (AZ826MIF or AZ726MIF).

12.3.4 Flip-chip and transfer

The flip-chip process widely finds applications in semiconductor industry and in the realization of quantum light sources. This approach has been utilized to transfer thin membranes of semiconductor to a specific substrate (a sample carrier or a piezoelectric PMN-PT). As an interesting example, rather than employing a 2D PhC fabricated in a slab of semiconductor surrounded by air, circular Bragg grating cavities have been developed (see Fig. 13.4(a), top and bottom, respectively), so that the fully etched semiconductor stands on a metallic mirror, which breaks the vertical symmetry and allows for more light to be extracted from the top surface (with respect to a symmetric PhC slab in which emitted light from the cavity propagates equally in the top and bottom directions). These cavities will be discussed in more detail in Chapter 14; here we provide a brief description of a possible process to realize such a CBG resonator (Fig. 12.8).

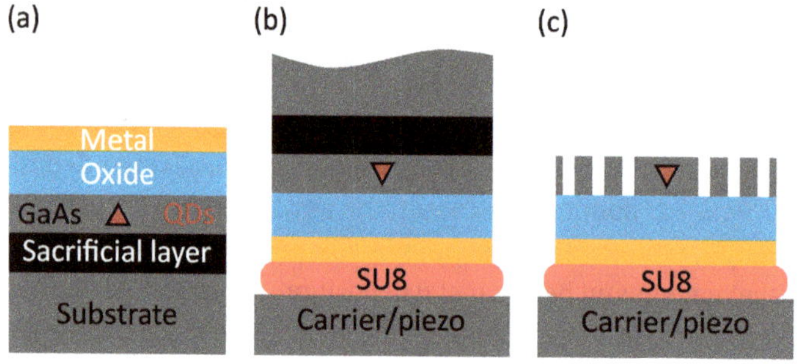

Figure 12.8: (a) Exemplary grown sample with sacrificial and semiconductor layer embedding quantum dots [100, 136] after oxide and metal layer deposition. (b) Flip-chip transfer of the sample on a carrier (exemplary glued utilizing SU8 resist), followed by substrate and sacrificial layer removal. (c) Final structure after e-beam and etching [98].

During epitaxial growth, a sacrificial layer is first deposited on the substrate, followed by the growth of the semiconductor layer embedding the emitters (exemplary QDs), which will constitute the semiconductor part of the CBG. After growth, an oxide layer is deposited, followed by metal deposition (for example, of a gold layer). The metal layer will act as a mirror in the finished cavity, whereas the oxide spacer will enable constructive interference of waves reflected at the various interfaces for a higher emission in the top direction. To continue with the fabrication, the top surface (i. e., the one with oxide and metal layer) is glued on the carrier, and the sample is flipped, resulting in the substrate now being on top, and the metal mirror constitutes the bottom of the sample. Mechanical polishing can be used to thin the substrate down (in case of GaAs substrates, the thickness is around 350 μm) to some tens of microns. Wet chemical etching can also be utilized to remove the substrate close to the sacrificial layer. This is then followed by a different wet chemical etchant that attacks the sacrificial layer, removing it completely together with eventual leftovers from the substrate. Spin coating, electron beam lithography, and etching are then employed to realize the circular photonic crystal (see [100] and references therein for more information).

12.3.5 Deterministic lithography

When operating with solid-state nanostructures as emitters of quantum light, a common challenge is related to their random position across the sample, as well as to the different emission wavelengths from emitter to emitter. In the case of semiconductor quantum dots, either deposited via Stranski–Krastanov growth or realized via droplet etching, they are randomly distributed on the sample surface (efforts are spent for site-controlled QD growth, but so far state-of-the-art optical properties are reached with Stranski–Krastanov growth or droplet etching). Furthermore, their size distribution results in emission at different wavelengths, having an inhomogeneous distributed Gaussian envelope of the PL emission collected from a large statistical number of emitters. Whereas this constitutes a challenge in multisource experiments and stimulates the development of single QD tuning mechanisms (see Chapter 9), the random spatial and spectral characteristics pose an even greater challenge in nanofabrication. Indeed, with the goal of employing cavity quantum electrodynamics for the modification of the emitter properties, spatial and spectral matching between the cavity mode and the emitter transition is a key (see Chapter 6 and Section 13.3). Even when employing geometrical effects for enhancing the light extraction (see Section 13.2), whereas the emission wavelength plays a minor role, the position of an emitter with respect to the photonic structure is key for the emission properties [62, 169]. For these reasons, deterministic fabrication approaches have been developed. Despite the various techniques employed, the main idea consists of determining precisely the spatial position of the emitter, as well as its emission wavelength, therefore fabricating around it a photonic resonator tailored to the preselected emitter. This is experimentally challenging since most of the highly developed nanostructures emit light at cryogenic temperature. Furthermore, the best performances are typically reached at liquid helium (around 4 K), so the spectral matching has to be ensured for such an operation condition. For this scope, four experimental techniques have been developed in the years and successfully employed. These techniques proved their capability of localizing a single emitter with a few nanometers accuracy, realizing devices with a deterministic positioning accuracy of a few tens of nanometers.

AFM position readout

This technique utilizes atomic force microscopy to determine the position of a shallow-capped QD with respect to predetermined metal markers, and it can be performed at room temperature. Further optical investigation, via low-temperature μ-PL, allowed determining the QD emission spectrum, tailoring the design of the cavity accordingly. So far, this technique has been employed for fabricating photonic crystal $L3$ cavities around the preselected QD [72]. $L3$ PhC slab cavities being rather sensitive to fabrication imperfections (below the achievable experimental tolerances), finding a QD-cavity system in perfect resonance becomes challenging (see also Chapter 14). Therefore local oxidation of a few nanometers of semiconductor and successive removal can be utilized for a controllable change of the diameters of the holes forming the PhC, resulting in a shift of the cavity resonance. This can be repeated until the cavity mode matches the QD transition [73].

Two-dimensional optical positioning

This approach starts with fabricating metallic arrays on the sample surface. The position of the emitters is then determined relatively to these marker arrays, which will be utilized for the successive alignment of the electron beam lithographer. The technique relies on making 2D images of the QD position and the array of markers (at low temperature). Typically, two images are acquired: for the first one, the sample is uniformly illuminated with light capable of exciting the emitters (for example, above the bandgap of the surrounding material). The luminescence signal is detected on a 2D array CCD after filtering the excitation laser. The second image collected is a reflection image from the sample surface to visualize on the detection CCD the geometry of the markers array. Practicality dictates that the reflectivity measurement is conducted employing light at the same wavelength as that of the emitter to avoid chromatic aberrations. Overlapping the two images allows for localizing the emitter's position with respect to markers. Alternatively, the excitation light source and the one employed for the reflectivity measurements can be utilized simultaneously, controlling the power levels to ensure simultaneous visualization of markers and emitters' luminescence in the same image (see Fig. 12.9). This would avoid the overlapping of two distinct images, increasing the

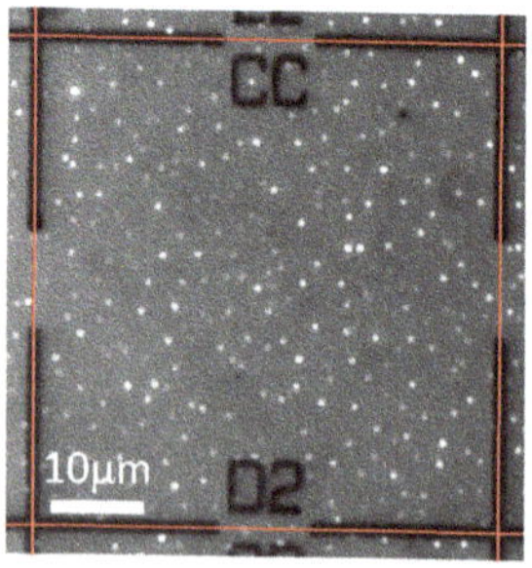

Figure 12.9: Exemplary two-dimensional optical positioning image: a uniform illumination allows for visualizing the alignment markers (lateral dark lines, letters, and numbers are further orientation references) as well as exciting the emitters. Quantum dots are indeed visible as bright circles. Their position can be recorded as a function as the surrounding markers, providing relative coordinates for the successive fabrication processing. Image courtesy of A. Rastelli and T. Krieger.

accuracy [167]. Once the emitter's position has been acquired, μ-PL can be conducted on the preselected emitters (which can be found thanks to the recorded coordinates relative to the markers), recording their emission wavelength. The cavity design parameters can be adjusted according to the desired wavelength and position with respect to the markers. The mask is employed in the successive electron beam lithography, aligning the machine to the firstly fabricated array of markers. E-beam lithographers can typically reach a few nanometers precision (called overlay accuracy) in overlapping the new (software defined) mask to previously fabricated markers. Resist development and etching are then utilized for the realization of the deterministically fabricated photonic structures [116].

Cathodoluminescence imaging and in situ electron beam lithography

This approach is based on low-temperature cathodoluminescence spectroscopy for the determination of the emitter's position and spectral features. Then the same e-beam can be employed for writing in the resist the desired structure around the precharacterized emitters. The challenge of this technique is mainly due to the fact that the sample needs to be spin-coated before starting the low-temperature cathodoluminescence spectroscopy (commercial resists are designed for room-temperature operations, and spin-coating parameters have to be adapted for a cryogenic environment). In addition, cathodoluminescence required for the observation of the emitter position and spectrum also exposes the resist. Therefore, for the actual mask exposure, the resist is overexposed, making it less soluble with respect to the rest of the resist (exposed during spectroscopy) [62]. The localization accuracy of this technique was proven to be down to a few nanometers. In principle, this approach can also be adapted for an operation principle similar to optical positioning, where cathodoluminescence maps are recorded, measuring the spatial position of the emitters with respect to prefabricated metal markers (so without need of resist). Nonetheless, optical positioning has the advantage that the localization of several emitters is performed in a single acquisition (whereas the spectra have to be collected one emitter at a time) instead of scanning a probing electron beam to create a cathodoluminescence 2D map.

Deterministic low-temperature optical lithography

Firstly demonstrated for the fabrication of micropillars operating at NIR [38], this technique has been highly developed in the past years and further interfaced with other nanofabrication setups [170]. This technique will be discussed in more detail in the following.

Deterministic low-temperature optical lithography utilizes three independent optical paths in a collinear setup for the emitter's precharacterization and for the deterministic lithography defining the photonic structure. As depicted in Fig. 12.10, one path is employed for the lithography laser, the second for the excitation laser used to inject car-

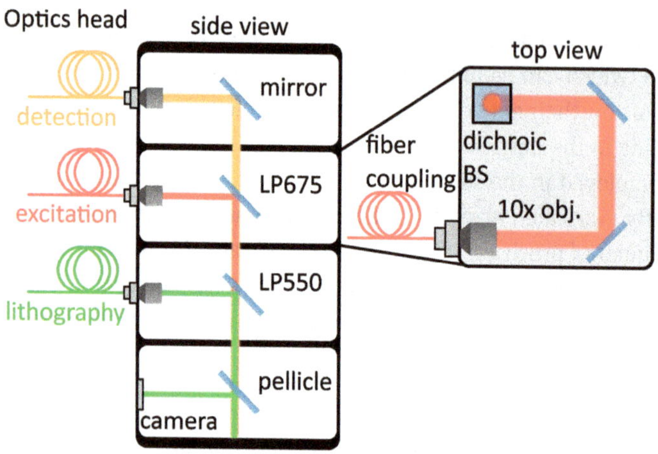

Figure 12.10: (left) Schematic side view of optics head with three optical paths utilized in deterministic optical lithography: detection, excitation, and lithography. One additional level (bottom) contains illumination and camera for sample visualization. Dichroic beamsplitters are employed to filter unwanted wavelength components in collection, ensuring that only the intended wavelength can be coupled back into the fiber (for alignment purposes; see text). (right) Exemplary top view of the excitation channel: an objective is utilized to collimate the laser light coming through the fiber (and coupling back the reflected light), whereas the two mirrors are used for ensuring the overlap of the beam paths via beam-walk procedure [168].

riers in the investigated nanostructures, and the third for the detection of the micro-PL emitted by the sample.

A first alignment has to be conducted to ensure the most efficient overlap between these channels: this has to be performed to ensure that excitation, detection, and lithography happen at the same coordinates. This is typically done in commercial systems by scanning repeatedly on areas where the reflectivity of the laser signal changes abruptly (as a sample edge or the edge between the sample and a metal marker). This results in a step-like reflectivity signal (see Fig. 12.11): measuring the reflected light sent through all three paths, lithography, excitation, and detection (where for alignment purposes, a laser at the same wavelength as the emitters under investigation is used) allow for a precise overlap (of the order of few nanometers accuracy) of the three paths. The figure exemplarily shows the alignment procedure in the x-direction, and it is repeated as well for the y-direction. Since the three paths are typically operated at distinct wavelengths, different focusing conditions due to chromatic aberration, are expected, and they result in a different steepness of the edge (see right part of Fig. 12.11). For the most effective alignment, the central point of the edge can be used as reference.

Before reaching low-temperature conditions, the sample is spin coated with resist (standard UV lithography resists are generally utilized). The choice of resist depends on the intended processing (eventual hard mask deposition, wet or dry chemical etching, etc.). Then the sample is placed into a cryostat and brought to cryogenic temperature. A commercial example of this setup relies on a bath cryostat, where the dipstick is filled

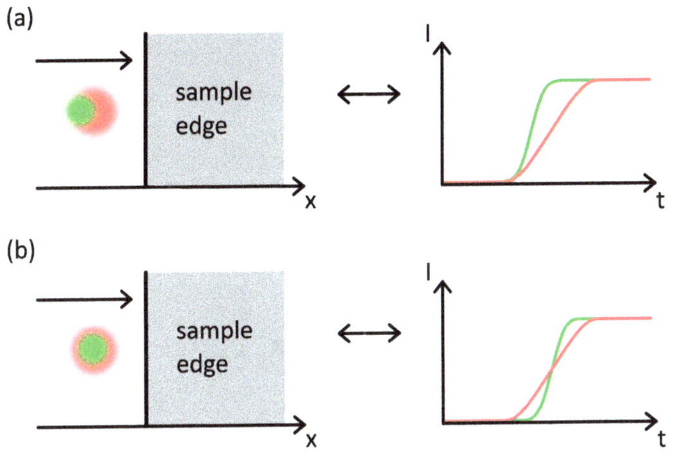

Figure 12.11: Schematic of the alignment procedure (exemplarily with two beams). The focused light is scanned over an edge having different reflectivity. (a) Nonoverlapped focused beams resulting in nonoverlapping intensity signals (plotted over time, repeatedly, as in an oscilloscope). (b) Well-overlapped focused beams. The intensity signal overlaps in the center between maximum and minimum intensities [168].

with helium as exchange gas and then dipped in liquid helium to reach 4 K (a window allows for optical access).

The sample is located on top of a stack of three positioners, furthermore having on top of the sample a large NA, high transmission cryogenic objective. Closed-loop positioning systems are employed, where the interferometric position readout (with subpicometric accuracy) is used as feedback for the compensation of any nonlinearity in the positioner motion. This setup allows for 2D mapping of the sample surface with high spatial resolution (Fig. 12.12). To do so, the excitation path is used to send a laser up to the sample to optically excite the emitters. In the case of semiconductor quantum dots, above bandgap excitation is utilized, making sure that the laser itself does not expose the resist. For this purpose, a laser at ~658 nm is employed. Indeed, most of the optical lithography resists are designed to be reactive to UV light, but they are transparent to red light (at least for the power levels necessary for the QD excitation). The detection path collects the emitter's signal. Scanning the sample in two directions allows for acquiring 2D micro-PL maps of the sample (detectors capable of detecting up to single-photon level are utilized for this purpose, shown as computer 1 in Fig. 12.12, where the spectrometer is set to zero-order sending all collected light to the APD). The collinearity of the two paths becomes a key to ensure that excitation and detection happen at the same spot. This, combined with the high-precision position readout, allows for the localization of the emitter position (i. e., relative coordinates set by the setup) with an accuracy below 2 nm. Several experimental aspects can have an impact on this accuracy: together with the position readout and the optical setup alignment, the signal-to-noise ratio for the detected μ-PL also comes into play. This accuracy can be best determined via a 2D fitting

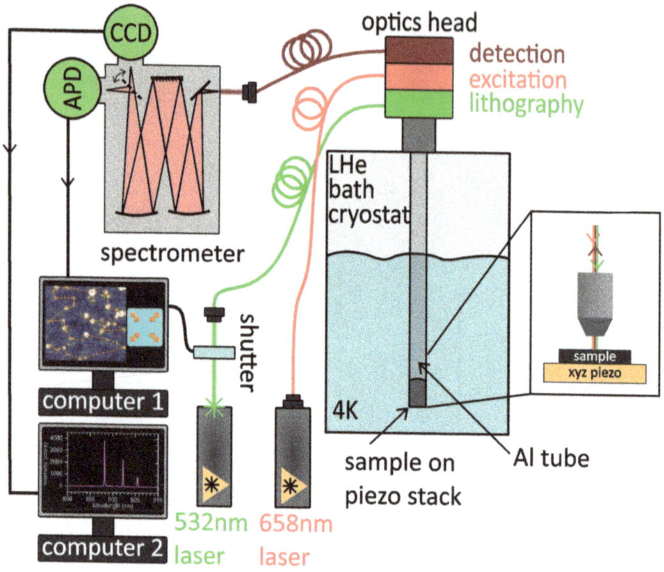

Figure 12.12: Full setup sketch. The head as in Fig. 12.10 is placed on the dipstick immersed in liquid helium. Inside the tube (filled with helium gas) the sample is placed on a piezo positioner stack (for *xyz* alignment and scanning). Excitation and lithography lasers (with computer-controlled liquid crystal shutter) are depicted. The detection channel (alignment laser not shown for clarity) is coupled to a spectrometer equipped with a CCD for spectral analysis and an avalanche photodiode (APD) for high signal-to-noise 2D maps acquisition. When used for micro-PL maps, the spectrometer is operated at the zero order. In some experiments the emission is spectrally filtered to acquire maps with light at specific wavelength ranges [168].

of the detected signal. The scanning excitation spot diffraction-limited diameter being of the order of a micron, and the emitter typically of a few tens of nanometer, the observed micro-PL map encompasses a Gaussian profile (i. e., Gaussian spot convoluted with a near-delta) of the order of a micron. The two-dimensional fit provides then the position of the emitter with respect to the system coordinates (the in-plane coordinates are the one sought-after). At this point, it is possible to use the subnanometric precision of the positioners to go back at the actual emitter's position and use the same setup for collecting the μ-PL spectrum (here using the spectrometer equipped with the CCD; see computer 2 in Fig. 12.12). Having now the spatial and spectral information of the emitter, the deterministic lithography can take place. For this purpose, the third optical path is employed. This is used to bring a green laser (~532 nm) to be focused on the sample surface: the resist is much less reactive at this wavelength, rather than its design one, but high powers still allow for exposure. Once more relying on the high accuracy and reproducibility of the positioners and on the collinearity of excitation, detection, and lithography paths, the location of the lithography spot can be precisely controlled, ensuring the writing of the software-defined mask in the aimed position. Similarly to e-beam lithography, the mask to be written is defined via a software. The laser path is opened and closed via a computer-controlled shutter, which allows the laser to hit the sample only during writing, blocking it while moving from spot to spot.

It is worth mentioning how the desired shape can be written in the resist. The software usually defines a shape, as the circle in Fig. 12.13, as a polygon. For large structures, the *continuous exposure mode* can be employed.

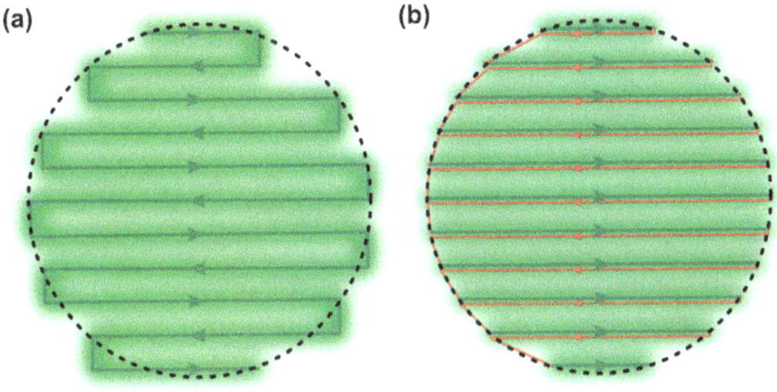

Figure 12.13: Schematic exposure of a circle. (a) Continuous writing mode. The laser is maintained on while moved across the sample (i. e., the laser is kept steady while the sample is scanned via the positioners). (b) Line writing mode. The laser reaches the sample only when scanned from left to write (green arrows), whereas it is shut during moving in the other direction (red arrows) [168].

In this mode the laser is scanned continuously on the sample surface (i. e., the sample moves via the positioners while the laser is kept at the same position to ensure its steady alignment in the center of the cryogenic objective). With a diffraction-limited spot size of around 500 nm, a distance between the horizontal line scan of around 100 nm is found to ensure the most uniform exposure of the central area. As depicted in Fig. 12.13(a), the top area of the circle results to be underestimated, whereas the bottom part is overestimated. Having the laser on while moving vertically further results in the overexposure of the horizontal dimension. Whereas this does not play a role for large structures (i. e., much larger than the spacing between the lines), small structures of a few microns should be written employing the *line writing mode*. With this approach, the laser is kept on only when moving from left to right, whereas the shutter closes the laser path during the movement right to left and subsequently down to the next line. Despite the exposure time is twice as long in the line writing, overexposure is avoided resulting in more symmetric structures. More details on the shape and writing parameters can be found in [168].

From an experimental point of view, a further calibration is necessary when employing such a kind of optical lithography. Indeed, distinct areas of the sample can result in different reflectivity and absorption of the employed green laser, which can modify the exposure conditions. To take this into account, calibration steps are first conducted, where the intended shapes are written on different areas of the sample with different writing parameters (for example, laser power). The ones closer to the intended shape are kept for the actual deterministic process. It is very important to mention that, first, different sample areas distant by a few hundreds of microns may already show different behavior upon green light illumination. Therefore the calibration should be performed in the neighboring area intended for the deterministic fabrication. Second, the optical lithography resists are designed for room temperature usage, and their reactivity upon exposure decreases with the temperature. On the one hand, this makes the resist less sensitive to laser power fluctuations, and on the other hand, the calibration has to be conducted in the same conditions as the deterministic lithography step. Interestingly, the resist itself can crack during cooling to low temperature because of the strain induced by the cooling (i. e., the resist is spun at room temperature, and it has a different expansion coefficient than the sample underneath). Despite unaffected areas of the sample can be large enough to conduct lithography, this impacts the reproducibility of the technique and the overall fabrication yield (since we cannot predict which areas will be usable after cooling down). One approach to circumvent this issue is to cover only small portions of the sample with resist. This can be done defining circles of a few hundreds of microns diameter via optical lithography, removing the rest of the resist during development. The reduced strain accumulated in the cooling down ensures the lack of crack in the resist at low temperature [37]. Alternatively, as studied in detail in [168], the choice of alternating resists with different viscosities can avoid the formation of cracks in the resist for areas of several squared millimeters.

Since this deterministic lithography approach is based on optical spectroscopy and lithography, the wavelength of the emitter can play a role in carrying out the described process. Indeed, the lithography laser wavelength (~532 nm) cannot be arbitrarily changed, since it must be absorbed by the resist itself to allow for the exposure. The wavelength of laser utilized for the above bandgap excitation is more flexible, but the closer to the lithography laser the less chromatic aberration in the employed optics will be observed. The longest wavelength is typically represented by the emitter's photons. Whereas this technique has been first employed for QDs emitting at around 900 nm, recently it has been extended up to telecom wavelength (~1310 nm), for which a careful optimization of the alignment procedure was necessary [169]. Interestingly, the same setup has also been employed for the realization of electrically driven single photon LEDs deterministically fabricated around one red-emitting quantum dot (InP) [171]. Having a source that can be excited electrically simplifies the deterministic optical lithography setup, since only two paths are required (detection and exposure), making it a valuable and versatile tool for the deterministic processes of future electrically driven quantum devices.

The high precision and reproducibility of this technique comes with the challenge that a visible laser is employed for the writing (similarly to laser lithography setups). This means that the diffraction limited spot focused on the sample is of the order of 500 nm. Thanks to fast liquid crystal shutters, a careful design of the writing procedure, and laser

Figure 12.14: SEM image of the University of Stuttgart, Institut für Halbleiteroptik und Funktionelle Grenzflächen (IHFG) logo realized on a GaAs wafer having then deposited 50 nm of SiO_2 followed by lift-off. In yellow, the polygon mask utilized in the writing is shown [168].

moving time (therefore exposure time), structures of this size (or slightly smaller) can be written in the resist (see Fig. 12.14, where the deposition of an oxide layer slightly enlarged the written structure, still showing details ≤500 nm).

Nevertheless, despite the writing size is small enough for defining the mask for micropillar or microlenses fabrication, it does not reach small enough spot sizes to be usable for photonic crystals. Therefore alternative strategies have to be found. First, it would be possible to utilize such a setup as for the above-described optical imaging, making 2D maps in emission and reflectivity of the emitters and alignment markers. Alternatively, the setup can be employed to write deterministic markers around preselected emitters, used for the successive alignment of other machines (as e-beam lithography).

This approach is appealing since it relies mostly on the capability of writing small markers precisely placed around the emitter (Fig. 12.15). The precision is easily achieved by the hardware itself (piezo positioners and precise position readout), whereas care must be taken when writing the markers to ensure the smallest structure is achieved. Since the written structure size is influenced by the diffraction-limited spot size of the lithography laser, it has been shown that markers made by two squares can serve the purpose. Indeed, to utilize these markers during room-temperature fabrication (therefore when the QD cannot be seen directly), the coordinate system set by the markers has to be precisely determined. For this purpose, the smaller the reference point, the more precise the alignment. Having this in mind, the two squares define in their touching point a valuable alignment spot, smaller than the lithography laser spot size (since it is not directly written but arises from the alignment of the two touching corners of the

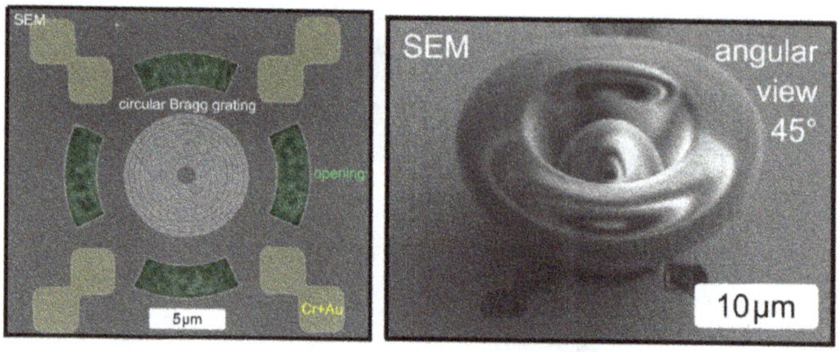

Figure 12.15: (left) SEM image of a circular Bragg grating cavity (suspender in air) deterministically fabricated around a preselected QD. Deposited markers (chromium and gold) are highlighted in yellow. In green, the openings employed for the underetching of the semiconductor layer [99]. (right) SEM angular view (with 45° tilt angle) of a Total Internal Reflection Solid Immersion Lens (TIR-SIL) deterministically 3D printed around a selected emitter (etched markers are visible at the bottom) [168, 172]. Figure on the left reprinted from [99] with the permission of AIP Publishing. Figure on the right from [172].

written squares). This alignment proved valid enough to effectively interface other machines as e-beam lithography (Fig. 12.15(left) [99]) or 3D printers (Fig. 12.15(right) [170, 172]), which can utilize the markers at room temperature to align on the preselected emitter.

12.4 Summary

- MOVPE and MBE represent valuable tools for the growth of semiconductor devices and nanostructures. Whereas the former allows for fast growth of high quality crystal (usable for example for DBR), the latter controls precisely the growth conditions, enabling highly pure environment for nanostructures.
- Strain plays a crucial role during growth; in particular, its engineering can be employed for depositing quantum dots.
- Resists can be employed to lithographically define patterns (via optical or e-beam lithography) on top of solid-state materials. These patterns can be then transferred in the material via etching.
- Deposition of oxide or metal layers can be extremely useful as a hard mask for sample surface protection, electrical isolation, or electrical contacting.
- Dry and wet chemical processes represent valid approaches for the etching of solid-state materials.
- Deterministic lithography techniques can be employed to tailor the fabrication of photonic structures around preselected quantum emitters. Thanks to the preselection process, the position and emission spectrum of the quantum emitter are characterized, allowing for the spatial and spectral matching of the photonic structure fabricated afterward.

- Deterministic low-temperature optical lithography has been further interfaced with other fabrication machines, demonstrating also its use for a broad range of wavelengths (from visible to telecom) and for electrically driven quantum light emitters.
- Deterministic fabrication techniques allow for localizing the position of quantum emitters with an accuracy of a few nanometers, together with an accuracy in the deterministic fabrication process of the device of a few tens of nanometers.

13 Semiconductor cavities and light extraction methods

13.1 Introductory remarks

Semiconductor nanostructures represent one of the most appealing platforms for the realization of high-performance sources of quantum light. Very high single-photon purity, near-unity photon indistinguishability, and highly entangled photon pair emission have been proven by several distinct studies, highlighting the maturity of the field. Together with the aforementioned properties, solid-state emitters reached unprecedented levels of brightness in regard to the emission of quantum light (see Section 5.2.4). Interestingly, after their deposition (see Chapter 12), these nanostructures are found embedded into a semiconductor material with a refractive index typically higher than the surrounding air or vacuum. As an example, indium arsenide (InAs) quantum dots in gallium arsenide (GaAs) find themselves embedded into a material with a refractive index of $n_{\mathrm{GaAs}} \approx 3.5$ over the surrounding $n_{\mathrm{air}} \approx 1$ (taking values at room temperature and for light of wavelength $\lambda \approx 900$ nm). Total internal reflection (TIR) at the semiconductor/air interface therefore limits the amount of light that can be extracted to a few percent of the total light emitted, severely limiting the maximum achievable source brightness (see Fig. 13.1). On the other hand, operating with semiconductor nanostructures is proven to be a valid approach to overcome this intrinsic limitation. Decades of fundamental research and industrial implementations of semiconductor devices provide a very large spectrum of micro- and nanofabrication techniques, which can be employed in combination with solid-state emitters (see Chapter 12). This means that semiconductor quantum dots can be employed in combination with optical resonators or photonic structures for overcoming the limit set by total internal reflection, drastically improving the achievable source brightness.

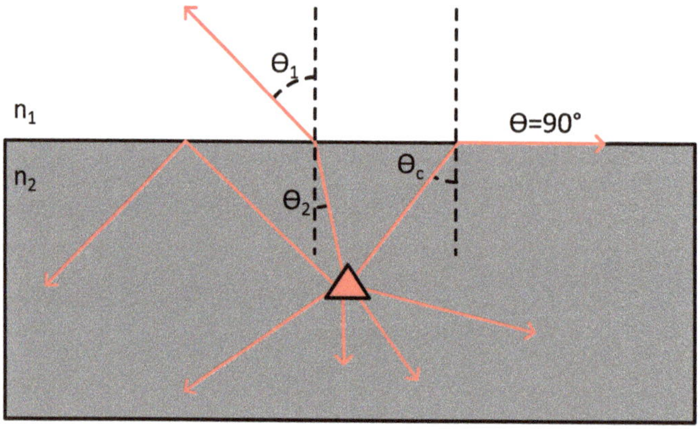

Figure 13.1: Total internal reflection (TIR) allows only the extraction of a few percent of the emitted light.

https://doi.org/10.1515/9783110703412-013

As it was preliminary discussed in Section 5.2.4, two approaches can be followed for reaching high photon extraction efficiencies: first, the shape of emitter's surrounding material can be modified to reduce the number of photons experiencing total internal reflection (geometrical approach). Second, the emitter can be embedded into photonic resonators to enhance the extraction via cavity quantum electrodynamics. These two approaches will be described in the following sections.

In both cases, following the line of arguments as in Eq. (5.10), a bright source does not only require that the light is efficiently extracted from the host material, but also that the out-coupling profile should have a shape that can be easily collected by the employed optics (i. e., the term η_{out} in the cited equation). To understand this statement, we can compare the two calculated far-field patterns shown in Fig. 13.2.

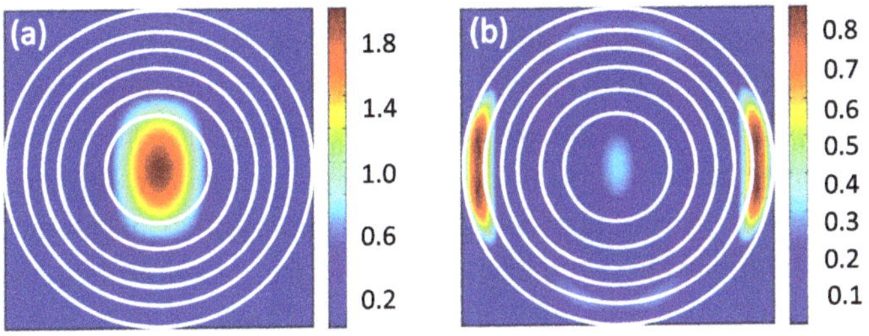

Figure 13.2: Exemplarily calculated far-field patterns (as the profile of the electric field intensity) for cavities with (a) and without (b) far-field optimization. The white circles represent angles of 20°, 30°, 40°, 50°, 60°, and 90°. The investigated structures are *L*3 photonic crystal cavities. Reprinted with permission from [152]. Copyright 2023 the Optical Society.

The pattern in Fig. 13.2(a) has a Gaussian-like profile, centered around small vertical angles, and hence it can be fully collected with relatively low numerical aperture optics. Conversely, the pattern in Fig. 13.2(b) shows emission at very large angles, making it experimentally difficult to collect. Furthermore, having a Gaussian-shaped far-field becomes highly advantageous to ensure good overlap with the mode sustained by single-mode fibers, and hence high fiber in-coupling can be achieved (see the term η_{FC} in Eq. (5.12)). This argument is also valid for the opposite case, i. e., when light needs to be efficiently coupled into the cavity: once again, an incoming Gaussian far-field profile (as that obtained from the collimation of light leaving a single-mode fiber) would allow efficient in-coupling of light spectrally resonant to the cavity mode (conversely, coupling to very wide *k*-vector components can be experimentally challenging). An easy access to the cavity mode would allow reducing the in-coupling power necessary to reach the enhanced light-matter interaction enabled by the resonator [68].

13.2 Geometrical effects

As schematically depicted in Fig. 13.1, a solid-state nanostructure could emit light in the full solid angle, but considering the refractive index contrast between the host semiconductor and the surrounding (air or vacuum), only few directions lie below the total internal reflection angle: this means that the large majority of photons will be reflected at the interface and will not leave the embedding material (~1–2 % can be collected by a microscope objective). The first approach to overcome this limitation is based on the modification of the TIR conditions: the host material surface is then shaped in a way that the emitted light will not impinge at the critical angle anymore. In addition, as explained in Eq. (5.9), the reduction of the decay rate Γ over all modes other than the selected mode employed for the light extraction can provide values of β close to one, an important step toward high source brightness (see Section 14.4.3). The main advantage of such geometric approaches relies on their broadband operation, since no optical resonances are employed (and only the dependence of the refractive index from the wavelength comes into play). High photon extraction can therefore be achieved on wavelength ranges much larger than the typical spectral separation between emission features of solid-state emitters. As an example, these geometric approaches can enhance the extraction efficiency of both biexciton and exciton photons from a semiconductor quantum dot (typically spectrally spaced by a few nanometers), allowing for use in entangled photon pair sources.

Various approaches have been followed to limit the impact of TIR, from the use of microlenses (Figs. 13.3(a) [62, 169] and (b) [170, 172]) to nanowires (Fig. 13.3(c) [25]) and trumpets (Fig. 13.3(d) [188]). Interestingly, shaping the host material does not only improve the extraction but can be also employed to modify the source emission profile.

One example of a modified surface geometry is based on the use of microlenses etched in the host material. These structures can be fabricated in various ways (see Chapter 12), for example, via optical lithography and wet chemical etching [169], enabling the control of the structure geometry together with achieving high surface quality (i. e., low surface roughness). Etching these microlenses directly in the host semiconductor material is not the only valid approach for modifying the emission pattern: 3D printing [170, 172] or alternatively transferring a thin semiconductor membrane embedding the emitters on the bottom of a macroscopic lens [24, 230] have been demonstrated as valid approaches to improve the light extraction and/or altering the emission far-field. At this point, it is necessary to remind that TIR conditions are mainly set by the refractive index mismatch between host and surrounding material: this means that employing an "externally fabricated" lens from a different material (as in Fig. 13.3(b), as opposed to the etched lens in (a)), should take into account the refractive indexes in play to achieve the best performances.

More complex structures, like nanowires and photonic trumpets, have been also subject of investigation, as they can guide the emitted light into a chosen direction, simultaneously modifying the light emitted pattern [25, 188].

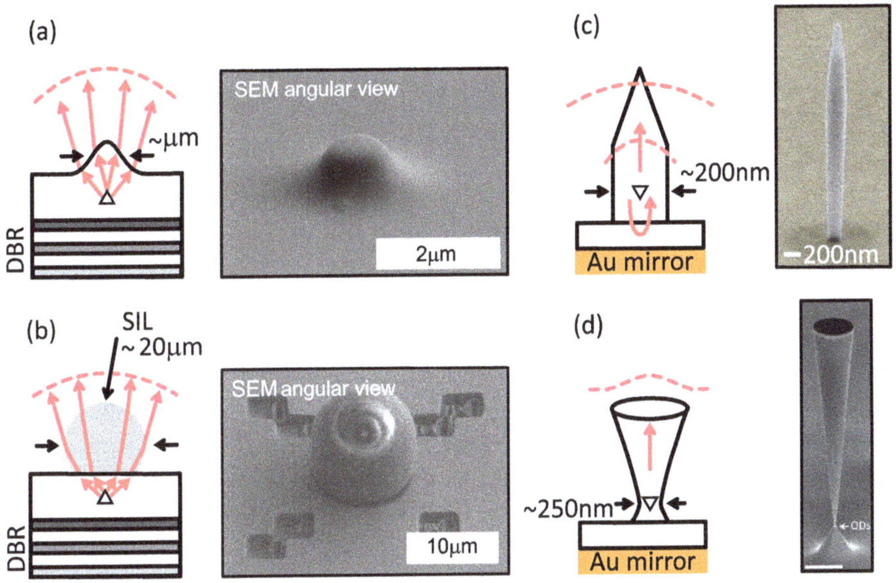

	(a)	(b)	(c)	(d)
Typical out-coupling eff.	~20%	~10%	~70%	~70%

Figure 13.3: Exemplary structures for enhancing light extraction via geometrical effects. (a) Gaussian-shaped microlens etched into the semiconductor (i. e., made of the same host material surrounding the emitter), as experimentally demonstrated for NIR quantum dots in [62] and for wet-chemically etched structures and telecom quantum dots in [169] (SEM picture reprinted from [169] with the permission of AIP Publishing). (b) Exemplary 3D-printed Weierstrass lens on top of a NIR quantum dot as in [172] (SEM picture from [172]). (c) Nanowire design as in [25] (SEM picture reprinted with permission from [46]. Copyright (2023) American Chemical Society) and (d) trumpet as in [188] (SEM picture reprinted from [188] with the permission of AIP Publishing. Scale bar represents 5 µm). Typically, a bottom reflective mirror (either semiconductor DBR or metallic) is employed to direct the photons emitted in the downward direction toward the top. Schematic wavefronts are also shown. Typical out-coupling efficiencies are listed in the table below. For external lenses as in (b), the extraction efficiency can be further improved employing materials with a refractive index closer to the host material (i. e., GaP could bring even more than 20% values.)

All the described approaches aimed at enhancing the amount of light that can be extracted from the host material, furthermore sending the photons into a specific direction (typically orthogonal to the sample surface) and having a suitable emission far-field (usually enabling the collection within the numerical aperture of the employed optics or reaching maximal coupling into a single-mode fiber). Nevertheless, controlling the direction of light emission is not only interesting for *off-chip* implementations, but also for all the realizations where the quantum light is guided into photonic structures directly embedded *on-chip*. Classical and quantum photonic integrated circuits (PIC) aim at realizing *on-chip* the optics functionalities and elements often performed off-chip with bulk optics: beamsplitters, phase shifters, photon detectors, and waveguides are only few examples of optical elements that benefit from a precise control of the emission direction of light from semiconductor emitters. A more complete description of these structures can be found in [76] and Section 14.4.

13.3 Cavity quantum electrodynamics: resonators and photonic crystals

The second approach for enhancing the light extraction of a semiconductor nanostructure from the host material is based on the exploitation of cavity quantum electrodynamics effects. As it has been discussed in Chapters 5 and 6, a two-level system (TLS) placed into an optical resonator (operating in the *weak coupling* regime) allows for increasing the probability of emitting light in the cavity mode over the other available decay channels. Purcell enhancement of the emission rate into the cavity Γ_c over the rate in bulk Γ_{bulk} enters in Eq. (5.7):

$$\beta = \frac{F_P}{F_P + \Gamma/\Gamma_{\text{bulk}}}, \tag{13.1}$$

where F_P is the Purcell factor defined in Eqs. (5.7) and (6.67) (and β as the fraction of the emission in the cavity mode versus all other modes), and Γ is the decay rate in modes other than the cavity. From the general expression in Eq. (6.68), the TLS has to spectrally and spatially match the cavity mode, having also its dipole oriented along the cavity field polarization. In this case the Purcell factor takes the form of Eq. (6.67):

$$F_P = \frac{\Gamma_c}{\Gamma_{\text{bulk}}} = \frac{3}{4\pi^2}\left(\frac{\lambda}{n}\right)^3\left(\frac{Q}{V_{\text{eff}}}\right) \tag{13.2}$$

with λ as the wavelength in free space, and n is the refractive index of the material. The cavity quality factor Q has been defined in terms of dissipation (see Section 6.4) as $Q = \omega_c/\kappa$, where ω_c is the frequency of the cavity mode, and κ the overall cavity decay. This leads to interpreting the factor Q as one of the following [89]:

- a dimensionless decay, where having the outgoing power P from the cavity and the electromagnetic energy U localized inside the resonator, we have $1/Q = P/\omega_c U$;
- a dimensionless lifetime, representing the time that the photon spends in the cavity or, in other words, the number of round-trips before the energy decays by a factor of $e^{-2\pi}$;
- the bandwidth of the cavity resonance having $1/Q$ proportional to the full width at half-maximum of the Lorentzian cavity resonance ($Q = \omega_c/\Delta\omega_c$).

The effective mode volume V_{eff} provides the measure of the spatial confinement of the electromagnetic field inside the cavity [89, 102, 204]:

$$V_{\mathrm{eff}} = \frac{\int P(\vec{r})d^3\vec{r}}{P_{\mathrm{max}}} = \frac{\int \varepsilon(\vec{r})|\vec{E}(\vec{r})|^2 d^3\vec{r}}{\max[\varepsilon(\vec{r})|\vec{E}(\vec{r})|^2]}. \tag{13.3}$$

From Eq. (13.2) it is evident that a high Purcell factor can be achieved for optical resonators having a large ratio Q/V. The search for strong Purcell enhancement and increased light-matter interaction has driven the efforts in designing and realizing optical resonators with high quality factors and small mode volumes. The first step is based on the theoretical design of the intended resonators, aiming for a certain mode volume and quality factor Q_{theor}. During the realization phase, losses due to fabrication imperfections can come into play, resulting in a quality factor defined to a first approximation as [55]

$$\frac{1}{Q_{\mathrm{exp}}} = \frac{1}{Q_{\mathrm{theor}}} + \frac{1}{Q_{\mathrm{loss}}}, \tag{13.4}$$

where any deviation of the experimentally observed quality factor from the theoretically achievable one is attributed to loss terms. Light absorption from the material typically being negligible, since the material is chosen to be transparent for the intended operation wavelength, losses are normally induced by fabrication imperfections, which add unwanted scattering channels allowing the light to escape from the cavity (see the definition of cavity decay rate κ in Section 6.4), or by a low-quality material itself. Together with the optimal cavity design, achieving high Purcell enhancement requires the TLS to spectrally and spatially match the cavity mode, having also its dipole oriented along the cavity field polarization (as in Eq. (6.68)). This can be ensured by employing deterministic fabrication techniques as described in Section 12.3.5.

Examples of semiconductor microcavities

In this section, we report a short overview of the most utilized types of cavities realized in semiconductor materials. We also provide a brief discussion on cavity resonance tuning mechanisms and fabrication tolerances.[1] The various resonators can be categorized in three subgroups [204]:

- *Photonic crystal-based cavities and circular Bragg grating resonators (Fig.* 13.4*(left))*: these systems are based on the formation of a defect in the otherwise perfectly periodic photonic crystal. As explained in the following, the light can exist only in the defect region, whereas the photonic bandgap prohibits the propagation within the PhC [101] (together with TIR for photonic crystal slabs). These cavities are widely

1 Optical resonators have been and still are a subject of intense research. Even restricting to only semiconductor-based systems, a very large number of studies pushed forward the development of the field, and several papers represent milestones in their respective field (i. e., silicon or III–V materials, laser-based cavities, optical filters, and modulators). Here the authors decided to report only a few examples for the readers to understand the topics at hand but still feel to acknowledge the numerous groups who played a key role in all these topics.

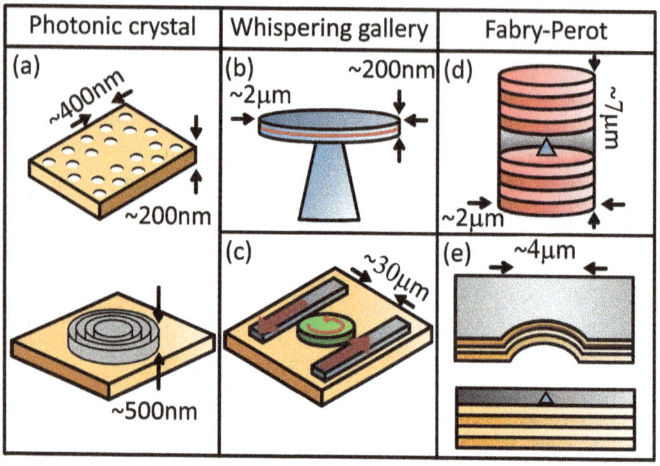

Figure 13.4: Exemplary photonic resonators geometries and respective footprint (actual values can vary depending on the employed material and operation wavelength). (left, top) Photonic crystal slab *L*3 cavity and (bottom) circular Bragg grating (CBG) cavity. (center) Whispering gallery cavities: (b) a microdisk on top of a pedestal and (c) a microdisk implemented as add-drop filter (microring geometries can also be employed). (right) Fabry–Perot resonators: (d) a monolithic micropillar cavity (with an exemplary QD as the sketched triangle) and (e) a tunable open Fabry–Perot cavity type. For all structures, exemplary quality factors Q of the order of 10^5–10^6 and mode volumes V_{eff} of the order of $(\lambda/n)^3 - 10(\lambda/n)^3$ have been achieved. Exemplary (d) has a mode volume of the order of $5(\lambda/n)^3$, whereas an *L*3 PhC has a cavity of the order of $(\lambda/n)^3$. As reference, out-coupling efficiencies have achieved the order of 80% ((a), CBG) and 70% ((d), micropillar).

used and have been the subject of intense research in the last decades. Various materials have been employed, i. e., GaAs, silicon, etc., and strong efforts have been made to understand and tailor the light–matter interaction in these systems. Despite the very high performances reached when utilized as single-photon sources (see, for example, [115]), photonic crystals are sensitive to imperfections in their implementation (various studies tried to understand the impact of disorder on PhC cavities; see, for example, [151]) and require high-quality fabrication methods. Furthermore, upscaling the experimental complexity may require multiple sources emitting at the exact same wavelength, which is challenging to implement using photonic crystal cavities since the wavelength of their narrow-band (fundamental) mode is highly sensitive with respect to deviations of the PhC geometry, deviations that are typically below nanofabrication tolerances. Various methods have been employed for achieving the tuning of PhC cavity modes, such as local nanooxidation [73], photochromic tuning [23], gas condensation [130], and advanced nanofabrication techniques, realizing photonic crystal membranes, which can be mechanically reconfigured [149]. Nevertheless, the choice of the tuning method should be done taking into account the operation conditions (as in room or cryogenic temperature) and the potential impact of the tuning mechanisms on the cavity properties (as, for example, on the quality factor). It is worth mentioning a resonator design called circular Bragg grating (CBG), where circular tranches are etched in the semiconductor (see Section 14.2.2) [33]. This kind of photonic cavity is technologically very appealing for the implementation of bright sources of quantum light. As discussed in Section 14.2, their low quality factor and small mode volume allow for the formation of a broad mode and a high Purcell factor, which enables the simultaneous enhancement of multiple, spectrally separated transitions. Together with the broadband increased extraction efficiency, CBGs have been used for the implementation of bright sources of single and polarization-entangled photon pairs [116, 136, 167, 218].

- *Whispering gallery cavities (Fig.* 13.4*(center))*: these structures are based on continuous total internal reflection of the light, which results to be guided around the circular resonator. They can be realized on pedestals, forming the so-called microdisks (Fig. 13.4(b)) [54] or microrings etched in semiconductor material (Fig. 13.4(c)) [204]. Also for these resonators, various approaches have been followed to tune their resonance to the desired one, for example, temperature tuning (varying the ring/disk refractive index) [158] or mechanical strain [233]. These structures find a variety of applications, as filters and switches, and have been employed in nonclassical photon sources as well [123]. Interestingly, it has been proven that surface treatments can sensibly enhance the achievable quality factor for GaAs-based microrings [63], showing once again the necessity of high-quality fabrication processes to achieve high-performance optical microcavities.
- *Fabry–Perot cavities (Fig.* 13.4*(right))*: as introduced in Section 5.2.2, optical resonators as well can be fabricated following a Fabry–Perot approach, i. e., using two mirrors facing each other. One widely used example is the micropillar cavity (Fig. 13.4(d)): there two mirrors are formed via two 1D photonic crystals (two distributed Bragg reflectors), whereas the etched sidewall surfaces ensure lateral optical confinement. This structure therefore allows for a 3D confinement of the light [56]. As discussed in the following, optimization of the cavity quality factor has also been shown. Interestingly, the light that leaves the cavity, propagating from the fundamental mode to the far-field, has a profile very close to a Gaussian one: this makes it very appealing for achieving high out-coupling from the cavity and in-coupling into single-mode fibers (and vice versa, good in-coupling in the cavity itself, considering an incoming Gaussian field profile). More advanced micropillar designs enabled the use of diode structures to stabilize and control the emission properties of the embedded emitters (see Section 9.4 and [186]). Another example of Fabry–Perot-based resonators is given by open tunable cavities (Fig. 13.4(e)): in most of the systems operating with semiconductor quantum dots, the cavity design includes a bottom semiconductor DBR mirror (on top of which the QDs are grown), combined with an external mirror (typically dielectric DBR, deposited on glass substrates or fibers [78]). These cavities are intrinsically tunable, since the distance between the two mirrors can be adjusted to spectrally match the investigated emitter's transition, as well as the lateral position, which enables

forming the cavity at controlled coordinates (therefore on each emitter present on the sample). Recent results showed impressive performances when the systems were driven in the strong coupling [135] or in the weak coupling regime [196]. In the latter case, the end-to-end efficiency of 53 % ± 3 % has been demonstrated.[2] Interestingly, the fundamental cavity mode has also a Gaussian profile, making it suitable for efficient fiber in-coupling. Recently, a similar design has been achieved in a full-semiconductor monolithic structure [43]: this approach allows for creating resonators with similar properties to micropillars but without the presence of lateral sidewalls for the 3D light confinement (hence avoiding the potential presence of surface defects). In principle, these resonators should be fully compatible with strain tuning, whereas micropillars face more challenges to be compatible with this tuning mechanism [129].

Basics of photonic crystals

In this section, we introduce the basic concepts of photonic crystals (PhC), since several optical resonators make use of these structures in their design and realization. For a more detailed reading, we refer to [89]. In simple terms, a photonic crystal is a material in which the dielectric constant is periodic in one (1D, Fig. 13.5(a)), two (2D, Fig. 13.5(b)), or three directions (3D, Fig. 13.5(c)). This periodicity plays a crucial role in the light propagation as described below. Furthermore, a widely used structure is represented by the photonic crystal slab, as depicted in Fig. 13.5(d).

For simplicity, let us consider here the case of a 1D crystal constituted by a periodic (infinite) stacking of alternating materials. The light propagation is now assumed along the direction of periodicity of the crystal itself.[3] In the case of a uniform material (i. e..

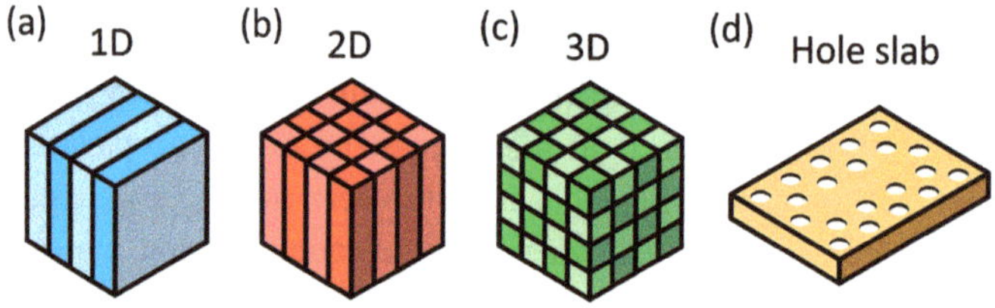

Figure 13.5: Exemplary sketch of (a) 1D, (b) 2D, and (c) 3D photonic crystals with the respective periodicities of refractive index. (d) A photonic crystal slab is also depicted as a triangular array of holes etched into the slab of host material.

2 The concept of end-to-end efficiency takes into account all experimental elements from the source to setup and detection. It can be compared as in Eq. (5.13) per excitation pulse (i. e., without the excitation rate term).

3 In case of off-axis propagation the light perceives a lower effective periodicity. In the simple description here introduced, only on-axis propagation is considered, since it constitutes one of the most interesting case studies. A full description can be found in [89].

a block of GaAs) with refractive index $n = \sqrt{\varepsilon}$, the light propagation is described by

$$\omega(k) = \frac{ck}{\sqrt{\varepsilon}}, \tag{13.5}$$

where ω is the frequency of the radiation, c is the speed of light in vacuum, and k is the corresponding wave vector. In strict similarity to the "empty lattice" model used in electronic crystals, the material being uniform, it is possible to assign it an arbitrary periodicity a, so that the material can be seen as in Fig. 13.6(a), having alternating layers with the same refractive index. The periodicity present in a real space reflects then in the reciprocal space.

The light line describing the propagation of the photons (Eq. (13.5)) folds in the so-called first Brillouin zone (BZ) defined by wave vectors between $[\pi/a, -\pi/a]$ (see Fig. 13.6(a)). The light propagation behavior changes sensibly when two materials with different refractive indexes are periodically stacked: at the edges of the BZ, there appears a frequency region where no modes exist independently of the wave vector (Fig. 13.6(b)). This region is known as the *photonic bandgap* or *photonic stop band*: the origin of the name lays on the fact that within this region, light cannot exist, and photon

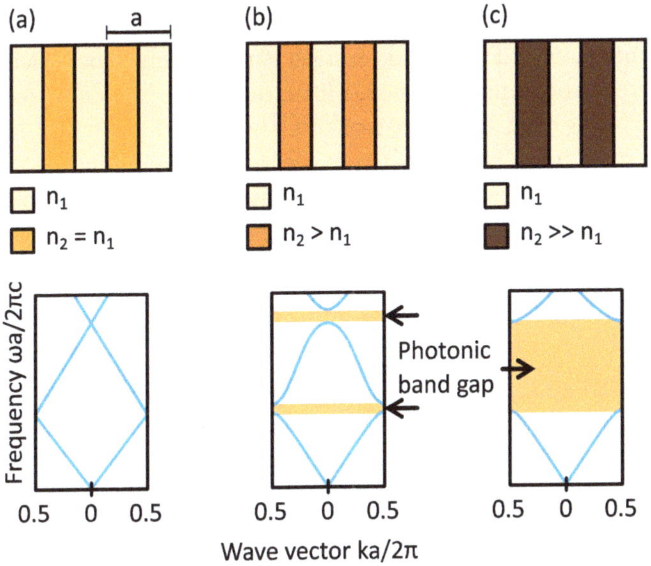

Figure 13.6: Schematic of 1D photonic crystal and respective band structure. (a) Case of the same material with assigned arbitrary periodicity (refractive index is the same for all repeating layers). Each layer thickness is equal to 0.5*a* where *a* is the lattice period. The respective band structure (bottom) can be folded in the first Brillouin zone. (b) Case study with periodic alternation of layers with different (but still close) refractive indexes. In the band structure (bottom), a photonic bandgap appears. An example of this structure can be GaAs/GaAlAs multilayer [89]. (c) Case study with periodic alternation of layers with different refractive indexes (i. e., GaAs/air). The bandgap in the band structure (bottom) increases.

propagation is forbidden. Figure 13.6(c) shows the enlarging of the photonic bandgap with the increase of the refractive index contrast. For photons with frequency within the photonic stop band, the crystal acts as a perfect mirror. This property makes the photonic crystal interesting and useful within the semiconductor platform. Indeed, there are several ways to implement materials with periodic refractive indexes and use them as high-quality mirrors for photonic resonators (see Chapter 12). As a few examples, the distributed Bragg reflector (DBR, as in Figs. 13.6(b) and (c)) constitutes a case of 1D photonic crystal. Woodpile [113], Yablonovite [229], Opals and inverse Opals [210] represent cases of 3D photonic crystals, for which a complete bandgap in all three dimensions can be expected. These photonic crystals and their ability of controlling the propagation of light in three directions make them appealing, although very challenging to fabricate. For this reason, other structures are experimentally employed as the photonic crystal slab (see Fig. 13.5(d)). This structure combines i) the photonic bandgap acting in two directions (thanks to a 2D periodicity of the refractive index) and ii) the total internal reflection used to confine the light in the third dimension (i. e., the one perpendicular to the periodicity plane of the PhC). For its implementation, the photonic crystal is fabricated into a *slab* of material that has a refractive index higher than the surrounding *cladding* (i. e., GaAs in air, silicon in air, or silica).[4] Because of the presence of total internal reflection, there exists an ensemble of states that are not confined in the crystal (since they violate the TIR conditions): for this reason, the so-called light-cone appears, depicted as azure color in Fig. 13.7 (for detailed band-structure calculations and description of the geometrical parameters, see [89]). Guided modes are present below the light line, and taking as an exemplary structure a hole slab (i. e., a triangular lattice of air holes fabricated into a slab of material), we can observe a bandgap for TE-like modes. Despite the bandgap being *incomplete*, because it refers only to guided modes in the slab and not above the light line, these structures allow for realizing a 3D confinement (thanks to the 2D photonic bandgap and the TIR in the perpendicular directions) and can be reliably fabricated in various solid-state materials.

Defects in photonic crystals: optical microcavities and quality factor optimization

Thanks to the previous discussion, it is now clear that photonic crystals can be employed to realize high-reflectivity mirrors. Interestingly, a local break of the periodicity results in the formation of *allowed states* in the photonic bandgap: this means that light with frequency matching the allowed state in the stop band can exist only in the defect region, whereas the surrounding photonic bandgap prevents its propagation in the rest

4 When designing such a photonic crystal, the slab thickness becomes an important parameter. Whereas for too thin slabs, the modes are weakly localized making the periodicity irrelevant, too thick slabs approach the case of a 2D PhC. In this case the formation of new states closes the bandgap. Therefore the best compromise is given by a slab thickness close to half of the wavelength in the dielectric. Further details can be found in [89].

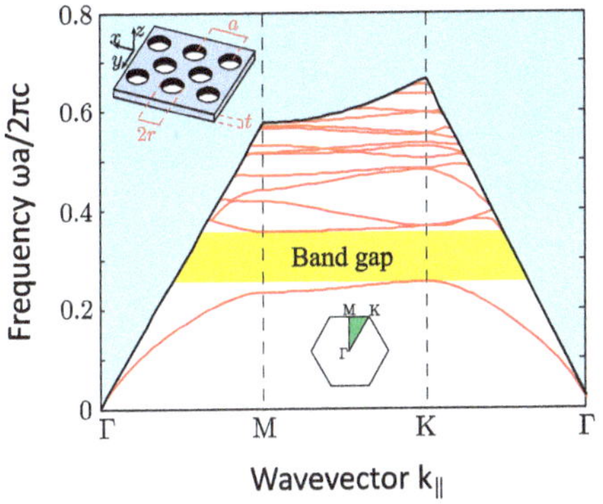

Figure 13.7: Band diagram. The wave vector $k_{||}$ lies in the plane of the PhC. Azure-colored area represents the light cone, whereas the lines below depict the guided modes (for TE-like modes). The modes above the light cone are extended in the surrounding material, and the red line accounts for modes in the slab. A bandgap for TE-like modes is observed [89]. Calculations have been performed for a triangular lattice as in the inset made of GaAs (the refractive index 3.5, the hole radius per lattice constant $r/a = 1/3$, and the thickness of the slab $t = 2a/3$). Reprinted with permission from [117]. Copyright (2023) by the American Physical Society.

of the PhC, where the periodicity is intact. The formation of a *point defect* in the periodicity can then be employed to design and implement optical resonators. As a case study, let us consider one of the most popular photonic crystal slab resonators, the $L3$ cavity. It is realized by three missing holes (hence the number 3) along a line (therefore the letter L) in the ΓK direction for a triangular lattice of holes (see Fig. 13.8(a)). This break of the PhC periodicity creates allowed states in the defect region. It is important to mention that since the introduced defect is spatially localized, each allowed mode at a certain frequency $\omega(k)$ will also have wave vector components lying above the light line: these k-vectors are coupled to the extended modes in air (i. e., in the low-index material surrounding the slab) and can leak in the free-space continuum.[5] This means that even without material absorption or fabrication imperfections, the lifetime of the photons inside the cavity (i. e., the cavity quality factor, as discussed earlier) is limited by this loss of light in the out-of-plane (vertical) direction. In more detail, for a cavity size comparably close to the radiation wavelength, the light confined in the cavity region can be seen as a combination of plane waves with various k-vector components for which the confinement conditions must hold. Conversely, all k-vector components lying inside the leaky region (i. e., where the in-plane wave vector violates TIR conditions) are not confined and escape from the cavity. The simple model introduced in [2] allows us to understand that the presence of these components within the leaky region can be seen as due to an abrupt change of the electric field at the edges of the cavity: with this simple

5 Conversely, guided modes outside the bandgap cannot couple to extended states and are infinitely extended in the slab (they decay exponentially outside the cladding).

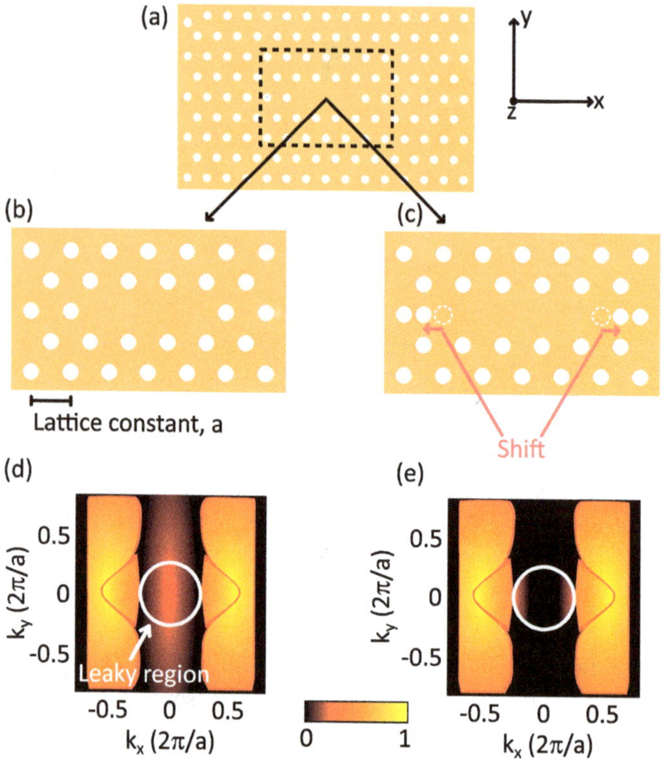

Figure 13.8: (a) Schematic of a photonic crystal slab, as a triangular lattice of holes, where an *L*3 cavity has been formed. (b) View of the central defect region, highlighting the PhC lattice constant *a*. All holes are periodically spaced. (c) Local geometry optimization performed via the shift of the two nearby holes forming the *L*3 cavity (original positions are marked as dotted circles). (d) and (e) Exemplary sketches of the Fourier transform of the typical electric field profile (E_y) inside the cavities in (b) and (c), respectively. A reduced number of *k*-vector components would be found in the leaky region. Reprinted and adapted with permission of Springer Nature from [2].

argument, the field in the cavity is seen as the product of a sinusoidal wave and an envelope function. For standard *L*3 cavities, the photonic crystal bandgap at the edges of the cavity creates an abrupt change of the envelope function, resulting in the appearance of wave vector components inside the leaky region. This can be avoided finding a more "gentle" envelope function instead.

In [2], this has been proposed and demonstrated via the optimization of the two nearby holes delimiting the cavity (see Fig. 13.8) by shifting them from their original position (at a distance *a* set by the lattice constant) to the design position or by changing their size with respect to the rest of the PhC holes (or both; see [55]). This break of the PhC symmetry results in a decrease of the reflectivity at the cavity edges and therefore in a larger penetration of the electric field within the photonic crystal lattice: whereas the electric field profile inside the cavity does not alter sensibly, its Fourier transformation in case of a locally optimized cavity shows that very little k-vector components are now present within the leaky region. This proved experimentally that the cavity Q-factor can be sensibly increased from that observed with a standard cavity design (for example, for $L3$ cavities, Q-factors from a few thousands were increased up to hundreds of thousands.). The milestone work of [2] stimulated the research field in the search for resonators with increased ratio Q/V, reaching extraordinary high quality factor values even for relatively small mode volumes (theoretically described for different cavity types in [128] and realized in cavities where more than only the two neighboring holes have been optimized [107]). Similar schemes have also been applied to other resonators as micropillar cavities [111], where Bloch-wave engineering has been employed to sensibly enhance the Q-factor.[6]

Defects in photonic crystals: optical microcavities and far-field optimization

In the previous section, we discussed the strategy for the improvement of the cavity quality factor. Nonetheless, when it comes to designing a resonator for the implementation of bright light sources, the far-field pattern and light extraction from the cavity also play a crucial role. After the optimization of the cavity Q-factor (performed via both shifting and shrinking the two holes forming the cavity, as in Fig. 13.9(a)), most of the k-vector components are found at the edge of the Brillouin zone, whereas very few lie above the light line (i. e., in the leaky region; see schematic in Fig. 13.9(c) and near-field in (e)). An idea was proposed and experimentally demonstrated [152, 197] based on the folding of some wave vector components above the light line and around the vertical direction (i. e., around $k_{||}$ equal to zero). This can be achieved by superimposing a PhC lattice variation with periodicity doubled with respect to that of the original lattice (with lattice period a and therefore periodicity $2a$; see Figs. 13.9(b) and 13.10(a)). This results in the folding of k-vector components around the point $k = \pi/2a$, and hence from the edge of the BZ to k equal to zero (as schematized in Fig. 13.9(d)). The simulated near-field distribution after far-field optimization shows indeed the wave vector components in the leaky region around $k = 0$ as per design (Fig. 13.9(f)). This far-field optimization creates leak channels in the vertical direction: this therefore increases the cavity out-coupling as shown in Fig. 13.10(b) (and the in-coupling as well if light from the outside is sent on the resonator) and renders the far-field narrower in the vertical direction (mostly Gaussian within small angles). This represents a clear improvement for a light source in comparison with $L3$ cavities with only Q-factor optimization, where the limited photon

6 The discussion within this section follows for the large majority the arguments of [2], whereas it can also be interpreted via matching the Bloch mode profile (see [173]). It is worth mentioning that the physical mechanisms behind the light confinement in cavities stimulated a lively discussion, and the work of P. Lalanne and coworkers helped shading light on this topic, further allowing for the optimization of optical resonators. For this reason, we would like to give credit to all scientists who helped the development and understanding of this branch of photonic crystal cavities.

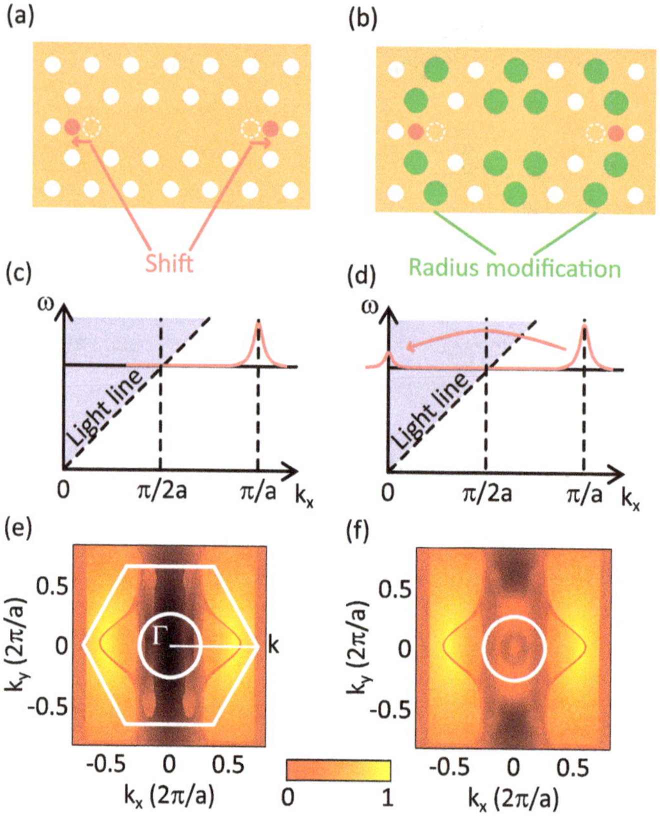

Figure 13.9: (a) $L3$ photonic crystal cavity with Q-factor optimization via shift (and/or additional shrinkage) of the two nearby holes forming the cavity. (b) Sketch of a resonator with both quality factor and far-field optimization. A second periodicity pattern with lattice constant $2a$ has been superimposed to the PhC lattice (with lattice constant a), depicted with green holes. (c) Schematic of the band structure after Q-factor optimization: most of the k-vectors lie at the edge of the BZ. (d) Folding of some k-vector components around the point $\pi/2a$, resulting in wave vectors at around zero (i. e., emission in the vertical direction). (e) and (f) Exemplary sketch of the near field distributions of the fields for cavities in (a) and (b). After far-field optimization, more wave vectors are found around the Γ point (i. e., vertical direction). Edges of the BZ and the light cones are depicted in white. Reprinted and adapted with permission from [197]. Copyright (2023) by the American Physical Society. Design in (b) follows [152].

loss happens on very large angles, which are difficult to collect experimentally (compare Fig. 13.10(c), 4 and 1). Having a Gaussian far-field is also advantageous for achieving high coupling to single-mode fibers, since Gaussian modes are here required. Nonetheless, we must take into account that adding decay channels to the cavity lowers the quality factors (as shown in Fig. 13.10(b)). This means that a trade-off has to be found between outcoupling and quality factor, depending on the intended utilization of the device. More details on the cavity design, simulations, and implementations for $L3$, $L5$, and $L7$ cavities can be found in [152].

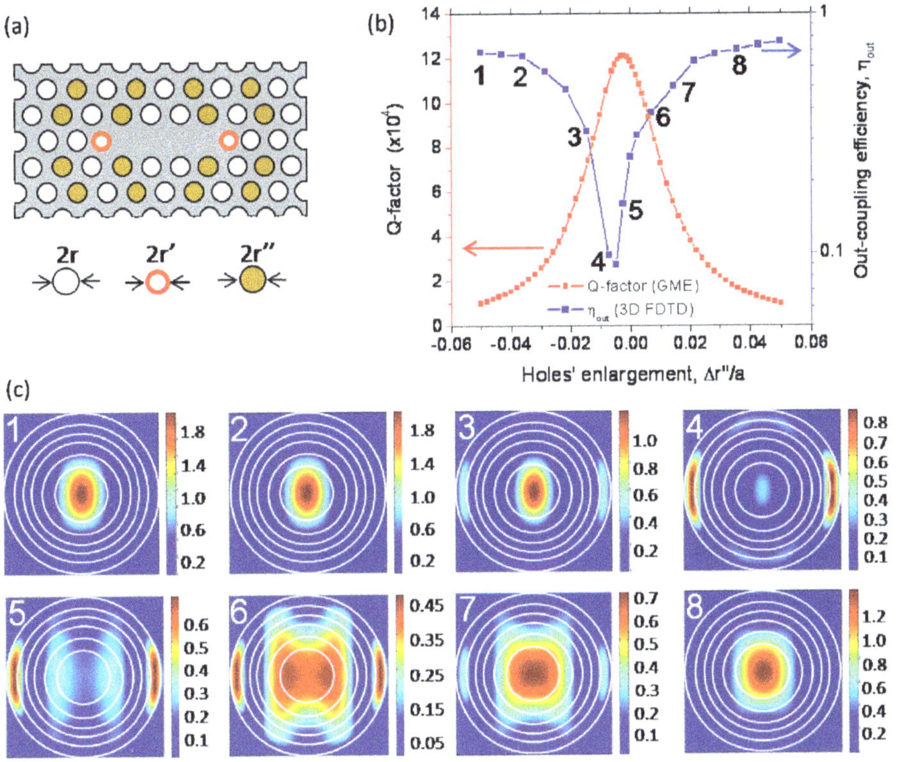

Figure 13.10: (a) Schematic of an $L3$ cavity with Q-factor and far-field optimization. White holes are the unmodified ones, red circled holes are shifted and shrunk for Q-factor optimization, yellow-highlighted are modified in diameter for far-field modification. (b) Behavior of the Q-factor and coupling efficiency over the hole enlargement (positive values of $\Delta r''/a$) or reduction (negative, respectively). PhC parameters can be found in [152]. Increasing the out-coupling efficiency via far-field optimization also results in a Q-factor reduction. (c) Calculated far-field patterns corresponding to the numbers in (b). Concentric circles represent angles of $\theta = 20°, 30°, 40°, 50°, 60°, 90°$. Intensities are normalized over the total power emitted in the vertical half-space. Reprinted with permission from [152]. Copyright 2023 the Optical Society.

13.4 Characterizing photonic cavities: micro-PL, reflectivity, resonant scattering

Several different techniques have been developed to characterize the cavity properties discussed in the previous section. Choosing which experimental approach to follow depends on the specific resonator under investigation. If the photonic cavity is realized with optically active materials (that emit at the frequencies of the cavity resonances), microphotoluminescence can be directly employed (see Section 5.2.1). After having excited the optically active material (i. e., QDs present in the material), its emission can feed the cavity mode making it visible directly on the spectrometer. From the observed mode resonance it is possible to deduct its wavelength and quality factor (as defined in

Section 13.3). This is easy to understand and interpret when the material emission can be assimilated to a white source of light, i. e., its emission intensity is the same for all wavelengths across the cavity mode. When this is not the case (for example, when considering an inhomogeneous emission distribution of an ensemble of QDs peaked at a different wavelength with respect to the mode), care must be taken since the spectrum of the cavity resonance may be altered by the nonequal distribution of the emitted light. Furthermore, limitations may arise from optical absorption or pump-induced losses, which can limit the maximum measurable quality factor [106].

Other techniques can be then employed when a very precise characterization is experimentally required. In particular, we can utilize methods that are also employed for resonators realized in optically passive material (i. e., silicon). For this purpose, various techniques have been developed along the years, both in room and low-temperature environments. One example is based on the use of waveguide coupling (as in the case of a channel-drop filter; see [89]), and it has been employed for the investigation of cavity quality factor and mode wavelength (Fig. 13.11(a)). For this scope, a waveguide is fabricated near the investigated cavity [2]. The injected light can propagate along the waveguide and be evanescently coupled to the PhC cavity, once in resonance with the cavity mode. Then the light can be emitted from the cavity leaking in the vertical direction (i. e., orthogonal to the PhC plane) or coupled back in the waveguide and detected collinearly to the injection direction. This is a very powerful experimental procedure even though we have to consider that the presence of the waveguide induces one additional loss channel, resulting in a lower observed quality factor. Therefore this loading effect induced by the presence of the waveguide needs to be taken into account. Fabricating the waveguide further apart from the cavity will reduce the evanescence coupling and hence this additional loss, but also will result in a lower light in-coupling in the PhC cavity.

An alternative strategy is based on the use of tapered fibers brought close to the resonator to be characterized (Fig. 13.11(b) [187]). The area with a reduced fiber diameter results in a lower light confinement, and the evanescent field can be coupled to the cavity itself if the distance is small enough. This technique has been used for planar photonic crystal cavities and for microdisks as well, proving its validity, even though also in this case the loading effect has to be taken into account.

Another technique employed to characterize photonic crystal resonators, even with very high quality factors, is based on cross-polarized light scattering [52, 121]. This technique, called *resonant scattering* (RS), relies on the modification that the incident light polarization undergoes upon interaction with the resonators. In more detail, light from a continuous wave, narrowband tunable laser, is firstly linearly polarized with a polarizer P (Fig. 13.11(c)). Then it is focused on the sample by means of a high-NA objective. The light reflected from the cavity is collected with a beamsplitter and sent through a cross-polarized analyzer A (i. e., a second polarizer oriented 90° with respect to P). When the laser wavelength is off-resonance with respect to the cavity modes, the light is simply reflected by the material maintaining its incoming polarization, therefore filtered by the analyzer. Conversely, when the laser frequency matches the modes, a signal can

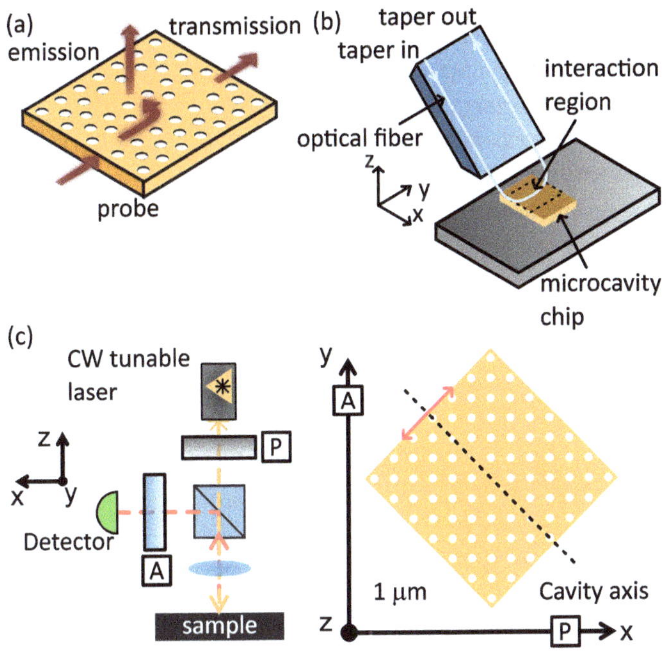

Figure 13.11: (a) Cavity characterization via a coupled waveguide: light propagating can evanescently couple to the cavity (here an exemplary *L*3). Emission from the cavity in the vertical direction (i. e., perpendicular to the PhC plane) can be detected as the transmission signal through the waveguide (therefore after eventual coupling to the cavity). [2] (b) Light coupling to a microcavity using a tapered fiber. The fiber is tapered reaching the minimal diameter next to the interaction region. The relative position between microcavity chip and fiber can be precisely controlled [187]. (c) Schematic resonant scattering setup (left) and respective orientation of the cavity axis with respect to analyzer and polarizer (right). A continuous wave laser is linearly polarized along the *x*-axis (P), whereas the reflected signal is collected with a beamsplitter and analyzed orthogonally (along the *y*-axis, A). Standard amplified photodiodes can be utilized as detectors. In [52], lock-in detection was employed to improve the signal-to-noise ratio, having the CW laser chopped optically at low frequency (few hundreds Hertz, set as reference signal for the lock-in). The cavity mode polarization is marked as a red arrow (for the present case of an *L*3 resonator).

be measured on the detector. The physical reason can be understood considering the respective orientation of cavity mode and polarizer/analyzer (Fig. 13.11(c), right): with the exemplary case of an *L*3 PhC resonator, the fundamental mode has a symmetry perpendicular to the cavity axis (marked in the figure as a red arrow). This means that incident light polarized along the X-axis has also a component parallel to the cavity mode, and hence it can couple to it when laser and mode frequencies match. After the in-coupling, light is out-coupled from the cavity having its polarization oriented as the cavity mode and therefore having components parallel to the Y-axis. The light component parallel to the analyzer is not filtered and can reach the detector. This means that while the laser which is not interacting with the cavity mode is filtered by the analyzer, the light interacting with the cavity can be detected, giving rise to a resonant scattering signal. This technique is ultimately limited by the laser linewidth and therefore can be suitable for measuring very high-Q cavities. Having now a clear picture of the origin of the signal, it also becomes clear why the cavity orientation with respect to polarizer and analyzer matters. Indeed, having the polarizer parallel to the mode polarization maximizes the in-coupling, but the light leaving the cavity will be fully perpendicular to the analyzer and therefore filtered out. Conversely, an incoming polarization orthogonal to the cavity mode will not couple to it, having no signal interacting with the cavity. This accounts for the necessity of the 45° orientation previously discussed. Nevertheless, it is worth noting that in case of large quality factors, another mechanism becomes relevant and gives additional signal to the RS. Indeed, light confined in the resonator region can also be subjected to out-of-plane scattering, leaving the cavity randomly polarized. This mechanism depends only on the in-coupled field and may become predominant for ultra-high Q-factors. Resonant scattering has been used with various cavity types from Ln [151, 152] to dispersion-adapted cavities [224].

Cavity mode and far-field profile

In some cases, measuring the cavity mode profile may be interesting to evaluate the cavity properties and the mode symmetry. Various techniques have been employed from microphotoluminescence imaging to scanning near-field microscopy (SNOM) [209], aiming at reaching the highest spatial resolution.

Another technique, also based on microphotoluminescence, is shown in Fig. 13.12. This approach relies on collecting the μ-PL in a collinear geometry (Fig. 13.12(a)) while scanning the excitation/collection spot in the xy-plane (i. e., the plane parallel to the sample surface) around the cavity location (Fig. 13.12(b)). As a case study, monolithic "truncated Gaussian microcavities" have been investigated [43]. Exemplary spectra for various positions are shown in Fig. 13.12(c). To study each single mode and its symmetry, the μ-PL signal of each resonance is transmitted while filtering all the others (see the spectral filtering sketched in Fig. 13.12(a)). Plotting the intensity per each xy position provides 2D maps resulting in the mode profiles in Fig. 13.12(d) (numbers correspond to each mode in (c)), and they agree very well with simulations (Fig. 13.12(e)). From these measurements we can see that the fundamental mode has a Gaussian-like profile, having then a clear signature of Hermite–Gaussian modes for the higher-order transverse modes. Further details can be found in the original publication [43].

Finally it is worth mentioning that when light in-coupling into a single mode fiber is targeted, a Gaussian mode profile is advantageous. In particular, it becomes important how it overlaps with the mode that can be sustained by the single-mode fiber: indeed, being a single-mode element, only one mode can propagate through the core. In [138] the far-field pattern is investigated using an infinity-corrected objective, which allows observing the angular dependence of the emitted field pattern as a spatial intensity distribution, simplifying the experimental implementation and data analysis with respect to back focal plane imaging (see, for example, [216]).

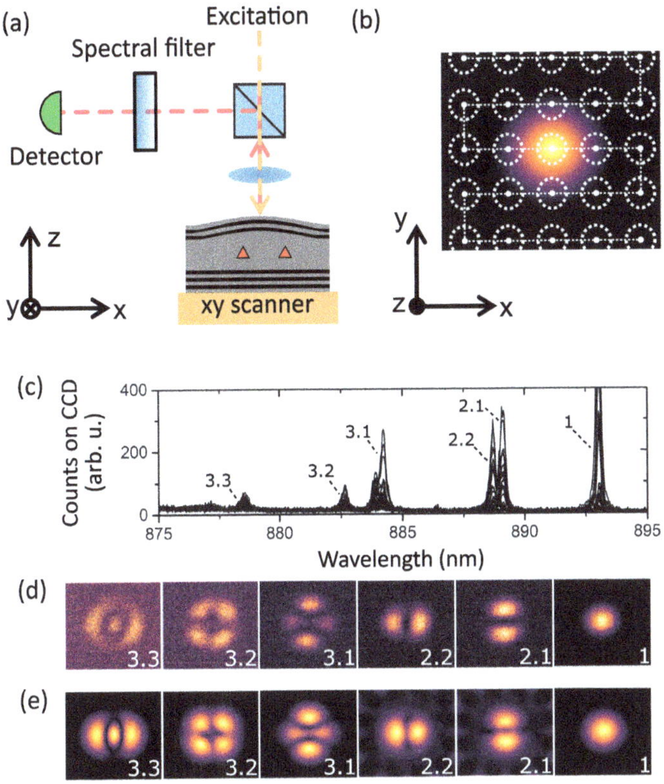

Figure 13.12: (a) Simplified sketch of the collinear μ-PL setup employed for acquiring the maps in (d): the sample exhibiting "truncated Gaussian microcavities" is scanned in the *xy*-plane, whereas the photoluminescence is spectrally filtered around each single mode. (b) Exemplary scan direction, where few collection spots are shown (dotted circles). The excitation/collection is scanned around the cavity having the spot moving across the mode profile (for clarity, the fundamental mode profile is shown underneath). (c) Exemplary μ-PL spectra acquired collecting the emission at various positions in the *xy*-plane. (d) Measured and (e) simulated profiles of the modes. The numbers are used to label each investigated mode. Reprinted from [43] with the permission of AIP Publishing.

13.5 Summary

- Extraction of light emitted by a nanostructure embedded into a solid-state material can be severely limited by the high refractive index contrast between host material and surrounding air or vacuum. This can be circumvented with two approaches that can sensibly increase the extraction efficiency. The first one is based on geometric effects, which modify the total internal reflection conditions, enabling larger extractions, or the coupling to a waveguiding mode. The second one is based on the use of cavity quantum electrodynamics, where a photonic resonator can be utilized to increase the coupling of the emitter to the cavity mode to enhance the extraction and collection efficiencies.

- A photonic resonator can be defined via parameters as, mode wavelength, quality factor Q, and mode volume V. The last two enter in the definition of the Purcell factor, where the term related to the cavity itself depends on the ratio Q/V.
- Various optical cavities make use of photonic crystals to realize high-reflectivity mirrors. PhC are materials with a refractive index, which is periodic in one, two, or three directions. This periodicity results in the formation of a photonic bandgap, respectively, in one, two, or three directions. Within this frequency region, light propagation is forbidden, and the crystal acts as a perfect mirror.
- A defect into a PhC can break the periodicity of the refractive index, resulting in the formation of allowed states in the band gap. In this case, light can only exist in the defect region, whereas the propagation is not allowed in other regions where the periodicity is intact because of the bandgap.
- PhC cavity geometry can be optimized to increase the quality factor and to narrow the emission far-field even making it more Gaussian. Although the former results in a better light confinement, the latter increases the light extraction, so a compromise between the two optimizations has to be made depending on the experimental conditions.
- Microphotoluminescence, reflectivity, and resonant scattering are techniques that can be employed for the characterization of cavity parameters as the mode wavelength and quality factor.
- The cavity mode profile can be characterized, for example, via microphotoluminescence imaging or scanning near-field microscopy.

14 Single- and entangled-photon sources, integrated quantum photonics

14.1 Introduction

Photonic quantum technologies, from fundamental science to more applied implementations, require sources of quantum light with performances as close as possible to ideal. Although most of the realizations benefit from high brightness and pure single-photon generation, some implementations may further require high photon indistinguishability, high degree of entanglement, and also the possibility to actively tune the source emission wavelength (see the discussion in Chapter 5). As discussed in Chapter 13, the strategies to increase the extraction of light generated by a nanostructure from its host solid-state environment can be categorized into two approaches: the first one is based on geometrical effects, which modify the total internal reflection conditions to limit the amount of light impinging at the critical angle at the solid-state/air interface; the second approach is based on the use of cavity quantum electrodynamics, in particular, in the weak coupling regime (see also Chapter 6). Both approaches can indeed sensibly enhance the source brightness as discussed in Chapter 5.

Geometrical approaches only require a precise spatial positioning of the emitter in the photonic device, whereas cavity quantum electrodynamics also require spectral matching between the resonator mode and the emitter's transition. Nonetheless, the latter approach allows for exploiting cQED effects. Particularly appealing for the realization of light sources is the weak coupling regime, where Purcell enhancement can increase the emission probability in the chosen optical mode. Interestingly, shortening the lifetime of a transition can be beneficial for reducing the impact of eventually present dephasing or spectral diffusion mechanisms. In addition, a shorter lifetime can also enable higher excitation rates, resulting in a larger number of available photons (see Eq. (5.13)). It is worth mentioning that exploitation of cavity quantum electrodynamics effects can be more challenging when bright sources of polarization-entangled photon pairs need to be realized. With the case study of a quantum dot as source of entangled photons, the photonic device needs to operate for two distinct transitions, typically spectrally separated by a few nanometers. Although this does not constitute a problem for geometrical approaches, which can be considered broadband for this purpose, cavity quantum electrodynamics require more advanced resonators designs.

In this chapter, we will focus on the description of the most commonly utilized photonic structure geometries for single- and entangled-photon sources. The cQED approach being in some sense more demanding than the geometrical one (it requires spatial and spectral matching of the cavity–emitter system, whereas geometric approaches are broadband for the purposes here described), we will mostly focus on this first strategy. We will also describe the approaches for realizing bright sources of entangled photon pairs, which may require the use of two distinct modes to enable the use of cQED on

https://doi.org/10.1515/9783110703412-014

the two transitions generating entangled photon pairs or a spectrally broad mode to simultaneously enhance both transitions. Most of the discussion will focus on realizations with semiconductor materials, since the growth and fabrication techniques are highly developed. Still, most of the discussion can be easily generalized to other solid-state systems.

Finally, it is worth noting that the enhanced light matter interaction due to the presence of a photonic cavity can be employed to sensibly reduce the amount of laser utilized for the optical excitation of the transition: this is particularly relevant for resonant pumping, as both the laser and the investigated QD transition are resonant with the cavity mode [68]. This aspect is strongly relevant in experimental implementations since it reduces the laser scattering and the pump induced dephasing, an important effect in resonant excitation schemes.

14.2 Single-photon sources

Employing cQED effects for the realization of bright sources of quantum light requires the spatial and spectral matching between the quantum emitter and the optical resonator. Operating in the weak coupling regime can allow the Purcell enhancement of the transition, hence increasing the source brightness by raising the β factor as in Eqs. (5.7) and (13.1). To ensure the required spatial and spectral matching (see Eq. (6.67) and following), deterministic fabrication techniques (see Section 12.3.5) can be applied, sensibly increasing the fabrication yield with respect to nondeterministic approaches. Here high Purcell factors (ultimately depending on the cavity properties via the quality factor Q over mode volume V ratio, as in Eq. (13.2)) can be achieved by increasing the factor Q. This results in a spectrally narrow cavity mode, which is anyhow still broader than typical emitter's transitions (even when not Fourier limited). Relatively large mode volumes can help reaching the spatial matching condition, the relative position between cavity mode and emitter being less sensitive to fabrication imperfections. Nonetheless, a too large mode volume would result in a lower achievable Purcell enhancement (and hence β factor, as discussed in the following).

Particular care will be given in describing optically pumped sources of single- and entangled-photon pairs, since this pumping strategy is the most developed for high-quality photon generation. Nonetheless, for most source types, electrical pumping will also be discussed.

14.2.1 Micropillar cavity-based sources

One of the first and most utilized cavity geometries is based on the realization of micropillar resonators. This was also firstly employed when deterministic low-temperature optical lithography was developed. To realize such a resonator, the following growth and fabrication steps are usually followed:

- Semiconductor growth of a planar cavity: a first bottom-distributed Bragg reflector (DBR) is deposited (with alternating $\lambda/4$ layers, acting as a 1D photonic crystal), followed by a cavity layer, typically a λ-cavity with the emitters sitting in the antinode of the fundamental cavity mode. MOVPE and MBE allows for a precise and accurate control of the vertical position of the emitters (normally around a few nanometers), therefore ensuring the correct vertical spatial position of the emitter over the cavity. The structure is completed via the deposition of a top DBR. To ensure light extraction through the vertical direction, the bottom DBR is designed for having a higher reflectivity than the top one (i. e., it consists of more DBR pairs). This structure would be characterized by the planar cavity quality factor Q_{pl}.
- Exposure: having now the substrate available for the fabrication, resist spin-coating followed by lithography (optical or e-beam, better if deterministically aligned on preselected emitters) can be conducted. A circle-shaped mask will be defined, which has then to be transferred in the semiconductor layers. The size of the micropillar being relatively large (around 1–2 μm), the mask can be defined effectively also with optical lithography. Eventually, a hard mask can be deposited, depending on the resist and etching recipe (see Section 12.3).
- Etching: dry etching techniques (i. e., ICP-RIE) can be employed to etch the processed sample. The circle defined in the resist, ideally centered on the top of the preselected emitter, will protect this area of material from etching, whereas the rest will be removed. The etching procedure must ensure verticality of the sidewalls, together with low surface roughness (which plays a role as explained in the following). This results in the cavity as seen in Fig. 14.1. Now the Fabry–Perot cavity confining the light in all three spatial directions, as described in Section 13.3, is ready.

Different types of micropillar cavities have been realized and various optical excitation schemes employed (see Fig. 14.1(a–c)). In Fig. 14.1, various optical excitation schemes are depicted. In Fig. 14.1(a), side excitation and top collection are schematized: this approach helps reducing the amount of laser light collected with the emitted single photons. Figure 14.1(b) shows an experimental realization where the pillar is laterally connected to larger structures with defined electrical contact pads. A vertically oriented electric field can be applied, resulting in wavelength tuning of the embedded emitter thanks to the quantum-confined Stark effect. In addition, a static electric field can further stabilize the QD environment, improving the photon indistinguishability. For this cavity geometry, collinear excitation/collection in the vertical direction can be employed. The mechanical stability of the structure can be improved, enabling the lateral light confinement with oxidation (rather than etching). This can enable the direct gluing of single-mode fibers to the micropillar (see Fig. 14.1(c)). As discussed in [185], one fiber can be utilized for carrying the excitation (bottom one in the figure), whereas a second one can collect the emission (top fiber in the figure).

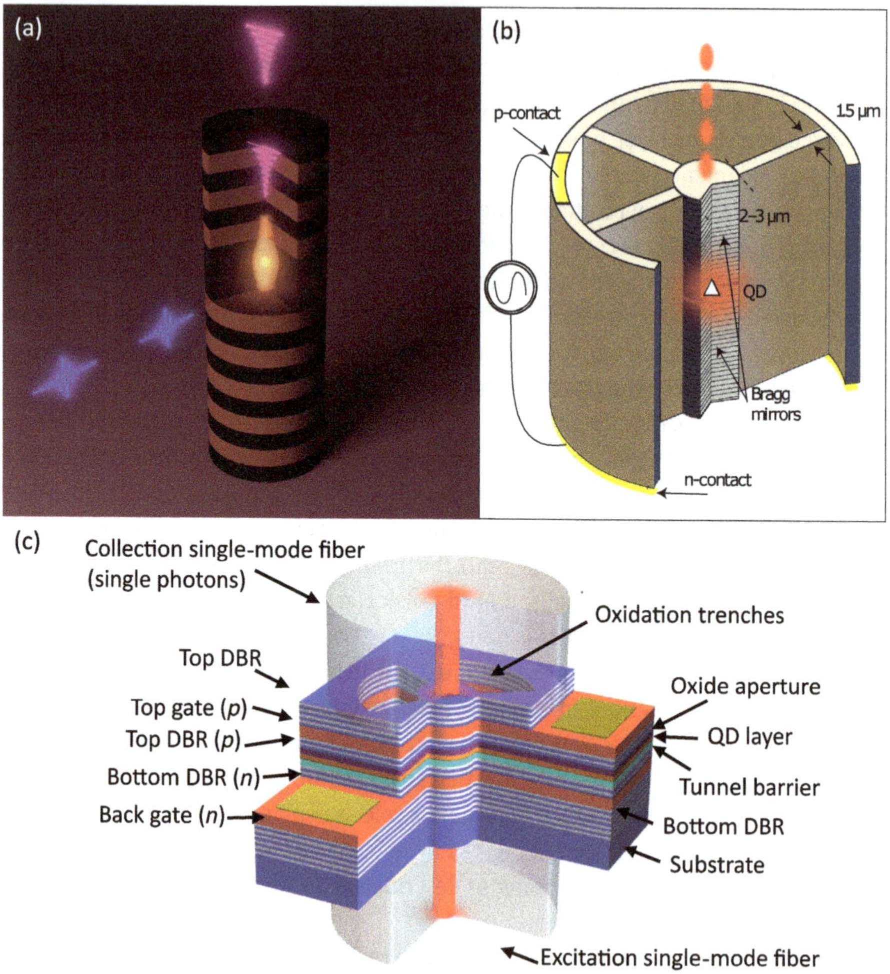

Figure 14.1: (a) Sketch of a micropillar cavity. The mode is exemplarily shown in the λ cavity, where one single emitter is located. Excitation photons (with Gaussian profile) reach the pillar from the side, having the photons generated by the emitter (with exponential profile) travel through the pillar in the vertical direction. Such excitation/collection geometry can be employed for effective laser suppression in resonant excitation schemes (see, for example, [68]). (b) Micropillar cavity with side trenches connecting to an outer structure: this enables the application of an external voltage for the wavelength tuning of the emitter via quantum-confined Stark effect, as well as the environment stabilization. Collinear top excitation/collection can be employed. Image reprinted with permission of Springer Nature from [186]. (c) Micropillar realized with lateral oxidation to define the three-dimensional confinement. Thanks to improved mechanical stability, fibers can be attached directly to the pillar. In the figure the bottom fiber carries the excitation laser, whereas the top one collects the emitted photons [185]. Image reprinted with permission from [185]. Copyright (2023) by the American Physical Society.

The design itself of the original planar cavity and respectively fabricated micropillar have to be carefully tailored to realize a bright source of light. Indeed, large pillar diameters would result in a large quality factor Q close to the planar cavity one Q_{pl}. Nevertheless, a large mode volume would result in low achievable Purcell factors which will have a detrimental effect on the achievable β, as in Eqs. (5.7) and (13.1). Smaller pillar diameters will allow for increasing the achievable Purcell factor (thanks to an improved light confinement) but resulting in a noticeable decrease of the quality factor because of losses induced by sidewalls roughness (see Fig. 14.2).

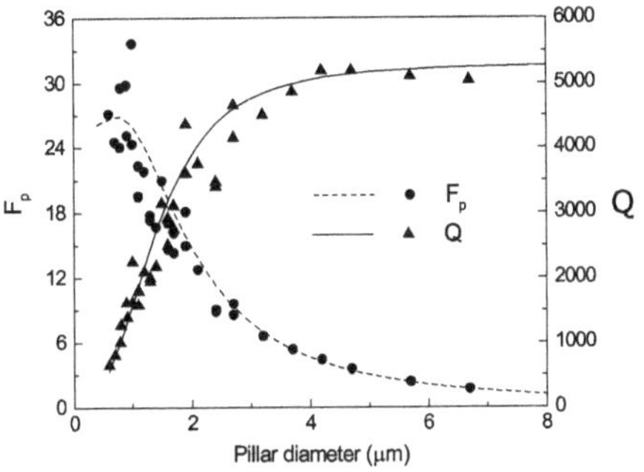

Figure 14.2: Quality and Purcell factors experimentally measured (symbols) and theoretically estimated (lines) for a series of micropillars (with $Q_{pl} = 5000$) [8]. Image reprinted with permission of Springer Nature from [8].

Indeed, for small pillars, the cavity mode will interact with the etched surfaces, and their eventual (and unavoidable) roughness will induce light scattering decreasing the lifetime of the photons inside the cavity (i. e., decreasing the factor Q). The actual micropillar quality factor can be written as follows:

$$\frac{1}{Q} = \frac{1}{Q_{pl}} + \frac{1}{Q_{scattering}}, \tag{14.1}$$

having a scattering term for encompassing the effect of sidewalls roughness. Therefore the impact of this loss channel on the spontaneously emitted photons that can be collected can be included in the β (taking Eq. (5.8)) as

$$\beta_{pillar} = \frac{Q}{Q_{pl}} \frac{F_P}{F_P + 1}. \tag{14.2}$$

A trade-off between high achievable Purcell and quality factor is then found [8], having in mind that typically intrinsic cavity losses are small (indeed, high Q_{pl} are routinely achieved). From a practical point of view, the pillar diameter can be adapted depending on the experimental requirements: for example, small pillars can ensure the presence of only one emitter in case of large emitter's density. In this case the planar quality factor can be modified, ensuring that scattering losses are small in comparison to intrinsic cavity losses [8].

Focusing on the use of micropillars in the weak coupling regime for the implementation of bright sources of quantum light, the high ratio Q/V allowed for observing high Purcell factors (as high as ~10 [153]). A relatively high quality factor (~10000) means a narrow resonance linewidth of the order of ~90 pm: therefore the spectral detuning between the transition employed for single-photon generation and the cavity resonance has to be smaller than the cavity linewidth to ensure high Purcell factor (and, as a consequence, high brightness). Deterministic fabrication techniques can help in the realization of a photonic structure with small detuning between the emitter transition and the cavity mode. Nonetheless, to ensure ideal spectral matching, further wavelength tuning of the emitter's transition may be necessary. As shown in Fig. 14.1, together with a deterministically fabricated micropillar (indeed, in (a) the emitter is in the spatial maximum of the cavity field), quantum-confined Stark tuning is experimentally employed to reach spectral matching between the QD transition and the cavity mode.

To employ the excitation scheme described in Section 10.2.3.1 (see, in particular, Fig. 10.5), elliptical micropillars can be easily realized [217]. This can be simply performed creating an elliptical instead of circular mask for the etching: the degree of ellipticity has to be chosen according to the aimed polarization mode splitting.

Finally, it is worth mentioning that having the emitter coupled to a narrowband cavity allows for reducing the impact of phonon dephasing on the emitted photons. Indeed, the cavity decreases the transition lifetime, making the photon less prone to pure dephasing. Furthermore, having the zero-phonon line (ZPL) spectrally resonant with the cavity mode further allows for the suppression of the phonon side bands (PSB, i. e., phonon-assisted emission). This funneling in the ZPL does not have the simple effect of spectral filtering, rather making the emission in the ZPL more probable over the emission in the PSB [58].

Micropillars could reach high photon indistinguishability (>95 %) with pure single-photon emission ($g^{(2)}(0) < 1\,\%$) and high extraction efficiencies (>60 %).

Electrical pumping of micropillars

The successful implementation of single-photon sources based on micropillars stimulated the search of analogous structures operated under electrical excitation. For this purpose, resonator geometries similar to those employed for electrical control (see Fig. 14.1(b,c)) can be utilized. A simple geometry, which enabled the generation of single photons with high rate [67] under electrical pumping, was based on the use of single-photon light-emitting diodes (SPLEDs; see Fig. 14.3(a)). Employing doped DBR mirrors, quantum dots have been embedded in the intrinsic region of a p–i–n diode. Optimized structures require the precise control of the current flow to avoid the excitation of multiple nearby emitters, which can impact the single-photon emission (see Section 10.3).

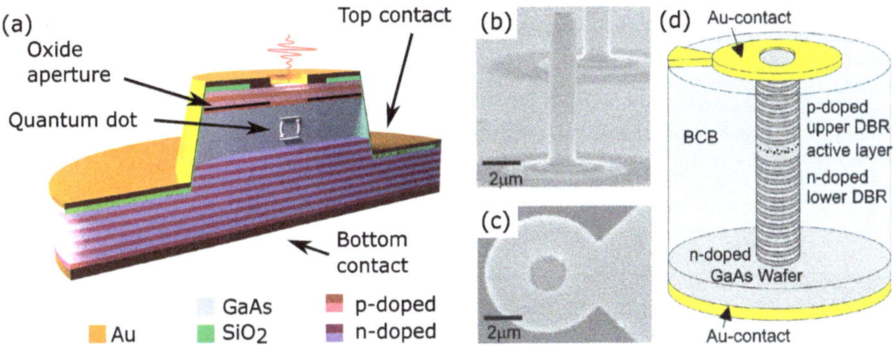

Figure 14.3: (a) Cross-sectional sketch of a SPLED. Doped top and bottom DBRs are utilized to enhance the light extraction. A metal layer (i. e., gold) serves as bottom contact, whereas the top one is defined around the mesa structure. An oxide layer (here SiO_2) isolates the top metal contact, which touches the semiconductor only around the top of the mesa, where an opening is created for the photon extraction. Current flow is further limited by the presence of an oxide aperture layer (dark color). (b) Angular SEM picture of a standard micropillar before the planarization step with BCB (benzocyclobutene). (c) SEM picture from the top of a fully processed sample: the opening in the gold contact corresponds to the pillar top. (d) Sketch of the fully fabricated device: the micropillar encompasses doped DBRs, together with a bottom and top (lithographically defined) metallic contacts. Images (b)–(d) reprinted from [15] with the permission of AIP Publishing.

Furthermore, current flowing into small volumes also results in smaller device capacity, key requirement to reach high excitation and emission rates. A commonly used structure for the realization of SPLEDs is based on the fabrication of a mesa structure. Furthermore, top and bottom DBR mirrors can be employed to increase the extraction efficiency and hence the source brightness. The etched mesa already helps in decreasing the device footprint, and to reduce even more the current flow, an oxide aperture is often employed. To do so, one specific layer is added next to the QD layer so that once the mesa has been fabricated (often via dry etching), oxidation can be triggered in controlled atmosphere (with an oxidation oven). The formation of a round opening (the symmetry and size are set by the mesa geometry and the oxidation time) allows for confining the current, which can only flow where no oxidation takes place. To reduce even more the footprint of the device, deterministic fabrication has been employed. Indeed, thanks to the improved accuracy in the spatial positioning of the device, smaller mesa can be employed, still ensuring the presence of one single emitter in the center of the oxide window (see Section 10.3 and [171]).

Although the described SPLED structure has the advantage of being rather broadband, electrically driven sources can also be realized employing cavities with higher Q-factors, as, for example, utilizing micropillars for the realization of electrically pumped QD-based light sources, further exhibiting Purcell enhancement (see Fig. 14.3(b–d)). An early study showed that indeed the Q-factor above 10000 can be achieved under electrical pumping, together with Purcell factors as high as 10 [15].

14.2.2 Circular Bragg grating cavity-based sources

An interesting photonic structure is constituted by the realization of a circularly symmetric Bragg gratings where the emitter is ideally located in the middle of a central region larger than the periodicity Λ (named circular Bragg gratings or also bull's eye; see Fig. 14.4).

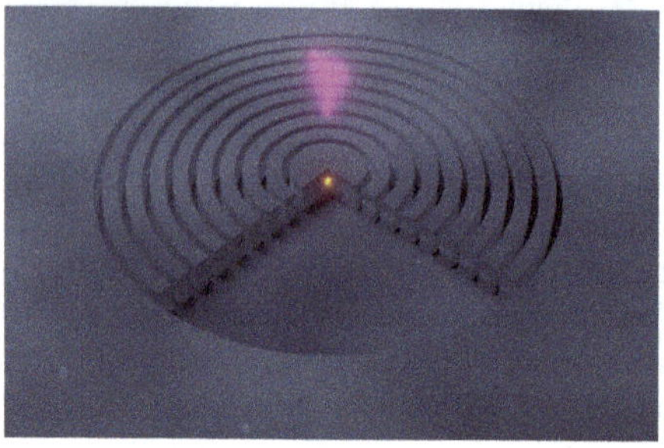

Figure 14.4: Sketch of a circular Bragg grating cavity. The structure is exemplarily depicted with a cut through to show the etching of the trenches fully through the semiconductor slab.

An interesting design, often used in recent studies, utilizes a membrane of semiconductor where trenches are fully etched through (Fig. 14.5(a,b)). Below the semiconductor material, an oxide layer serves as a spacer for precisely controlling the position of a metallic mirror. This reflector helps redirecting toward the top direction photons emitted toward the bottom (i. e., toward the substrate). In fact, a fully vertically symmetric cavity (as an $L3$ resonator surrounded by air) provides light emission in both directions perpendicular to the plane of the photonic crystal. In this case the presence of an oxide layer makes the emission toward the bottom direction even more probable (being the refractive index mismatch lower). In early designs, CBG was realized as a semiconductor layer fully surrounded by air (see Fig. 14.5(c)), where the trenches were not fully etched through, ensuring vertical asymmetry and higher emission toward the

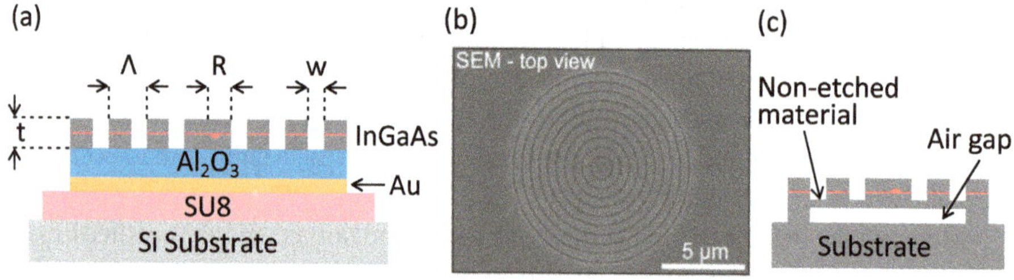

Figure 14.5: (a) Cross-sectional scheme of the fabricated CBG resonator; design parameters are marked. (b) SEM picture of a fully fabricated circular Bragg grating cavity. (c) Exemplary design of a CBG surrounded by air. The trenches are not fully etched. An exemplary SEM picture is shown in Fig. 12.15(a), where nearby openings for the wet-chemical underetching are shown. Image adapted from [98]. Image (b) reprinted with permission from [100]. Copyright (2023) American Chemical Society.

top direction. This design, despite effective, presents two practical challenges: firstly, stopping the dry processing at a precise point to ensure a precise partial etching can be challenging to be reproduced; secondly, membrane structures fully surrounded by air are more fragile, and mechanical bending has been observed (gradual, over time spans of a few months), resulting in the shift of the cavity mode from its design spectral position [99]. Fully etched semiconductor membranes realized on oxide spacer and metallic mirrors can counteract the aforementioned challenges: longer etching times only result in a minor removal of the oxide material, which has no severe impact on the photonic properties. In addition, no bending has been observed.

The realization of CBG as depicted in Fig. 14.4 can be performed as follows, aiming to reach the final device as shown in Fig. 14.5:

- Semiconductor growth of the layers: first, a sacrificial layer (with composition depending on the wet chemical process utilized, for example, AlAs) is deposited, followed by the growth of the semiconductor layer embedding the QDs (see Fig. 12.8(a)). This sacrificial layer plays an important role for the following fabrication. Then the semiconductor layer embedding the quantum emitters is deposited. The thickness has to be designed so that the semiconductor slab can support only the fundamental TE mode (and the fundamental TM mode, even though the quantum emitter with in-plane dipole moments will couple only to the supported TE mode). As a rule of thumb, the optical thickness of the semiconductor (i. e., the product of the material thickness with the refractive index) should be smaller than the resonance wavelength of the designed cavity.
- Deposition of oxide spacer layer and metal mirror (Fig. 12.8(a); see also Fig. 14.5). The reflector will be placed at the bottom of the structure after flip-chip, and it will reflect the light toward the top for being collected. The oxide spacer ensures the correct reflection conditions, where the light emitted toward the bottom (circa 80 % of the total emission) and then reflected, constructively interferes with the light emitted directly toward the top (circa 20 % of the total emission). This will ensure extraction efficiencies larger than 80 % for a broad wavelength range. Interestingly, whereas the cavity factor Q decreases for increasing oxide thickness, the extraction efficiency increases up to a saturation value of 80–90 % for increasing spacer layer thickness (whereas the mode volume and cavity resonance are barely modified by the oxide layer geometry). This means that a trade-off between the factor Q (and hence the Purcell factor) and extraction efficiency needs to be found. Typical oxide layer thickness of the order of 300 nm is chosen (for CBG operating at telecom wavelengths).[1]
- Flip-chip process: at this stage the sample is ready for the flip-chip as described in Section 12.3.4. This results in having the semiconductor membrane on the top, whereas the oxide layer and metal mirror lay on the bottom (see Fig. 12.8(b)).

1 More details on the design can be found, for example, in [98] and references therein.

– E-beam lithography and reactive ion etching: now the semiconductor membrane is ready for the fabrication of the resonators. Electron beam lithography and reactive ion etching can be employed to define and transfer in the semiconductor the designed trenches (see Fig. 12.8(c)). Here the design parameters include the central disk radius R, the trench width w, and the periodicity Λ (see Fig. 14.5(a)). Typical values for these parameters are: $R \sim 500$ nm, $w \sim 150$ nm, and $\Lambda \sim 500$ nm (for operation in the telecom O-band, slightly larger values can be considered for C-band operation).

Together with the periodicity of the refractive index creating the circular grating, a double periodicity can also be superimposed to enable efficient extraction of light toward the top direction (as explained in detail in Section 13.3). For CBG, this condition is always fulfilled as long as the grating periodicity Λ follows the condition $\Lambda = \lambda_{QD}/n_{eff}$, with λ_{QD} the wavelength of the quantum dot emission and n_{eff} the effective refractive index of the semiconductor slab [98].

Interestingly, the simulated far-field profile of the cavity shows that more than 90 % of the emitted light could be collected employing optics with a rather modest numerical aperture of 0.6. Furthermore, the profile of the emitted pattern shows almost 90 % overlap with a Gaussian mode, implying that a good coupling into single-mode fibers can be achieved. Despite the advantage of broadband extraction efficiency and relatively broadband achievable Purcell enhancement, the small mode volume makes the spatial alignment of the emitter with respect to the cavity mode much more challenging than for micropillars. According to simulations, displacing the emitter from the ideal central position by ~70 nm would result in a decrease of the achievable Purcell enhancement by a factor of 2. This could be circumvented employing deterministic fabrication techniques (see Section 12.3.5).

CBG resonators have rather modest quality factors (150–200), but also a small mode volume (of the order of 0.6–1.5 $(\lambda/n)^3$), resulting in large ratios Q/V despite the modes quality factor. This means that the Purcell enhancement >15 can be achieved. This can also be observed even for nonzero detuning between the spectral maximum of the cavity mode and the transition. We can see this in Fig. 14.6, where simulations have been conducted for a CBG with mode spectrally centered at ~1551.8 nm. Relatively high Purcell factors can be achieved even for transitions detuned a few nanometers (blue or red) from the maximum. Interestingly, collection efficiencies larger than 85 % (for $NA = 0.6$) can be expected in more than 20-nm-large wavelength range.

As for the case of micropillars, elliptical CBG have been realized, enabling the use of the excitation method as described in Section 10.2.3.1 [217].

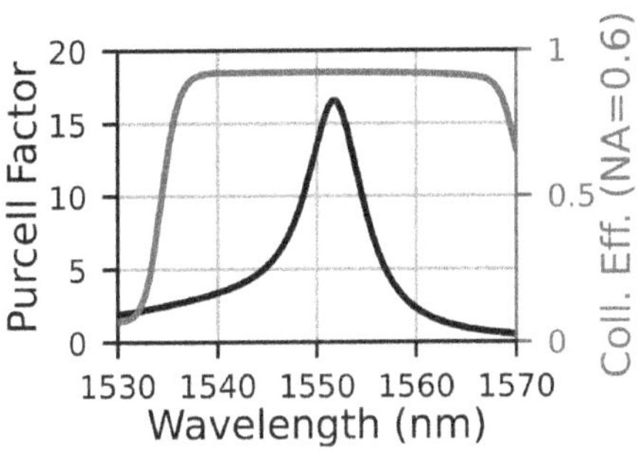

Figure 14.6: FDTD simulations of the achievable Purcell factor and collection efficiency (for a numerical aperture of 0.6) over wavelength. The maximum Purcell factor is achieved for perfect spectral resonance between the transition and the cavity mode. Image adapted from [136].

Electrical pumping of circular Bragg gratings

As discussed in the following section, photonic crystal cavities based on arrays of holes etched in the semiconductor show the advantage of having the cavity region conductively connected to the outer semiconductor. This makes it easily suitable for the implementation of electrical excitation schemes (as shown for classical light sources in [182] and references therein). Still, CBGs have also been adapted to become compatible with electrical control/excitation. For this purpose, bridges connecting the cavity region with the surrounding have been added (see Fig. 14.7). In this way, contact pads can be defined (in lateral or vertical geometry) enabling the application of an external voltage to the active layer.

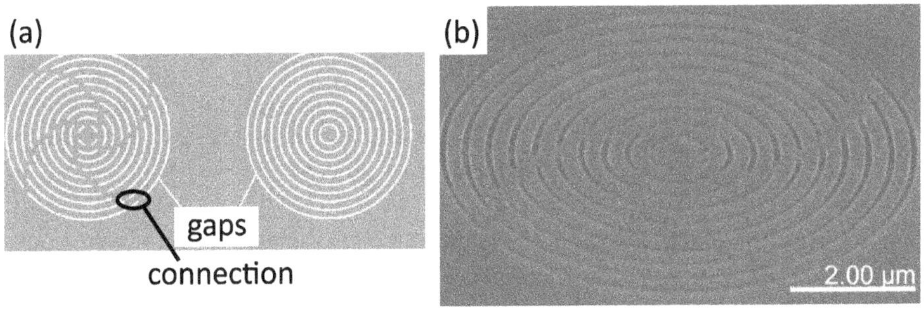

Figure 14.7: (a) Simplified schematic of a circular Bragg grating with (left) and without bridges (right) to ensure the electrical connection [98]. Since the symmetry of the cavity, and therefore its photonic properties, can be influenced by the addition of bridges, advanced designs have been proposed as sketched in (a) and realized (see the SEM angular view in (b)). Images (a) and (b) reprinted from [22] with the permission of AIP Publishing.

14.2.3 Photonic crystal cavity-based sources

Photonic crystal cavities have been deeply investigated and optimized in the past years. They have also been employed in the realization of sources of nonclassical light. Their design and the physics of the photonic crystal lattice have been discussed in Chapter 13.

The realization of a photonic crystal cavity-based single-photon source (an exemplary *L*3 one) based on a triangular lattice of holes etched in a semiconductor slab and surrounded by air (see Fig. 14.8) is described in the following.

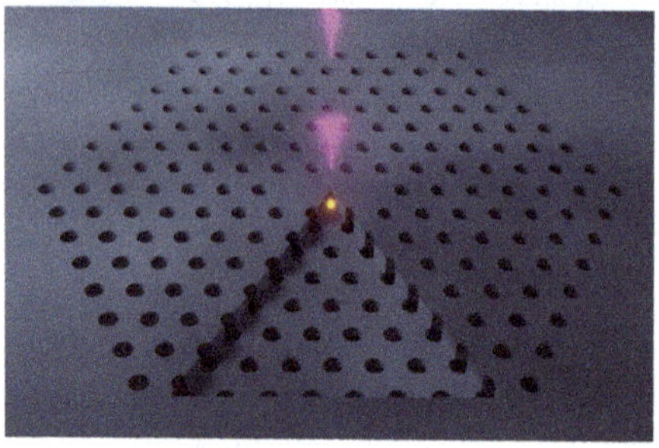

Figure 14.8: Exemplary sketch of an *L*3 PhC cavity. As depicted in the image opening, the holes are etched fully through the semiconductor.

- Semiconductor growth: similarly to CBG, the semiconductor slab containing the emitters has to be grown on top of a layer, which will be etched away to realize a PhC slab cavity surrounded by air. The slab thickness is also here an important design parameter. Indeed, for too thin slabs, the modes will be weakly localized, rendering the periodicity of the fabricated photonic crystal irrelevant. On the other hand, too thick slabs would resemble the case of a 2D PhC, where modes with a higher number of vertical nodes will have lower energy, and therefore the newly formed states close the photonic bandgap. As a rule of thumb, slab thickness of around half a wavelength (in semiconductor, for a triangular lattice of holes) represents a good compromise [89].
- E-beam and etching: as for CBG, electron beam lithography and dry etching can be employed to define the photonic crystal lattice and the chosen defect forming the resonator.
- Wet chemical underetch: as a last step, wet chemical etching is typically employed to remove the layer underneath the semiconductor PhC to create a symmetric slab surrounded by air. Depending on the material constituting this layer and the wet etching conditions, relying on the penetration of the chemicals through the previously defined holes may not be enough to ensure a homogeneous underetching. In some cases, additional opening could be realized (as in Fig. 12.15(a)).

Photonic crystal cavities can reach very large quality factors $>10^5$. Resonators with a small mode volume, like the *L*3 or the *H*1, can therefore enable reaching high Purcell factor values (for example, >40 in [115]). The narrow spectrum of the cavity mode makes them suitable for the implementation of optically pumped bright sources of single photons. Photonic crystal based structures further showed compatibility with electrical pumping [50] or electrical control of QDs (in PhC waveguides [202]).

14.2.4 Open tunable Fabry–Perot cavity-based sources

Open tunable Fabry–Perot cavities have been introduced in Section 13.3. In combination with semiconductor quantum dots, further embedded into p–i–n diode structures, these resonators achieved extremely high extraction efficiency values, furthermore with emission of indistinguishable single photons. These structures can be realized as follows:

- Semiconductor growth: a high-reflectivity DBR is firstly deposited. Then an n-doped layer is grown, followed by the deposition of the semiconductor QDs inside an undoped layer. Finally, the p-doped layer is deposited together with a capping layer. The design of layers thickness has to be chosen to enable the QDs to be placed in the antinode of the vacuum electric field [196]. The layers are further designed to ensure the doped layers to sit into nodes of the cavity mode (to limit free-carrier absorption).
- The p- and n-doped layers are made accessible from top via etching at different depths, and metal contacts can then be deposited to apply a voltage to the diode.
- The surface native oxide (forming on the first few nanometers of semiconductor surface because of interaction with air) is firstly wet-chemically removed. Then a passivation layer is deposited on the top of the semiconductor material to decrease the impact on the cavity performances of absorption due to surface states.
- Top mirror fabrication: for this purpose, CO_2 laser ablation (or focused ion beam milling) can be employed to form a curved surface in the glass (with typical curvature radii of ~10 μm).[2] Dielectric DBRs can be deposited to reach high reflectivity (still lower than the bottom DBR to ensure extraction from the top direction).
- The sample can be then placed in the cryostat and the top mirror aligned. Low-temperature operation has to be ensured, i. e., the position of the top mirror needs to be precisely controlled with respect to the bottom sample even at cryogenic temperatures.

The structure as in Fig. 14.9 combines:

- the possibility to spatially and spectrally match the emitter with the cavity mode thanks to the tunability of cavity position and mirror separation.

2 More information on the radius of curvature can be found in [196], supplementary material and related bibliography.

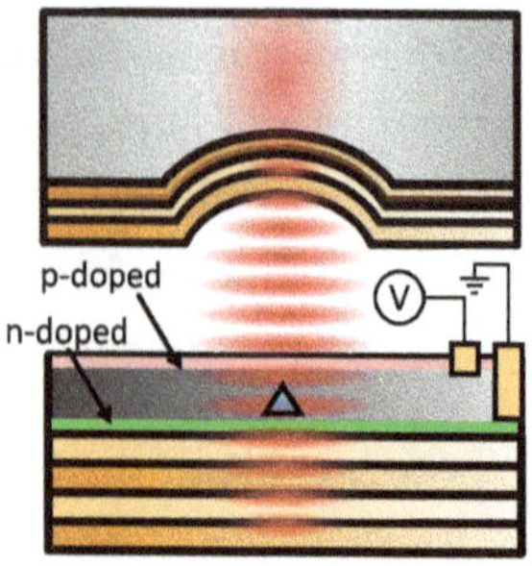

Figure 14.9: Sketch of an open, tunable Fabry–Perot cavity formed by a bottom semiconductor DBR and a top mirror realized via dielectric deposition on a machined glass substrate. The contacts to the embedding p–i–n structure are also shown. An exemplary mode structure is also depicted. The number of antinodes depends on the distance between the mirrors (i. e., which Fabry–Perot mode number is formed). Doped layers are by design in the cavity mode nodes, whereas the QD is in the antinode. Image designed from [196].

- an embedding diode, which stabilizes the emitter's electronic environment, strongly reducing the dephasing and spectral diffusion effects.
- a Gaussian-shaped output mode, which couples well into a single-mode fiber.

The cavity operation requires a precise control of the top mirror position and distance over the bottom mirror (on top of which the emitters are deposited). Together with that, also mechanical stability of the resonator needs to be achieved to avoid spectral jittering of the cavity mode induced by unwanted vertical displacements of the top mirror. Experimental realizations showed β factors larger than 85 %, with end-to-end efficiencies >50 %. Comparing with Eq. (5.13), recent results showed that countrates on the detector of ~40 MHz for 76 MHz excitation rate can be achieved.

14.2.5 Geometrical effects-based sources

As discussed in Section 13.2, the modification of the TIR conditions can be an effective way to enhance the extraction of light from a semiconductor nanostructure. This approach has the advantage that the structures have a broadband operation, therefore enhancing the extraction for several optical transitions also when spectrally separated (they have also been used for the generation of entangled-photon pairs; see the next section). Nanowires have been grown via vapor–liquid–solid mechanism into MOVPE reactors. The use of a metal catalyst particle then controls the position of the growth of the nanowire [159]. Alternatively, these structures can also be realized employing a flip-chip process followed by the removal of the substrate via a sacrificial layer and then by e-beam defined hard masks for the dry etching process defining the nanowires [25]. A similar approach has also been employed for the implementation of photonic trumpets [188], as also described in Section 13.2. Structures where the coupling to modes other than the investigated one (named cavity mode in Eq. (5.7)) is suppressed allow reaching high factors β even in the absence of Purcell enhancement (see Eq. (5.9)). Although very appealing because of the high extraction efficiency reached (>70 %), it becomes challenging to suppress the laser in pure resonant excitation, and Purcell enhancement is difficult to achieve.

An alternative structure employed to modify the TIR conditions is based on the use of microlenses (see Section 13.2). Focusing on the structures etched in the semiconductor material (either via dry etching [62] or via wet chemical etching [169]), they allow for a broadband enhancement of the optical transitions, even allowing for pulsed two-photon resonant excitation (supported by the increased focusing of the excitation laser because of the microlens itself). Finally, macroscopic (millimeter-sized) lenses have also been employed for enhancing the light extraction from QDs embedded in semiconductor membranes. In this case the layer embedding the emitters is transferred on the bottom of hemispheric lenses made with materials with as high as possible refractive index (for example, GaP for visible and NIR, silicon for telecom wavelengths). A metallic mirror deposited on the bottom of the semiconductor membrane and the lens at the top allow for the formation of an optical antenna with broadband operation [24]. The precise control of the distance between the semiconductor and the lens (adjusted via the glue thickness) allows for optimization of extraction efficiency and far-field for increased coupling in a single-mode fiber (>60 %) [138].

14.2.6 Waveguide-based sources for off-chip operation

In Section 14.4, on-chip realizations of photonic integrated circuits will be discussed. A key point will be the efficient coupling of the emitter's generated photons into the waveguide, where high factors β can be reached suppressing the emission in modes other than that sustained by the employed waveguides (even without Purcell enhancement; see Eq. (5.9)). Here it is worth mentioning the strategies followed to realize optically pumped bright sources of single photons for off-chip experiments, still employing waveguides for achieving large couplings to a specific single mode (i. e., large factors β). One exemplary system is based on the use of photonic crystal waveguides (PhCWG), for example, standard $W1$ structures (i. e., one row of missing holes). These systems demonstrated a coupling in the waveguide mode (β) larger than 95 %. Photonic crystal waveguides allow reaching high factors β thanks to i) Purcell enhancement (which is achieved without a resonator thanks to slow light modes, having a group velocity that increases toward the band edge of the waveguide mode; further details can be found in Section 14.4) and ii) inhibition of radiative emission in the plane of the semiconductor thanks to the photonic bandgap (whereas total internal reflection reduces out of plane radiative emission) [6]. It is important to mention that emission in a single-mode waveguide happens in both directions: if photons are needed in only one specific direction, then a reflector can be added in the waveguide.

To be used as single-photon sources for off-chip implementations, out-couplers can be employed following tapered mode adapters (Fig. 14.10(c)) or grating couplers (Fig. 14.10(a,b)). In the GaAs platform, these structures have shown out-coupling efficiencies into small numerical apertures (~0.6) greater than 60 %, which can then make PhCWG a valid approach as single-photon sources also for off-chip applications (since >90 % coupling in the WG mode can be achieved). Even higher out-couplings have been achieved in the silicon platform (where light has been injected externally), up to almost 90 % [9, 82], providing the perspective of finding near-unity out-coupling strategies for III–V platforms as well. Despite the need of designing and including an out-coupler for extracting the light off chip (which can therefore lower the overall source brightness), the advantages of PhCWGs lie on the broadband operation (over tens of nanometers) and also achievable (narrowband) Purcell enhancement (see also Section 14.4.3). An overview of the discussed single-photon sources and respective parameters is shown in Tab. 14.1.

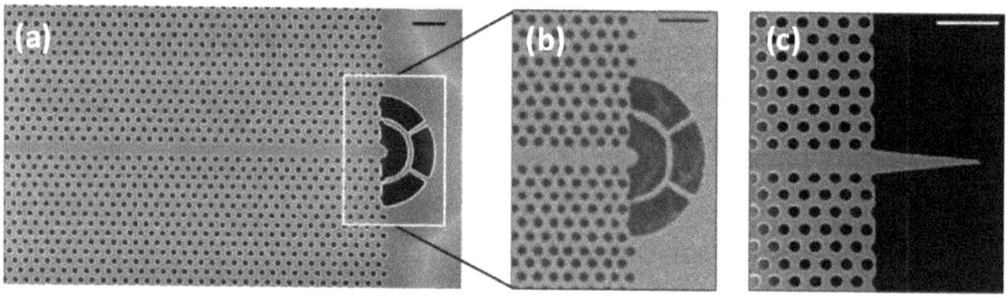

Figure 14.10: (a) SEM picture of a PhCWG with a grating coupler for off-chip photon extraction. (b) Close up image of the out-coupler. (c) Exemplary tapered mode adapter. Scale bars mark 1 μm. Reprinted with permission from [6]. Copyright (2023) by the American Physical Society.

Table 14.1: Exemplary optically pumped single-photon sources and respective typical parameters. In experimental realizations the numbers can vary depending on the specific optimization and design.

SPS	Spectrum	Mode Volume	F_P	Out-coupling
Pillar	Narrow (~90 pm)	2–200 ($(\lambda/n)^3$)	>10	>60 %
CBG	Broad (~10 nm)	≤1 ($(\lambda/n)^3$)	>15	~90 %
PhC ($L3$)	Narrow <90 pm	<1 ($(\lambda/n)^3$)	>40	>10 %
Open cavity	Narrow ~70 pm	≥10 μm^3	>10	>85 %
Nanowire	Very broad (geometrical)	–	–	~70 %
Trumpet	Very broad (geometrical)	–	–	~70 %
Microlens	Very broad (geometrical)	–	–	~20 %
Waveguide	Broad (extraction, tens of nm)	–*	~10	>60 %

*Emitter's position sensitive for more than 100-nm displacement.

14.3 Entangled-photon sources

Bright sources of entangled photon pairs find several applications in various aspects of photonic quantum technologies. Indeed, entanglement plays a crucial role in the currently ongoing second quantum revolution. This means that for these purposes, a bright source of entangled photons needs to be achieved. This requires i) the use of photonic structures capable of enhancing the light extraction for all entangled photons, without inducing polarization projection in the emission process, and ii) effectively zero fine-structure splitting when utilizing semiconductor quantum dots for the generation of polarization-entangled photon pairs (via the biexciton–exciton–ground cascade, as discussed in Section 7.3.2).

The challenge of enhancing the brightness with cavity quantum electrodynamics lies on the spectral separation between the two transitions (biexciton and exciton), typically of a few nanometers (4–5 for InGaAs QDs operating at telecom C-band): this requires a spectral resonance broad enough to accommodate both transitions or the use of two separate modes, each resonant with one QD transition (usually, in Fabry–Perot-like cavities, fundamental and first excited modes have a spectral separation larger than typical biexciton–exciton energy splitting). Furthermore, as explained for the single-photon source, the enhanced light matter interaction due to the presence of the cavity reduces the excitation laser intensity required in the experiment. Nonetheless, for a resonant excitation of the biexciton, two-photon excitation schemes are often employed (see Section 10.2.3.4), having then the laser spectrally in between the XX and X transitions. Having a third resonance matched with the laser makes the cavity implementation even more challenging. This could be overcome with photonic structures employing geometric effects, forgoing any advantage carried by the Purcell effect. This can be particularly limiting in terms of achievable indistinguishability as discussed in Section 7.3.3 and in the next section.

Zero fine-structure splitting can be achieved during growth (as, for example, with strain-free symmetric dots) or at the device stage, e. g. via piezo-tuning. This means that photonic structures compatible with piezo-tuning are attractive for the realization of highly entangled photon pairs.

14.3.1 Circular Bragg grating cavity-based sources

Circular Bragg gratings, as discussed in the previous section, are highly appealing for the realization of bright sources of entangled photons. Indeed, their low quality factor results in a broad mode, which can accommodate biexciton and exciton transitions and the excitation laser frequency (e. g., in TPE) simultaneously. For the exemplary case of semiconductor quantum dots, biexciton and exciton transitions employed for the generation of polarization-entangled photon pairs have a spectral separation of a few nanometers (~2.5 nm for NIR GaAs QDs and ~5 nm for telecom C-band In(Ga)As quantum dots), whereas typical CBG full widths at half-maximum are of the order of 5 to 10 nm. The

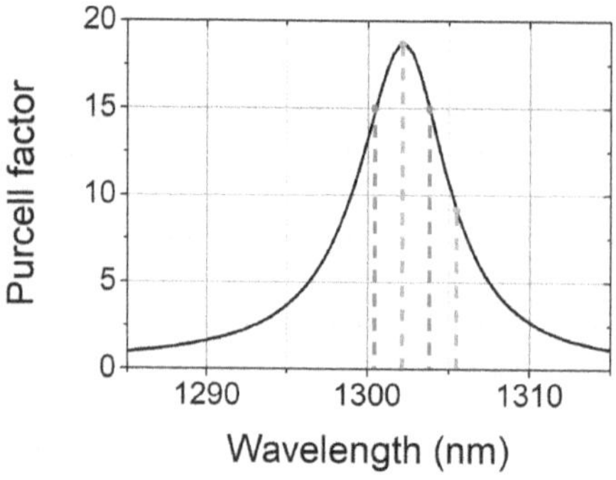

Figure 14.11: Exemplary FDTD calculation of Purcell enhancement over wavelength (for a CBG optimized for the telecom O-band, periodicity of 500 nm, central disk radius of 535 nm, trench width of 160 nm). Red dashed lines show the two potential spectral transitions (separated by ~3 nm), equally detuned from the cavity maximum to obtain similar Purcell enhancement (~15). Asymmetric configuration is shown by blue dashed lines, where one transition undergoes the maximum Purcell enhancement (~18), whereas the second, red detuned, reaches a lower Purcell enhancement (simulated around 9). Image adapted from [98].

small mode volume results in large ratios Q/V despite the modes quality factor, ensuring the possibility of reaching Purcell enhancement factors larger than 15, whereas smaller enhancement can still be reached for multiple QD transitions (see Fig. 14.11). In addition, the high extraction efficiency also results in large photon pair brightness. Experimentally, a photon pair extraction efficiency of ~65 % (i. e., circa 80 % for each photon) and an entanglement fidelity larger than 85 % have been demonstrated.

It is important to mention that maximal Purcell enhancement is reached for a transition fully in resonance with the cavity mode: this means that employing CBG for the acceleration of the spontaneous emission of spectrally separated transitions may result in different Purcell enhancement for each transition (see Fig. 14.11). This can become a very appealing feature since, as discussed in Section 7.3.3, the indistinguishability of photons generated from the radiative biexciton–exciton cascade can be limited depending on the lifetime ratio of these two transitions. This can be overcome by shortening the biexciton radiative lifetime more than the exciton: for a biexciton decay 10 times shorter than the exciton one, the indistinguishability could reach ~90 %.

Alternatively, a comparable Purcell factor can be achieved by having two transitions equally detuned from the mode maximum (i. e., one spectrally red and the other blue detuned from the maximum). High Purcell enhancement (>20) resulted in near-unity photon indistinguishability (~90 %), high single photon extraction efficiency (>80 %) [116, 218] and in the possibility of exciting the source with higher rates (thanks to the reduced transition decay time), reaching unprecedented single-photon emission rates even at telecommunication wavelength [136].

Finally, these structures can be fabricated on top of multiaxis piezo-actuators, which can be utilized for tuning to near-zero the exciton fine-structure splitting, reaching even higher values of entanglement.

14.3.2 Coupled cavities-based sources

Coupled cavities, as, for example, two micropillars, can be employed for the realization of bright sources of entangled photon pairs. Indeed, bringing two pillars with the same diameter (and therefore with equal spectral resonances) close together results in the hybridization of the two fundamental modes of each micropillar (Fig. 14.12). This results in the creation of supermodes, localized on both pillars, with a spacing between the fundamental and the excited modes much smaller than the spectral separation between the modes of a single pillar itself, i. e., the modes of the Fabry–Perot cavity. The geometry can be tuned so that the newly formed modes have a spectral separation of a few nanometers, each matching exciton and biexciton, respectively (Fig. 14.12(b)). Despite the interesting results achieved with such a structure, these geometries require a more complex fabrication and design. Indeed, it is necessary to achieve the spectral matching of two separate transitions, having therefore the cavity designed so that the newly formed supermodes can simultaneously spectrally match the transitions.

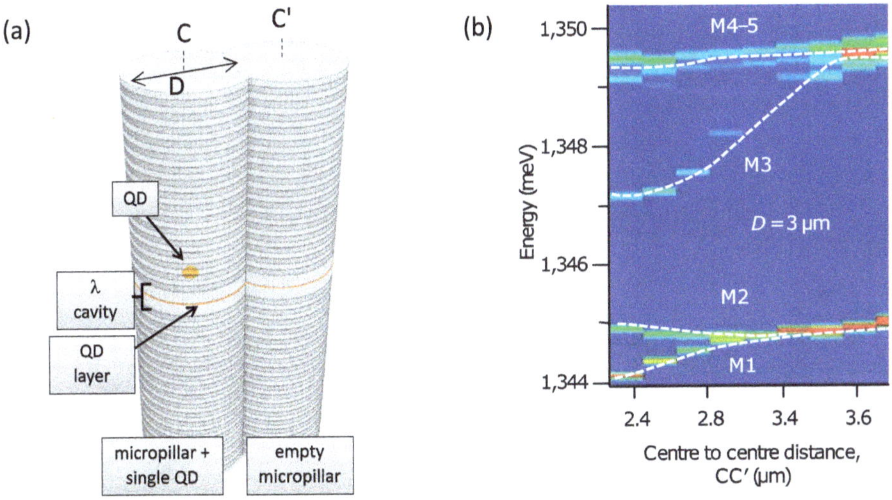

Figure 14.12: (a) Scheme of the realization of coupledmicropillars. Design parameters are: D is the diameter of the pillars forming the molecule, and C and C' are the centers of the micropillars. One QD is inserted in one pillar (vertically in the center of the λ cavity as depicted by the orange layer), whereas the other is empty. Image reprinted from [40] with permission of AIP Publishing. (b) Modes energies as a function of the distance between centers CC' for pillars of diameter 3 µm. Modes of the coupled cavity are labeled as $M1$ to $M5$. Figure reprinted with permission of Springer Nature from [39].

Successful implementation of coupled cavities have been achieved with micropillars [39], microdisks [14, 181], photonic crystals [7], stacked Fabry–Perot cavities [122], and tunable open cavities [77]. Particularly important in the realization of polarization-entangled photon pair sources is a cavity with modes with minimal to zero polarization splitting, otherwise bringing a limiting factor to the achievable entanglement. As exemplary discussed for micropillars, their diameter and the distance between the centers can be employed to control the energy of the modes together with the splitting of the newly formed hybridized modes. Ensuring the low-polarization splitting requires the pillars to be circular (or the quality factor to be low enough to ensure good spectral overlap between the modes of different linear polarizations).

Single micropillars have been fabricated on top of piezoelectric actuators for controlling the QD properties, but this approach still remains challenging to apply to high Q-factor pillar and coupled-pillar resonators. This is because it is nontrivial to transfer the strain induced by the actuator up to the height of the QD position.

14.3.3 Electrically pumped diode-based sources

As for single-photon sources, entangled photons generated under electrical excitation are technologically appealing since they can enable a reduction of the setup dimension (no laser or optical filtering setups would be required). Generation of entangled photon pairs under electrical excitation has been achieved. One exemplary device integrated telecom wavelength quantum dots into a p–i–n diode, further having the emitters embedded into a low-Q planar cavity (formed by a bottom DBR and a lower reflectivity top DBR) to enhance the extraction for both biexciton and exciton photons. Interestingly, a small device size with QDs embedded in a few micron-sized pillars (further electrically connected) allowed for fast electrical operation (reaching the GHz clock level; see Fig. 14.13) [183].

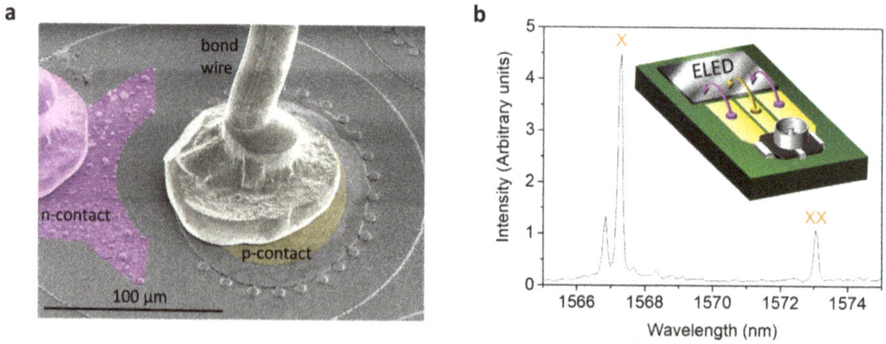

Figure 14.13: SEM picture (colorized) of a quantum dot-based LED operated under 1 GHz electrical excitation (pulsed). The ovally shaped mesa includes various connected pillars (small circular structures next to p-contact). Electrical isolation between pillar and enclosing wafer is ensured. P- and n-contacts are also shown. (b) Exciton and biexciton are visible in the luminescence spectrum under electrical pumping. In the inset the exemplary packaging for high-frequency driving is shown. The diode is bonded to tracks, which are impedance matched. Contacts are color coded as in (a). Reprinted with permission from [183]. Copyright 2023 the Optical Society.

14.3.4 Geometrical effect-based sources

Analogously as for single-photons sources and as discussed in Section 13.2, the modification of the TIR conditions can be an effective way to enhance the extraction of light of a semiconductor nanostructure. The appeal of this approach for the generation of polarization-entangled photon pairs resides in the broadband operation of such structures. Exemplary nanowires have been effectively employed for the generation of bright polarization-entangled photon pairs [207]. Also microlenses etched in the semiconductor showed their use as sources of entangled photon pairs, even under two-photon resonant excitation [16]. Nevertheless, as discussed in Section 13.2, the use of Purcell enhancement can be advantageous in several implementations, therefore requiring a photonic resonator as with the aforementioned sources.

14.4 Integrated quantum photonics

So far, the properties of single and entangled sources of light have been discussed, with particular attention on the possibility of extracting as much light as possible from the native semiconductor chip. Indeed, several experiments are designed for free-space operation, therefore having the quantum light propagating in collimated beams through free-space optics. Free-space quantum optical experiments reached long distances, even to satellites. Nevertheless, increasing the achievable distance on Earth implementations requires the use of optical fibers, which are currently the backbone of the ground telecommunication infrastructure. Light propagates in the high refractive index core and is confined thanks to the surrounding lower refractive index cladding. This operation principle will come at hand for understanding the physics behind the use of waveguides in on-chip applications. Indeed, upscaling the experimental complexity, having more and more optics involved, cannot be affordably done with bulky optics. Rather, as classical computing benefited from high complexity enabled by small device footprint, also on-chip quantum optics would take advantage of the high density of components available on semiconductor platforms. Even more appealing, Knill, Laflamme, and Milburn (KLM) proved in 2001 that photonic quantum computing can be efficiently performed employing only linear optical elements, i. e., single-photon sources, beamsplitters, phase shifters, and photo detectors [76]. This stimulated the field of quantum photonic integrated circuits (PIC) to demonstrate all required functionalities on the same sample. In addition, depending on the task at hand, additional devices may be required as filters, delay lines, or light switches operating at the single-photon level.

Once light has been generated, for example, employing a semiconductor quantum dot, it needs to be guided on chip employing single-mode elements (importantly with the lowest possible loss), which can further define a precise polarization. Guiding of light on chip can be achieved utilizing waveguides defined in the semiconductor material. For this purpose, structures are defined having a core material with high refrac-

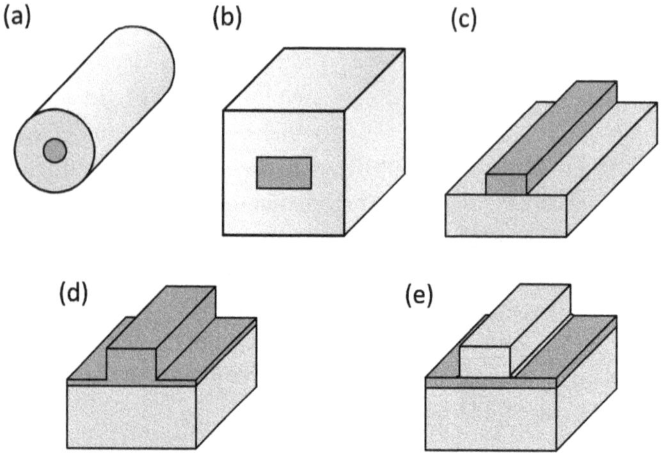

Figure 14.14: Exemplary sketches of typical waveguide designs. Higher refractive indexes are in darker colors. (a) Optical fiber with central core and surrounding cladding. (b) Channel waveguide. (c) Ridge waveguide having distinct materials on top (for example, air or vacuum) and bottom (typically, lower refractive index material) claddings. (d) Rib waveguide. It differs from the previous because of the presence of a slab of high refractive index material on the side of the waveguide. (e) Strip-loaded waveguide. Reprinted with permission of Springer Nature from [126], Chapter 13.

tive index, where the light is mostly propagating, and the surrounding lower refractive index cladding ensuring confinement. Several structures have been investigated, as exemplary summarized in Fig. 14.14.

14.4.1 Waveguides

Whereas Fig. 14.14(a) shows a sketch of a commonly utilized silica fiber (with a round core surrounded by a concentric cladding), Fig. 14.14(b) depicts a possible solid-state implementation of such a waveguide, where a central core (this time with rectangular cross-section) is surrounded by a lower refractive index cladding. This is often implemented in silicon photonics, where the core is made of Si, whereas the cladding is constituted of SiO_2. When a III–V material is employed, for example, GaAs, to enable the epitaxial integration of InAs quantum dots, a slight modification from the previous structure has been implemented: the WG material is indeed surrounded by air (or vacuum) instead of an oxide cladding, further increasing the refractive index mismatch between the waveguide and surrounding (often named nanobeam WG). To avoid the intrinsic fragility of a micrometric-sized WG suspended in air, alternative waveguide geometries can be employed. Ridge waveguides utilize a bottom low index material (which can be grown epitaxially before the deposition of the top semiconductor used to form the waveguide), whereas the WG after fabrication results to be surrounded from top and side by air (Fig. 14.14(c)). Alternatively, rib waveguides also posses an asymmetric vertical structure but, differently from the previous, present a small slab of material (same as the WG) below the waveguiding element (Fig. 14.14(d)). Finally, it is worth mentioning the geometry employed for the implementation of certain side-emitting laser structures: as shown in Fig. 14.14(e), the lateral confinement required for the waveguiding can also be achieved via the realization of a top low refractive index layer, which allows the confinement of light while it propagates in the waveguide layer underneath.

To understand the process of light guiding, Fig. 14.15 shows the mode profile for the fundamental TE mode of all waveguides described in Fig. 14.14.[3] Optimized fabrication, in particular, enabling low sidewalls roughness, is highly required to ensure low propagation losses (summarized in Table 14.2).

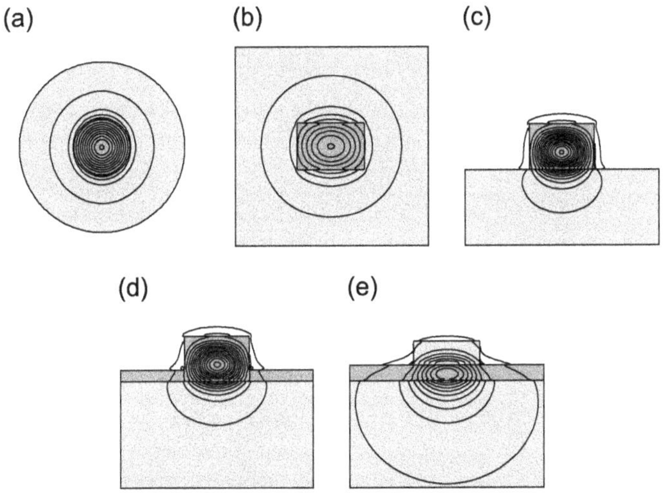

Figure 14.15: Profiles of the fundamental TE mode of the waveguides for all discussed geometries. MIT Photonic-Bands package was employed [126]. It is represented by the squared modulus of the electric field in linear scale. Lower symmetry waveguides also display a lower symmetry mode profile. For (c–e), calculations were performed assuming single-mode (TE) waveguides with layers formed by GaAs (dark gray) and $Al_{0.42}Ga_{0.58}As$ (light gray). Reprinted with permission of Springer Nature from [126], Chapter 13.

The use of mechanically stable waveguides, as ridge or rib WGs, simplifies the realization of further functionalities on chip (as, for example, the implementation of superconducting nanowire single-photon detectors as discussed in the following), since the

3 As discussed in [160], the modes in a WG are not fully transverse electromagnetic modes because of the coupling between all three field components upon interaction with the index step at the WG boundary. Nevertheless, transverse electric (TE) and magnetic (TM) still identify modes with the majority of the field in the respective direction.

Table 14.2: Exemplary waveguide types embedding semiconductor quantum dots and respective approximate performances. More details can be found, for example, in [76].

Waveguide type	Losses in dB/cm^{-1}	β factor	F_P
Ridge	0.2–10 (at 1550 nm and 900 nm)	~0.3	–
Ridge with Bragg cavity	N/A	~0.7	>3
DBR-based	~8 (at 900 nm)	~0.07	–
Nanobeam	~75 (at 950 nm)	~0.95	–
Nanobeam cavity	N/A	~1	~10
PhC $W1$	7 (at 1275 nm) >150 (at 900 nm)	~0.98	~27
PhC cavity with $W1$	>150 (at 900 nm)	>0.85	~40
PhC with ridge	~8 (at 900 nm)	0.95	>2

semiconductor photonic chip is more robust to further fabrication steps. The drawback of these systems, as, for example, GaAs/AlGaAs WGs, is based on the low refractive index mismatch between the waveguide and the bottom cladding: therefore an efficient coupling to radiative modes toward the bottom substrate impacts the achievable factors β (typically, around 10 % per direction). This can be a severe limiting factor when quantum emitters are embedded in the waveguide to realize on-chip sources of quantum light. Strategies to overcome this limitation are explained in the following sections. It is worth mentioning that once a WG structure has been chosen, a careful design of the photonic chip is required to ensure the minimization of eventual waveguide losses induced by bending.

14.4.2 Waveguide-based single-photon sources: Fabry–Perot resonators

One alternative to increase the coupling to the rib and ridge waveguide mode is employing distributed Bragg reflectors realized in the waveguide to form a Fabry–Perot-like resonator (see Fig. 14.16). In Fig. 14.16, the mirrors are exemplary defined by etching periodic air layers in the semiconductor constituting the waveguide.

14.4.3 Waveguide-based single-photon sources: photonic crystal waveguides

An interesting approach for realizing waveguides is based on photonic crystals. As explained in Section 13.3, a local breaking of the periodicity in the refractive index can bring to the formation of allowed states in the photonic bandgap: whereas light can be inhibited where the PhC is periodic, i. e., the bandgap is present, photons can exist in the defect region. Besides a PhC cavity, this behavior can also be utilized to form a waveguide.

As for PhC resonators, where a triangular lattice of holes is formed in a slab of semiconductor surrounded by air, waveguides can be realized in the same way: the photonic

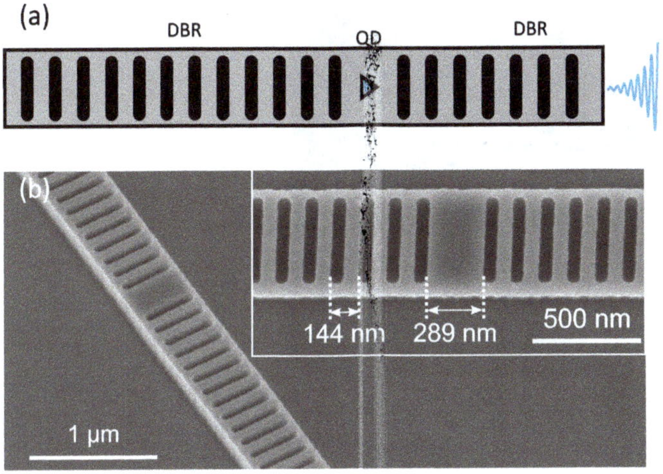

Figure 14.16: (a) Illustration of the Fabry–Perot cavity constituted by two distributed Bragg reflectors. An exemplary QD is sketched in the cavity region. Most of the photons leave the cavity toward the lower reflectivity mirror. (b) SEM picture of a resonator (utilizing two distinct Bragg mirrors) embedded into a ridge waveguide (~900-nm operation wavelength). The cavity region and respective key geometrical parameters are shown in the inset. Reprinted with permission from [74]. Copyright 2023 the Optical Society.

crystal bandgap prevents light propagation in the plane of the PhC, whereas total internal reflection results in confinement in the vertical direction, perpendicular to the plane of the semiconductor slab (see Fig. 14.17(a) for an exemplary $W1$ waveguide). As for resonators, also these waveguides present a light cone in the band structure, where total internal reflection conditions are violated (named radiation modes in Fig. 14.17(b)). Together with the confinement that comes from the PhC bandgap and the TIR conditions, which provide high achievable factors β, PhCWG can also enable Purcell enhancement of the transition of a two-level system placed in the waveguide. This effect, despite the lack of a photonic resonator, can be attributed to the modification of the local density of states due to the presence of the waveguide: Purcell enhancement is here enabled by the induced slow-light, therefore proportional to the group index $n_g = c/v_g$, where c is the speed of light in vacuum, and v_g is the group velocity. The latter is the slope of the band of the waveguide, which, as shown in Fig. 14.17(b), decreases at the band edge [6]. Such waveguide structures have shown broadband enhancement of the factor β (>90 % over several tens of nanometers) together with Purcell enhancement (over a few nm; see also the source description in the previous sections). The controlled and enhanced light-matter coupling between the light propagating in the WG and a TLS placed in it shows interesting perspective for the implementation of on-chip functionalities as transistors and single-photon gates, appealing for the implementation of on-chip information processing.

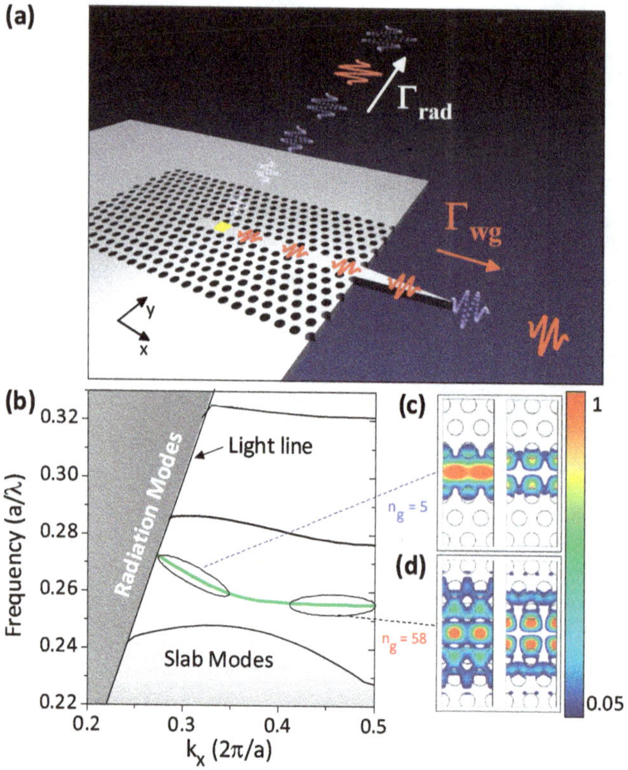

Figure 14.17: (a) Sketch of a photonic crystal *W*1 waveguide, further marking the decay rate in the waveguide over the decay in other radiation modes (Γ_{wg} and Γ_{rad}, respectively, in the picture; these correspond to the nomenclature utilized as Γ_c and Γ). (b) TE band structure of the waveguide: the even and odd waveguide modes are marked in green and black, respectively. The continuum of radiation modes are represented by the gray-shaded area. (c) and (d) show the electric field profiles (*x* components on the left, *y* components on the right) for the two regions of interests in the even mode. Reprinted with permission from [6]. Copyright (2023) by the American Physical Society.

14.4.4 Beamsplitters

Having now defined the waveguide geometries, which, together with an embedded semiconductor nanostructure, can constitute the source of quantum light on chip, it is important to discuss how to realize beamsplitters and how to tune their splitting ratio. Indeed, on-chip BS can be employed for realizing two-photon interference and key elements for creating superposition states (as, for example, Hadamard gates). Considering the discussed WG geometries, two beamsplitter designs can be exploited. The first one is based on evanescent coupling. Two single-mode waveguides are brought together, close enough so that the evanescent fields of the WGs, as seen in Fig. 14.15, can interact with each other (see Fig. 14.18(a,b)). The splitting ratio can be controlled by altering the evanescent coupling region length or the WG optical distance (for a specific wave-

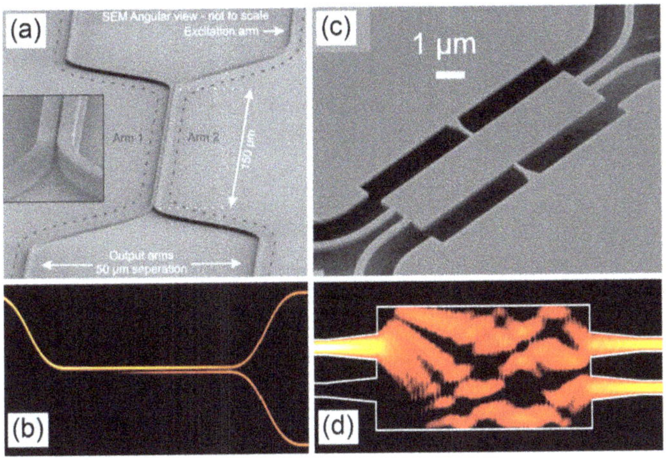

Figure 14.18: (a) SEM picture of a directional coupler (~900-nm operation wavelength). The ridge waveguides have a length of 150 µm in the coupling region. (b) Electric field profile simulated for a directional coupler showing 50:50 splitting ratio. (c) SEM picture of a beamsplitter based on MMI geometry, realized with nanobeam waveguides. (d) Electric field profile simulated for an MMI coupler showing 50:50 splitting ratio. Image (a) reprinted from [161] with permission of AIP Publishing. Image (c) reprinted with permission from [127]. Copyright 2023 the Optical Society. Images (b) and (d) reprinted with permission of Wiley Company from [76].

length) [161]. This design is referred to as a directional coupler. An alternative design is based on the use of a portion of a multimode waveguide element, where the self-imaging principle allows for creating splitting patterns, which result in certain splitting ratios of the beamsplitter (Fig. 14.18(c,d)). These elements are often refer to as multimode interference (MMI) couplers and have the advantage of being more fabrication tolerant with respect to the evanescent counterpart (at the expense of a slightly increased losses due to fabrication imperfections).

14.4.5 Phase shifters

These elements play a central role in quantum photonic integrated circuits since, in combination with a Mach–Zehnder interferometer, arbitrary unitary operations on single qubits are possible. Phase shifters are often realized on silicon PIC employing the thermo-optic effect. One heater positioned on one arm of an MZI can modify the refractive index via temperature change, which transforms in a change of the propagation constant (and of the phase) in that specific arm. The phase change takes the form $\Delta\Phi = \frac{2\pi L}{\lambda}\frac{\mathrm{d}n}{\mathrm{d}T}\Delta T$, where L is the device length, λ is the operation wavelength, and $\frac{\mathrm{d}n}{\mathrm{d}T}$ (~$2.35 \times 10^{-4}\,\mathrm{K}^{-1}$ for GaAs at 300 K) is the thermo-optic coefficient. For cryogenic applications, the use of heaters on the chip needs to be typically avoided to maintain the required temperature. Noncentrosymmetric materials (GaAs, $LiNbO_3$, InP) can exploit

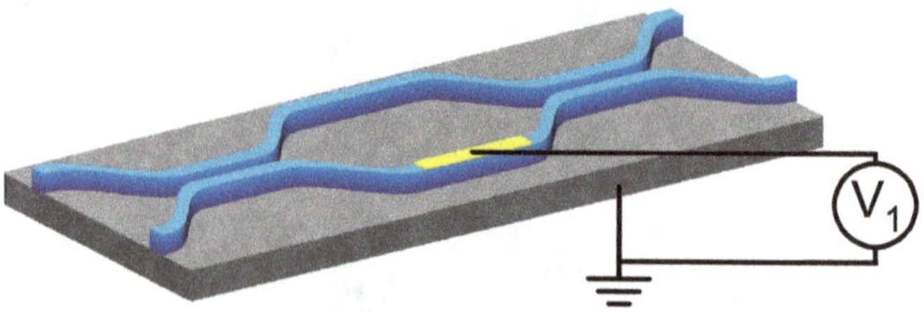

Figure 14.19: Sketch of a Mach–Zehnder interferometer with one arm controlled with a phase shifter. It is implemented with a metal contact enabling refractive index modulation via electro-optical effect. Reprinted and adapted with permission of Springer Nature from [126], Chapter 13.

electro-optic (see Fig. 14.19) or Pockels effect. Despite the refractive index change can be modest, for light propagations much longer than the wavelength (in the material), it can become substantial. For modulators, an important parameter is given by the half-wave voltage, i. e., the voltage that provides a π phase shift [76].

14.4.6 Single-photon detectors

For the effective scalability of functionalities on-chip, the last fundamental building block is given by single-photon detectors. Similarly to classical detectors (as semiconductor-based photodiodes), the performances of a single-photon detector can be characterized by detection efficiency, time resolution, operation wavelength, dead time, dark counts, and maximum measurable intensity (or count rate). Particularly appealing for the task are superconducting nanowire single-photon detectors (SNSPDs). These systems have some appealing advantages with respect to their semiconductor counterpart (where typically single-photon sensitivity is reached in avalanche photodiodes, where the carriers created by photon absorption are accelerated so that they can generate secondary electrons): they can reach near-unity detection efficiencies (typically, in a wavelength range of a few tens of nanometers, although they can still detect with lower efficiencies in a very broad range even between UV and mid-infrared), together with time resolutions of a few picoseconds and dark counts as low as few Hz, and they can detect count rates higher than a few MHz. The drawback is that SNSPDs require cryogenic temperatures (typically of a few degree Kelvin or below) for optimal operation. As their name suggests, SNSPDs are realized via the deposition of a high-quality film of superconducting material, followed by the definition of a pattern via e-beam lithography and etching, resulting in a meander or spiral design of a superconducting nanowire (see Fig. 14.20). Without entering into the details of the physics behind the detection principle (which is still in discussion), the operation principle can be schematized as follows [76]:

– The wire is kept below its critical temperature T_c, and a current I_b is applied, being below, albeit close, to the critical current I_c.

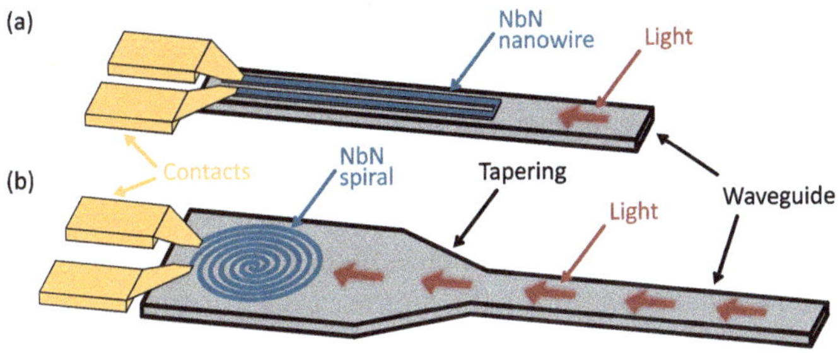

Figure 14.20: (a) Schematic of a waveguide including an NbN superconducting nanowire single-photon detector realized on top. The light travels inside the waveguide, and it is evanescently coupled to the SNSPD. Because of the small size, this design can enable high time resolution. Exemplary realization can be found in [90]. (b) Exemplary spiral design of a nanowire detector realized at the end side of a tapering in the waveguide, aiming at high detection efficiencies thanks to the large area. Double Archimedean spirals were exemplarily employed in [178]. Metallic contacts for the signal readout are schematized.

- Once a photon gets absorbed, a photo-excited electron thermally relaxes leading to formation of quasiparticles (e. g., broken Cooper pairs) and phonons. The formation of this cloud of quasiparticles locally suppresses the superconductivity, resulting in a "hot-spot."
- A local highly resistive spot creates current crowding in the area surrounding this nonsuperconductive location. When the current density exceeds the switching current, a resistive barrier is created.
- The resulting increased resistance is responsible for the current to be redirected from the nanowire to the detection electronics: this results in a detected voltage spike, indicating the photon detection event.
- The heat in the system is then dissipated, restoring the superconductivity, and the SNSPD is ready for another detection event.

More elaborate models are currently in discussion to understand the detection mechanisms (for example, including magnetic vortices; see [44] and references therein).

SNSPDs have been realized with various superconducting materials, such as niobium nitride (NbN; see Fig. 14.20), niobium titanium nitride (NbTiN), tantalum nitride (TaN), tungsten silicide (WSi), and molybdenum silicide (MoSi). In particular, detectors have been fabricated on several material platforms as gallium arsenide, lithium niobate, alluminium nitride, silicon, and silicon nitride. Some materials show better performances, and others higher operation temperatures (as for the nitride-systems, 4–11 K operation has been demonstrated). Nonetheless, it is appealing to realize such a system in combination with a developed quantum photonic integrated circuit. This complicates the realization of effective SNSPDs, since their performances are not forgiving to small fabrication tolerances (for example, a nonuniform deposited film). Still, the demonstration of all required functionalities on the same chip stimulated the research to allow fabricating such single-photon detectors on photonic chips.

Furthermore, operating SNSPDs on a photonic chip bears another challenge: optimal performance of semiconductor light sources is nowadays achieved via optical pumping. The excitation laser needs then to be strongly suppressed before it reaches the detector. Although for nonresonant pumping schemes, filters can be realized on the photonic chip (as DBR, cavities, microrings, etc.), this approach is more challenging for resonant pumping, since the excitation laser cannot be discriminated in wavelength.

Whereas the recent results on SNSPD realization on PIC and their respective performances can be found, for example, in [76], it is worth mentioning the first implementation of a fully on-chip Hanbury Brown–Twiss experiment on a monolithic semiconductor-superconductor platform [178]. In the work, an InAs quantum dot was embedded in a GaAs single-mode ridge waveguide. The resonant excitation laser has been shaped to make the pulse spectrally narrower [177] to improve the signal-to-laser background in the waveguide (this further aided by the laser polarization chosen parallel to the WG to sensibly diminish the amount of light that can couple to the waveguide mode). Two-single mode waveguides were employed to form a 50:50 BS as a directional coupler, followed by two SNSPDs fabricated on top of the waveguide to allow for the evanescent field to interact (and be absorbed) by the detector (Fig. 14.21). This study showed the simultaneous realization of a WG-coupled light source, even resonantly excited, a beamsplitter, and two working SNSPDs. Further technological details on this work can be directly found in the original paper [178].

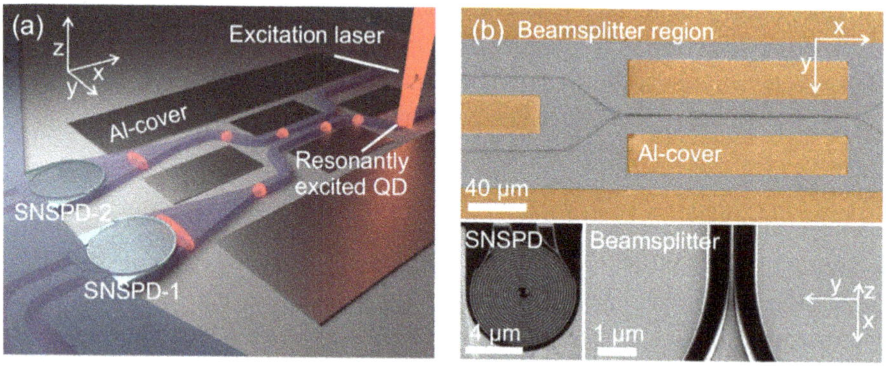

Figure 14.21: (a) Scheme of the device: one resonantly excited QD is embedded into a single-mode ridge waveguide, and a directional coupler is realized with two SNSPDs at the output ports. AlN/Al layers are utilized as covers for reducing the laser scattering (not shown on the SNSPDs for clarity). (b) SEM pictures (colorized) after fabricating the device and depositing the Al-covers (in gold color). Various elements are visible: the beamsplitter (top view) with the Al-covers, one SNSPD (bottom, left, top view), and a close image of the BS region, where the two WGs are coming close (bottom, right, angular view). Reprinted with permission from [178]. Copyright (2023) American Chemical Society.

14.5 Quantum dot emitters: properties in a nutshell

In the following, we will summarize a few properties of semiconductor quantum dots. Rather than focusing on the performances of light emission that can depend on the device utilized (and already discussed along the chapter), here the emission wavelength, operation temperature (for which $g^{(2)}(0) < 0.5$ was observed), and measured decay time will be reported. These values are summarized in Table 14.3. More information can be found in [5] and references therein.

Table 14.3: Summary of semiconductor quantum dots properties. Most of the numbers are obtained from [5] and references therein.

Material system (QD/surrounding)	Emission wavelength (nm)	Decay time exciton or trion (ns)	Max. operating temperature (K)
InAs/InP	~1550	~0.5–2	~80
InAs/GaAs	~870–1550	~1–2	~120
GaAs/AlGaAs	~780–900	~0.3–0.5	~80
GaAs/GaAsP	~750	~0.3	~160
InP/AlInGaP	~650–680	~2	~80
CdSe/ZnSe	~520–570	~0.3	~300
CdSe/ZnSSe	~500–550	~0.9	~300
InGaN/GaN	~400–600	~1	~280
GaN/AlGaN (wz)	~280–450	–	~350
GaN/AlGaN (zb)	~330	~0.3–3	~100
CdTe/ZnTe	~550	~0.3	–

14.6 Summary

- Bright sources of single- and entangled-photons can be realized utilizing photonic resonators or geometric effects. Although the latter only requires spatial matching (being practically broadband for the wavelength range of interest), the former requires also spectral matching between the emitter's transition (or transitions) and the cavity mode (or modes).
- Micropillar cavities have the advantage of high extraction efficiency (>60 %) and Purcell enhancement (>10), with a relatively narrow mode (~90 pm). Electric fields can be applied to these resonators with a proper epitaxial and nanofabrication-realized geometry. High single-photon purity and near-unity indistinguishability under optical pumping have been shown. Even electrical pumping with high excitation rates can be realized.
- Circular Bragg grating cavities display very high extraction efficiencies over a broad spectral range (~90 % for several tens of nanometers). They operate with a rather broad spectral mode (~10 nm), but the small mode volume allows for achieving high

Purcell factors (>15). Near-unity indistinguishability and high single-photon purity have been demonstrated (optical pumping). Thanks to the spectrally broad mode, several different emitter's transitions can be coupled to the cavity. As sources of entangled photon pairs, high degree of entanglement has been achieved under optical excitation. Novel designs have been investigated to realize electrically controlled and, in future, electrically pumped CBGs.

- Open Fabry–Perot cavities reached high values of out-coupling efficiencies (>85 %), together with very high coupling to single mode fibers, which enabled unprecedented source efficiencies. Large Purcell factors (>10) have been demonstrated (for a narrow mode ~70 pm) with near-unity indistinguishability and high single-photon purity.
- Linear optics photonic quantum computing can be realized with only a few optical elements: single-mode waveguides, beamsplitters, phase shifters, and single-photon detectors. These elements have been demonstrated for on-chip operation. Still, realizing them all in a scalable manner on the same chip is a technological challenge.
- Optical resonators and photonic crystal waveguides can be employed to achieve, on-chip, high coupling to the waveguide mode together with Purcell enhancement.

Bibliography

[1] M. Abramowitz and I. A. Stegun, editors. *Handbook of Mathematical Functions: With Formulas, Graphs, and Mathematical Tables*. Dover books on advanced mathematics. Dover Publications, New York, 1964. 255, 256

[2] Y. Akahane, T. Asano, B.-S. Song, and S. Noda. High-Q photonic nanocavity in a two-dimensional photonic crystal. *Nature*, 425(6961):944–947, 2003. 299, 300, 301, 304, 305

[3] J. B. Altepeter, E. R. Jeffrey, and P. G. Kwiat. Photonic state tomography. Volume 52 of *Advances In Atomic, Molecular, and Optical Physics*, pages 105–159. Academic Press, 2005. 115, 118

[4] C. Antonelli, M. Shtaif, and M. Brodsky. Sudden death of entanglement induced by polarization mode dispersion. *Physical Review Letters*, 106:080404, 2011. 119

[5] Y. Arakawa and M. J. Holmes. Progress in quantum-dot single photon sources for quantum information technologies: A broad spectrum overview. *Applied Physics Reviews*, 7(2):021309, 2020. 203, 339

[6] M. Arcari, I. Söllner, A. Javadi, S. Lindskov Hansen, S. Mahmoodian, J. Liu, H. Thyrrestrup, E. H. Lee, J. D. Song, S. Stobbe, and P. Lodahl. Near-unity coupling efficiency of a quantum emitter to a photonic crystal waveguide. *Physical Review Letters*, 113(9):093603, 2014. https://doi.org/10.1103/PhysRevLett.113.093603. 323, 324, 333, 334

[7] K. A. Atlasov, K. F. Karlsson, A. Rudra, B. Dwir, and E. Kapon. Wavelength and loss splitting in directly coupled photonic-crystal defect microcavities. *Optics Express*, 16(20):16255–16264, 2008. 327

[8] W. L. Barnes, G. Björk, J. M. Gérard, P. Jonsson, J. A. E. Wasey, P. T. Worthing, and V. Zwiller. Solid-state single photon sources: Light collection strategies. *European Physical Journal D*, 18:197–210, 2002. 313, 314

[9] S. Bauer, D. Wang, N. Hoppe, C. Nawrath, J. Fischer, N. Witz, M. Kaschel, C. Schweikert, M. Jetter, S. L. Portalupi, M. Berroth, and P. Michler. Achieving stable fiber coupling of quantum dot telecom C-band single-photons to an SOI photonic device. *Applied Physics Letters*, 119(21):211101, 2021. 324

[10] L. Béguin, J. P. Jahn, J. Wolters, M. Reindl, Y. Huo, R. Trotta, A. Rastelli, F. Ding, O. G. Schmidt, P. Treutlein, and R. J. Warburton. On-demand semiconductor source of 780-nm single photons with controlled temporal wave packets. *Physical Review B*, 97:205304, 2018. https://doi.org/10.1103/PhysRevB.97.205304. 238, 239

[11] J. S. Bell. On the einstein podolsky rosen paradox. *Physics Physique Fizika*, 1:195–200, 1964. 117

[12] C. H. Bennett, G. Brassard, and N. D. Mermin. Quantum cryptography without bell's theorem. *Physical Review Letters*, 68:557–559, 1992. 118

[13] C. H. Bennett, D. P. DiVincenzo, J. A. Smolin, and W. K. Wootters. Mixed-state entanglement and quantum error correction. *Physical Review A*, 54:3824–3851, 1996. 114

[14] M. Benyoucef, J.-B. Shim, J. Wiersig, and O. G. Schmidt. Quality-factor enhancement of supermodes in coupled microdisks. *Optics Letters*, 36(8):1317–1319, 2011. 327

[15] C. Böckler, S. Reitzenstein, C. Kistner, R. Debusmann, A. Löffler, T. Kida, S. Höfling, A. Forchel, L. Grenouillet, J. Claudon, and J. M. Gérard. Electrically driven high-Q quantum dot-micropillar cavities. *Applied Physics Letters*, 92(9):091107, 2008. 315

[16] S. Bounouar, C. de la Haye, M. Strauß, P. Schnauber, A. Thoma, M. Gschrey, J.-H. Schulze, A. Strittmatter, S. Rodt, and S. Reitzenstein. Generation of maximally entangled states and coherent control in quantum dot microlenses. Applied Physics Letters, 112(15):153107, 2018. 329

[17] S. Bounouar, M. Elouneg-Jamroz, M. Den Hertog, C. Morchutt, E. Bellet-Amalric, R. André, C. Bougerol, Y. Genuist, J. P. Poizat, S. Tatarenko, and K. Kheng. Ultrafast room temperature single-photon source from nanowire-quantum dots. *Nano Letters*, 12(6):2977–2981, 2012. 203

[18] S. Bounouar, M. Müller, A. M. Barth, M. Glässl, V. M. Axt, and P. Michler. Phonon-assisted robust and deterministic two-photon biexciton preparation in a quantum dot. *Physical Review B*, 91:161302, 2015. https://doi.org/10.1103/PhysRevB.91.161302. 242, 243, 244

https://doi.org/10.1515/9783110703412-015

[19] A. J. Brash, J. Iles-Smith, C. L. Phillips, D. P. S. McCutcheon, J. O'Hara, E. Clarke, B. Royall, L. R. Wilson, J. Mørk, M. S. Skolnick, A. M. Fox, and A. Nazir. Light scattering from solid-state quantum emitters: beyond the atomic picture. *Physical Review Letters*, 123(16):167403, 2019. https://doi.org/10.1103/PhysRevLett.123.167403. 181

[20] X. Brokmann, M. Bawendi, L. Coolen, and J.-P. Hermier. Photon-correlation Fourier spectroscopy. *Optics Express*, 14(13):6333, 2006. 89, 91, 94

[21] R. Hanbury Brown and R. Q. Twiss. Correlation between photons in two coherent beams of light. *Nature*, 177(4497):27–29, 1956. 77

[22] Q. Buchinger, S. Betzold, S. Höfling, and T. Huber-Loyola. Optical properties of circular Bragg gratings with labyrinth geometry to enable electrical contacts. *Applied Physics Letters*, 122(11):111110, 2023. 319

[23] T. Cai, R. Bose, G. S. Solomon, and E. Waks. Controlled coupling of photonic crystal cavities using photochromic tuning. *Applied Physics Letters*, 102(14):141118, 2013. 294

[24] Y. Chen, M. Zopf, R. Keil, F. Ding, and O. G. Schmidt. Highly-efficient extraction of entangled photons from quantum dots using a broadband optical antenna. *Nature Communications*, 9(1):2994, 2018. 290, 323

[25] J. Claudon, J. Bleuse, N. S. Malik, M. Bazin, P. Jaffrennou, N. Gregersen, C. Sauvan, P. Lalanne, and J.-M. Gérard. A highly efficient single-photon source based on a quantum dot in a photonic nanowire. *Nature Photonics*, 4(3):174–177, 2010. 290, 291, 322

[26] J. F. Clauser, M. A. Horne, A. Shimony, and R. A. Holt. Proposed experiment to test local hidden-variable theories. *Physical Review Letters*, 23:880–884, 1969. 117

[27] V. Coffman, J. Kundu, and W. K. Wootters. Distributed entanglement. *Physical Review A*, 61:052306, 2000. 114

[28] D. Cogan, Z.-E. Su, O. Kenneth, and D. Gershoni. Deterministic generation of indistinguishable photons in a cluster state. *Nature Photonics*, 17(4):324–329, 2023. 174

[29] N. Coste, D. A. Fioretto, N. Belabas, S. C. Wein, P. Hilaire, R. Frantzeskakis, M. Gundin, B. Goes, N. Somaschi, M. Morassi, A. Lemaître, I. Sagnes, A. Harouri, S. E. Economou, A. Auffeves, O. Krebs, L. Lanco, and P. Senellart. High-rate entanglement between a semiconductor spin and indistinguishable photons. *Nature Photonics*, 17(7):582–587, 2023. 174

[30] S. F. Covre Da Silva, G. Undeutsch, B. Lehner, S. Manna, T. M. Krieger, M. Reindl, C. Schimpf, R. Trotta, and A. Rastelli. GaAs quantum dots grown by droplet etching epitaxy as quantum light sources. *Applied Physics Letters*, 119(12):120502, 2021. 268

[31] G. E. Cragg and A. L. Efros. Suppression of auger processes in confined structures. *Nano Letters*, 10(1):313–317, 2010. 192

[32] A. C. Dada, T. S. Santana, R. N. E. Malein, A. Koutroumanis, Y. Ma, J. M. Zajac, J. Y. Lim, J. D. Song, and B. D. Gerardot. Indistinguishable single photons with flexible electronic triggering. *Optica*, 3(5):493, 2016. 167

[33] M. Davanço, M. T. Rakher, D. Schuh, A. Badolato, and K. Srinivasan. A circular dielectric grating for vertical extraction of single quantum dot emission. *Applied Physics Letters*, 99(4):041102, 2011. https://doi.org/10.1063/1.3615051. 295

[34] K. De Greve, L. Yu, P. L. McMahon, J. S. Pelc, C. M. Natarajan, N. Y. Kim, E. Abe, S. Maier, C. Schneider, M. Kamp, S. Höfling, R. H. Hadfield, A. Forchel, M. M. Fejer, and Y. Yamamoto. Quantum-dot spin-photon entanglement via frequency downconversion to telecom wavelength. *Nature*, 491(7424):421–425, 2012. 171

[35] E. V. Denning, J. Iles-Smith, N. Gregersen, and J. Mork. Phonon effects in quantum dot single-photon sources. *Optical Materials Express*, 10(1):222, 2020. 181

[36] C. Diederichs. *Habilitation à diriger de recherches: The resonance fluorescence of single semiconductor quantum dots for the generation of indistinguishable photons*. Université Pierre et Marie Curie (UPMC), 2016. tel-01416901. 145, 147

[37] J. E. Dixon. *Towards Integrated Scalable Nanophotonic Circuits*. PhD thesis, The University of Sheffield, Faculty of Science, Department of Physics and Astronomy, 2017. 284

[38] A. Dousse, L. Lanco, J. Suffczyński, E. Semenova, A. Miard, A. Lemaître, I. Sagnes, C. Roblin, J. Bloch, and P. Senellart. Controlled light-matter coupling for a single quantum dot embedded in a pillar microcavity using far-field optical lithography. *Physical Review Letters*, 101:267404, 2008. 279

[39] A. Dousse, J. Suffczyński, A. Beveratos, O. Krebs, A. Lemaître, I. Sagnes, J. Bloch, P. Voisin, and P. Senellart. Ultrabright source of entangled photon pairs. *Nature*, 466(7303):217–220, 2010. 327

[40] A. Dousse, J. Suffczyński, O. Krebs, A. Beveratos, A. Lemaître, I. Sagnes, J. Bloch, P. Voisin, and P. Senellart. A quantum dot based bright source of entangled photon pairs operating at 53 K. *Applied Physics Letters*, 97(8):081104, 2010. 327

[41] A. Einstein, B. Podolsky, and N. Rosen. Can quantum-mechanical description of physical reality be considered complete? *Physical Review*, 47(10):777–780, 1935. 109

[42] D. J. P. Ellis, A. J. Bennett, A. J. Shields, P. Atkinson, and D. A. Ritchie. Electrically addressing a single self-assembled quantum dot. *Applied Physics Letters*, 88(13):133509, 2006. 247

[43] L. Engel, S. Kolatschek, T. Herzog, S. Vollmer, M. Jetter, S. L. Portalupi, and P. Michler. Purcell enhanced single-photon emission from a quantum dot coupled to a truncated Gaussian microcavity. *Applied Physics Letters*, 122(4):043503, 2023. 273, 274, 296, 306, 307

[44] I. Esmaeil Zadeh, J. Chang, J. W. N. Los, S. Gyger, A. W. Elshaari, S. Steinhauer, S. N. Dorenbos, and V. Zwiller. Superconducting nanowire single-photon detectors: A perspective on evolution, state-of-the-art, future developments, and applications. *Applied Physics Letters*, 118(19):190502, 2021. 337

[45] K. A. Fischer, K. Müller, K. G. Lagoudakis, and J. Vučković. Dynamical modeling of pulsed two-photon interference. *New Journal of Physics*, 18(11):113053, 2016. 83, 167

[46] R. Fons, A. D. Osterkryger, P. Stepanov, E. Gautier, J. Bleuse, J.-M. Gérard, N. Gregersen, and J. Claudon. All-optical mapping of the position of quantum dots embedded in a nanowire antenna. *Nano Letters*, 18(10):6434–6440, 2018. 291

[47] A. M. Fox. *Optical Properties of Solids*. Oxford master series in condensed matter physics. Oxford University Press, 2001. 64

[48] M. Fox. *Quantum Optics*. Oxford Master Series in Physics. Oxford University Press, 2006. 37, 128, 131

[49] F. C. Frank, J. H. van der Merwe, and N. F. Mott. One-dimensional dislocations. i. static theory. *Proceedings of the Royal Society of London. Series A, Mathematical and Physical Sciences*, 198(1053):205–216, 1949. 267

[50] M. Francardi, L. Balet, A. Gerardino, N. Chauvin, D. Bitauld, L. H. Li, B. Alloing, and A. Fiore. Enhanced spontaneous emission in a photonic-crystal light-emitting diode. *Applied Physics Letters*, 93(14):143102, 2008. https://doi.org/10.1063/1.2964186. 321

[51] B. Gaál, M. A. Jacobsen, L. Vannucci, J. Claudon, J.-M. Gérard, and N. Gregersen. Near-unity efficiency and photon indistinguishability for the "hourglass" single-photon source using suppression of the background emission, *Applied Physics Letters*, 121(17):170501, 2022. https://doi.org/10.1063/5.0107624. 105

[52] M. Galli, S. L. Portalupi, M. Belotti, L. C. Andreani, L. O'Faolain, and T. F. Krauss. Light scattering and Fano resonances in high-Q photonic crystal nanocavities. *Applied Physics Letters*, 94(7):071101, 2009. 304, 305

[53] W. B. Gao, P. Fallahi, E. Togan, J. Miguel-Sanchez, and A. Imamoglu. Observation of entanglement between a quantum dot spin and a single photon. *Nature*, 491(7424):426–430, 2012. 172

[54] B. Gayral, J. M. Gérard, A. Lemaître, C. Dupuis, L. Manin, and J. L. Pelouard. High- Q wet-etched GaAs microdisks containing InAs quantum boxes. *Applied Physics Letters*, 75(13):1908–1910, 1999. 295

[55] D. Gerace and L. C. Andreani. Effects of disorder on propagation losses and cavity Q-factors in photonic crystal slabs. *Photonics and Nanostructures – Fundamentals and Applications*, 3(2):120–128, 2005. 293, 300

[56] J. M. Gérard, B. Sermage, B. Gayral, B. Legrand, E. Costard, and V. Thierry-Mieg. Enhanced spontaneous emission by quantum boxes in a monolithic optical microcavity. *Physical Review Letters*, 81:1110–1113, 1998. 295

[57] J. P. Gordon and H. Kogelnik. PMD fundamentals: Polarization mode dispersion in optical fibers. *Proceedings of the National Academy of Sciences*, 97(9):4541–4550, 2000. 119

[58] T. Grange, N. Somaschi, C. Antón, L. De Santis, G. Coppola, V. Giesz, A. Lemaître, I. Sagnes, A. Auffèves, and P. Senellart. Reducing phonon-induced decoherence in solid-state single-photon sources with cavity quantum electrodynamics. *Physical Review Letters*, 118:253602, 2017. https://doi.org/10.1103/PhysRevLett.118.253602. 182, 314

[59] N. Gregersen, P. Kaer, and J. Mork. Modeling and design of high-efficiency single-photon sources. *IEEE Journal of Selected Topics in Quantum Electronics*, 19(5):1–16, 2013. 72, 73, 74

[60] J. Q. Grim, A. S. Bracker, M. Zalalutdinov, S. G. Carter, A. C. Kozen, M. Kim, C. S. Kim, J. T. Mlack, M. Yakes, B. Lee, and D. Gammon. Scalable in operando strain tuning in nanophotonic waveguides enabling three-quantum-dot superradiance. *Nature Materials*, 18(9):963–969, 2019. 224, 225, 226

[61] M. Grundmann. *The Physics of Semiconductors*. Springer, 2006. 145, 152, 215, 216

[62] M. Gschrey, A. Thoma, P. Schnauber, M. Seifried, R. Schmidt, B. Wohlfeil, L. Krüger, J.-H. Schulze, T. Heindel, S. Burger, F. Schmidt, A. Strittmatter, S. Rodt, and S. Reitzenstein. Highly indistinguishable photons from deterministic quantum-dot microlenses utilizing three-dimensional in situ electron-beam lithography. *Nature Communications*, 6:7662, 2015. 277, 279, 290, 291, 323

[63] B. Guha, F. Marsault, F. Cadiz, L. Morgenroth, V. Ulin, V. Berkovitz, A. Lemaître, C. Gomez, A. Amo, S. Combrié, B. Gérard, G. Leo, and I. Favero. Surface-enhanced gallium arsenide photonic resonator with quality factor of 6×10^6. *Optica*, 4(2):218–221, 2017. 295

[64] R. Hafenbrak. *Tuning the exciton fine structure of single (In,Ga)As/GaAs quantum dots to realize a triggered entangled photon source*. PhD thesis, Institut für Halbleiteroptik und Funktionelle Grenzflächen, Universität Stuttgart, 2011. Verlag Dr. Hut, München. 114, 115, 116

[65] R. Hafenbrak, S. M. Ulrich, P. Michler, L. Wang, A. Rastelli, and O. G. Schmidt. Triggered polarization-entangled photon pairs from a single quantum dot up to 30 K. *New Journal of Physics*, 9(9):315, 2007. 113

[66] L. Hanschke, K. A. Fischer, S. Appel, D. Lukin, J. Wierzbowski, S. Sun, R. Trivedi, J. Vučković, J. J. Finley, and K. Müller. Quantum dot single-photon sources with ultra-low multi-photon probability. *npj Quantum Information*, 4(1):43, 2018. 168

[67] F. Hargart, C. A. Kessler, T. Schwarzbäck, E. Koroknay, S. Weidenfeld, M. Jetter, and P. Michler. Electrically driven quantum dot single-photon source at 2 GHz excitation repetition rate with ultra-low emission time jitter. *Applied Physics Letters*, 102(1):011126, 2013. 249, 314

[68] F. Hargart, M. Müller, K. Roy-Choudhury, S. L. Portalupi, C. Schneider, S. Höfling, M. Kamp, S. Hughes, and P. Michler. Cavity-enhanced simultaneous dressing of quantum dot exciton and biexciton states. *Physical Review B*, 93(11):115308, 2016. 289, 310, 312

[69] Y.-M. He, Y. He, Y.-J. Wei, D. Wu, M. Atatüre, C. Schneider, S. Höfling, M. Kamp, C.-Y. Lu, and J.-W. Pan. On-demand semiconductor single-photon source with near-unity indistinguishability. *Nature Nanotechnology*, 8(3):213–217, 2013. 62

[70] Y. M. He, H. Wang, C. Wang, M. C. Chen, X. Ding, J. Qin, Z. C. Duan, S. Chen, J. P. Li, R. Z. Liu, C. Schneider, M. Atatüre, S. Höfling, C. Y. Lu, and J. W. Pan. Coherently driving a single quantum two-level system with dichromatic laser pulses. *Nature Physics*, 15(9):941–946, 2019. 232

[71] J. N. Helbert. *Handbook of VLSI Microlithography*. Materials science and process technology series: Electronic materials and process technology. Elsevier Science, 2001. 268

[72] K. Hennessy, A. Badolato, M. Winger, D. Gerace, M. Atatüre, S. Gulde, S. Fält, E. L. Hu, and A. Imamoğlu. Quantum nature of a strongly coupled single quantum dot-cavity system. *Nature*, 445(7130):896–899, 2007. 278

[73] K. Hennessy, C. Högerle, E. Hu, A. Badolato, and A. Imamoğlu. Tuning photonic nanocavities by atomic force microscope nano-oxidation. *Applied Physics Letters*, 89(4):041118, 2006. 278, 294

[74] S. Hepp, S. Bauer, F. Hornung, M. Schwartz, S. L. Portalupi, M. Jetter, and P. Michler. Bragg grating cavities embedded into nano-photonic waveguides for Purcell enhanced quantum dot emission. *Optics Express*, 26(23):30614, 2018. 333
[75] S. Hepp, F. Hornung, S. Bauer, E. Hesselmeier, X. Yuan, M. Jetter, S. L. Portalupi, A. Rastelli, and P. Michler. Purcell-enhanced single-photon emission from a strain-tunable quantum dot in a cavity-waveguide device. *Applied Physics Letters*, 117(25):254002, 2020. 217, 218
[76] S. Hepp, M. Jetter, S. L. Portalupi, and P. Michler. Semiconductor quantum dots for integrated quantum photonics. *Advanced Quantum Technologies*, 2(9):1900020, 2019. 292, 329, 332, 335, 336, 338
[77] T. Herzog, S. Böhrkircher, S. Both, M. Fischer, R. Sittig, M. Jetter, S. L. Portalupi, T. Weiss, and P. Michler. Realization of a tunable fiber-based double cavity system. *Physical Review B*, 102(23):235306, 2020. 327
[78] T. Herzog, M. Sartison, S. Kolatschek, S. Hepp, A. Bommer, C. Pauly, F. Mücklich, C. Becher, M. Jetter, S. L. Portalupi, and P. Michler. Pure single-photon emission from In(Ga)As QDs in a tunable fiber-based external mirror microcavity. *Quantum Science and Technology*, 3(3):034009, 2018. 295
[79] S. Hill and W. K. Wootters. Entanglement of a pair of quantum bits. *Physical Review Letters*, 78:5022–5025, 1997. 114
[80] M. J. Holmes, S. Kako, K. Choi, M. Arita, and Y. Arakawa. Single photons from a hot solid-state emitter at 350 K. *ACS Photonics*, 3(4):543–546, 2016. 203
[81] C. K. Hong, Z. Y. Ou, and L. Mandel. Measurement of subpicosecond time intervals between two photons by interference. *Physical Review Letters*, 59:2044–2046, 1987. 48
[82] N. Hoppe, W. S. Zaoui, L. Rathgeber, Y. Wang, R. H. Klenk, W. Vogel, M. Kaschel, S. L. Portalupi, J. Burghartz, and M. Berroth. Ultra-efficient silicon-on-insulator grating couplers with backside metal mirrors. *IEEE Journal of Selected Topics in Quantum Electronics*, 26(2):1–6, 2020. 324
[83] M. Horodecki, P. Horodecki, and R. Horodecki. Separability of mixed states: necessary and sufficient conditions. *Physics Letters A*, 223(1):1–8, 1996. 114
[84] D. Huber, M. Reindl, S. F. Covre Da Silva, C. Schimpf, J. Martín-Sánchez, H. Huang, G. Piredda, J. Edlinger, A. Rastelli, and R. Trotta. Strain-tunable GaAs quantum dot: A nearly dephasing-free source of entangled photon pairs on demand. *Physical Review Letters*, 121(3):033902, 2018. 241
[85] A. J. Hudson, R. M. Stevenson, A. J. Bennett, R. J. Young, C. A. Nicoll, P. Atkinson, K. Cooper, D. A. Ritchie, and A. J. Shields. Coherence of an entangled exciton-photon state. *Physical Review Letters*, 99:266802, 2007. 116
[86] J. Iles-Smith, D. P. S. McCutcheon, A. Nazir, and J. Mørk. Phonon scattering inhibits simultaneous near-unity efficiency and indistinguishability in semiconductor single-photon sources. *Nature Photonics*, 11(8):521–526, 2017. 183, 186
[87] D. F. V. James, P. G. Kwiat, W. J. Munro, and A. G. White. Measurement of qubits. *Physical Review A – Atomic, Molecular, and Optical Physics*, 64:052312, 2001. 117, 118, 123
[88] H. Jayakumar, A. Predojević, T. Kauten, T. Huber, G. S. Solomon, and G. Weihs. Time-bin entangled photons from a quantum dot. *Nature Communications*, 5:4251, 2014. 124
[89] J. D. Joannopoulos, S. G. Johnson, J. N. Winn, and R. D. Meade. *Photonic Crystals: Molding the Flow of Light*. Second edition, Princeton University Press, 2011. 292, 296, 297, 298, 299, 304, 320
[90] O. Kahl, S. Ferrari, V. Kovalyuk, G. N. Goltsman, A. Korneev, and W. H. P. Pernice. Waveguide integrated superconducting single-photon detectors with high internal quantum efficiency at telecom wavelengths. Scientific Reports, 5(1):10941, 2015. 337
[91] B. Kambs and C. Becher. Limitations on the indistinguishability of photons from remote solid state sources. *New Journal of Physics*, 20(11):115003, 2018. https://doi.org/10.1088/1367-2630/aaea99. 253, 254, 256, 257
[92] C. Kammerer, C. Voisin, G. Cassabois, C. Delalande, P. Roussignol, F. Klopf, J. P. Reithmaier, A. Forchel, and J. M. Gérard. Line narrowing in single semiconductor quantum dots: Toward the control of environment effects. *Physical Review B*, 66:041306, 2002. 183

[93] Y. Karli, F. Kappe, V. Remesh, T. K. Bracht, J. Münzberg, S. Covre da Silva, T. Seidelmann, V. M. Axt, A. Rastelli, D. E. Reiter, and G. Weihs. SUPER scheme in action: experimental demonstration of red-detuned excitation of a quantum emitter. *Nano Letters*, 22(16):6567–6572, 2022. 232

[94] K. Karrai, R. J. Warburton, C. Schulhauser, A. Högele, B. Urbaszek, E. J. McGhee, A. O. Govorav, J. M. Garcia, B. D. Gerardot, and P. M. Petroff. Hybridization of electronic states in quantum dots through photon emission. *Nature*, 427(6970):135–138, 2004. 199

[95] A. Kavokin, J. J. Baumberg, G. Malpuech, and F. P. Laussy. *Microcavities*. Series on Semiconductor Science and Technology. Oxford University Press, Oxford, 2007. 139, 140

[96] J. Kettler. *Telecom-wavelength nonclassical light from single In(As)GaAs quantum dots*. PhD thesis, Institut für Halbleiteroptik und Funktionelle Grenzflächen, Universität Stuttgart, 2017. Verlag Dr. Hut. 193, 194, 196

[97] J. Kim, O. Benson, H. Kan, and Y. Yamamoto. A single-photon turnstile device. *Nature*, 397:500–503, 1999. 245

[98] S. Kolatschek. *Efficient single-photon emission from semiconductor quantum dots in photonic structures*. PhD thesis, Institut für Halbleiteroptik und Funktionelle Grenzflächen, Universität Stuttgart, 2023. 276, 316, 317, 318, 319, 326

[99] S. Kolatschek, S. Hepp, M. Sartison, M. Jetter, P. Michler, and S. L. Portalupi. Deterministic fabrication of circular Bragg gratings coupled to single quantum emitters via the combination of in-situ optical lithography and electron-beam lithography. *Journal of Applied Physics*, 125(4):045701, 2019. 286, 317

[100] S. Kolatschek, C. Nawrath, S. Bauer, J. Huang, J. Fischer, R. Sittig, M. Jetter, S. L. Portalupi, and P. Michler. *Bright purcell enhanced single-photon source in the telecom O-band based on a quantum dot in a circular Bragg grating*. Nano Letters, 21(18):7740–7745, 2021. 276, 277, 316

[101] T. F. Krauss, R. M. De La Rue, and S. Brand. Two-dimensional photonic-bandgap structures operating at near-infrared wavelengths. *Nature*, 383(6602):699–702, 1996. 293

[102] P. T. Kristensen, C. Van Vlack, and S. Hughes. Generalized effective mode volume for leaky optical cavities. *Optics Letters*, 37(10):1649, 2012. 292

[103] B. Krummheuer, V. M. Axt, and T. Kuhn. Theory of pure dephasing and the resulting absorption line shape in semiconductor quantum dots. *Physical Review B*, 65:195313, 2002. https://doi.org/10.1103/PhysRevB.65.195313. 178, 179

[104] A. V. Kuhlmann, J. Houel, D. Brunner, A. Ludwig, D. Reuter, A. D. Wieck, and R. J. Warburton. A dark-field microscope for background-free detection of resonance fluorescence from single semiconductor quantum dots operating in a set-and-forget mode. *Review of Scientific Instruments*, 84(7):073905, 2013. 62

[105] S. Kumar, R. Trotta, E. Zallo, J. D. Plumhof, P. Atkinson, A. Rastelli, and O. G. Schmidt. Strain-induced tuning of the emission wavelength of high quality GaAs/AlGaAs quantum dots in the spectral range of the 87Rb D 2 lines. *Applied Physics Letters*, 99(16):161118, 2011. 217

[106] M. El Kurdi, X. Checoury, S. David, T. P. Ngo, N. Zerounian, P. Boucaud, O. Kermarrec, Y. Campidelli, and D. Bensahel. Quality factor of Si-based photonic crystal L3 nanocavities probed with an internal source. *Optics Express*, 16(12):8780–8791, 2008. 304

[107] Y. Lai, S. Pirotta, G. Urbinati, D. Gerace, M. Minkov, V. Savona, A. Badolato, and M. Galli. Genetically designed L3 photonic crystal nanocavities with measured quality factor exceeding one million. *Applied Physics Letters*, 104(24):241101, 2014. 301

[108] P. Lambropoulos and D. Petrosyan. *Fundamentals of Quantum Optics and Quantum Information*. Springer, 2007. 29, 110, 111, 127, 128, 130, 131, 133, 134, 135, 136, 137, 138, 139

[109] Y. Léger, L. Besombes, L. Maingault, and H. Mariette. Valence-band mixing in neutral, charged, and Mn-doped self-assembled quantum dots. *Physical Review B*, 76:045331, 2007. https://doi.org/10.1103/PhysRevB.76.045331. 214

[110] T. Legero, T. Wilk, A. Kuhn, and G. Rempe. Time-resolved two-photon quantum interference. *Applied Physics. B, Lasers and Optics*, 77(8):797–802, 2003. 254

[111] M. Lermer, N. Gregersen, F. Dunzer, S. Reitzenstein, S. Höfling, J. Mørk, L. Worschech, M. Kamp, and A. Forchel. Bloch-wave engineering of quantum dot micropillars for cavity quantum electrodynamics experiments. *Physical Review Letters*, 108(5):057402, 2012. 301

[112] S.-K. Liao, H.-L. Yong, C. Liu, G.-L. Shentu, D.-D. Li, J. Lin, H. Dai, S.-Q. Zhao, B. Li, J.-Y. Guan, W. Chen, Y.-H. Gong, Y. Li, Z.-H. Lin, G.-S. Pan, J. S. Pelc, M. M. Fejer, W.-Z. Zhang, W.-Y. Liu, J. Yin, J.-G. Ren, X.-B. Wang, Q. Zhang, C.-Z. Peng, and J.-W. Pan. Long-distance free-space quantum key distribution in daylight towards inter-satellite communication. *Nature Photonics*, 11(8):509–513, 2017. 119

[113] S. Y. Lin, J. G. Fleming, D. L. Hetherington, B. K. Smith, R. Biswas, K. M. Ho, M. M. Sigalas, W. Zubrzycki, S. R. Kurtz, and J. Bur. A three-dimensional photonic crystal operating at infrared wavelengths. *Nature*, 394(6690):251–253, 1998. 298

[114] N. H. Lindner and T. Rudolph. Proposal for pulsed On-demand sources of photonic cluster state strings. *Physical Review Letters*, 103:113602, 2009. 172

[115] F. Liu, A. J. Brash, J. O'Hara, L. M. P. P. Martins, C. L. Phillips, R. J. Coles, B. Royall, E. Clarke, C. Bentham, N. Prtljaga, I. E. Itskevich, L. R. Wilson, M. S. Skolnick, and A. M. Fox. High Purcell factor generation of indistinguishable on-chip single photons. Nature Nanotechnology, 13(9):835–840, 2018. 294, 321

[116] J. Liu, R. Su, Y. Wei, B. Yao, S. F. Covre Da Silva, Y. Yu, J. Iles-Smith, K. Srinivasan, A. Rastelli, J. Li, and X. Wang. A solid-state source of strongly entangled photon pairs with high brightness and indistinguishability. *Nature Nanotechnology*, 14(6):586–593, 2019. 241, 279, 295, 326

[117] P. Lodahl, S. Mahmoodian, and S. Stobbe. Interfacing single photons and single quantum dots with photonic nanostructures. *Reviews of Modern Physics*, 87:347–400, 2015. https://doi.org/10.1103/RevModPhys.87.347. 299

[118] R. Loudon. *The Quantum Theory of Light*. Oxford Science Publications, 2001. 3, 6, 7, 10, 14, 17, 29, 33, 36, 37, 46, 50, 51, 52

[119] I. Marcikic, H. de Riedmatten, W. Tittel, H. Zbinden, M. Legré, and N. Gisin. Distribution of time-bin entangled qubits over 50 km of optical fiber. *Physical Review Letters*, 93:180502, 2004. 122

[120] J. Martín-Sánchez, R. Trotta, A. Mariscal, R. Serna, G. Piredda, S. Stroj, J. Edlinger, C. Schimpf, J. Aberl, T. Lettner, J. Wildmann, H. Huang, X. Yuan, D. Ziss, J. Stangl, and A. Rastelli. Strain-tuning of the optical properties of semiconductor nanomaterials by integration onto piezoelectric actuators. *Semiconductor Science and Technology*, 33(1):013001, 2018. https://doi.org/10.1088/1361-6641/aa9b53. 215, 216, 217, 219, 221, 222, 223, 224

[121] M. W. McCutcheon, G. W. Rieger, I. W. Cheung, J. F. Young, D. Dalacu, S. Frédérick, P. J. Poole, G. C. Aers, and R. L. Williams. Resonant scattering and second-harmonic spectroscopy of planar photonic crystal microcavities. *Applied Physics Letters*, 87(22):221110, 2005. 304

[122] P. Michler, M. Hilpert, and G. Reiner. Dynamics of dual-wavelength emission from a coupled semiconductor microcavity laser. *Applied Physics Letters*, 70(16):2073–2075, 1997. 327

[123] P. Michler, A. Kiraz, C. Becher, W. V. Schoenfeld, P. M. Petroff, L. Zhang, E. Hu, and A. Imamoglu. A quantum dot single-photon turnstile device. *Science*, 290(5500):2282–2285, 2000. 81, 295

[124] P. Michler, editor. *Single Quantum Dots*. Springer, 2003. 145, 157, 213, 214

[125] P. Michler, editor. *Single Semiconductor Quantum Dots*. Springer, 2009. 117, 155, 159

[126] P. Michler, editor. *Quantum Dots for Quantum Information Technologies*. Springer, 2017. 72, 89, 105, 119, 145, 153, 163, 165, 171, 184, 186, 187, 191, 196, 197, 198, 200, 213, 228, 242, 259, 260, 330, 331, 336

[127] L. Midolo, S. L. Hansen, W. Zhang, C. Papon, R. Schott, A. Ludwig, A. D. Wieck, P. Lodahl, and S. Stobbe. Electro-optic routing of photons from a single quantum dot in photonic integrated circuits. *Optics Express*, 25(26):33514–33526, Dec 2017. 335

[128] M. Minkov and V. Savona. Automated optimization of photonic crystal slab cavities. *Scientific Reports*, 4(1):5124, 2014. 301

[129] M. Moczała-Dusanowska, Ł. Dusanowski, S. Gerhardt, Y. M. He, M. Reindl, A. Rastelli, R. Trotta, N. Gregersen, S. Höfling, and C. Schneider. Strain-tunable single-photon source based on a quantum dot-micropillar system. *ACS Photonics*, 6(8):2025–2031, 2019. 296

[130] S. Mosor, J. Hendrickson, B. C. Richards, J. Sweet, G. Khitrova, H. M. Gibbs, T. Yoshie, A. Scherer, O. B. Shchekin, and D. G. Deppe. Scanning a photonic crystal slab nanocavity by condensation of xenon. *Applied Physics Letters*, 87(14):141105, 2005. 294

[131] J. R. A. Müller, R. M. Stevenson, J. Skiba-Szymanska, G. Shooter, J. Huwer, I. Farrer, D. A. Ritchie, and A. J. Shields. Active reset of a radiative cascade for entangled-photon generation beyond the continuous-driving limit. *Physical Review Research*, 2:043292, 2020. 249

[132] M. Müller. *Generation of Indistinguishable and Entangled Photons from Semiconductor QuantumDots*. PhD thesis, Institut für Halbleiteroptik und Funktionelle Grenzflächen, Universität Stuttgart, 2017. Verlag Dr. Hut, München. 115, 116, 117, 120, 122, 123, 124

[133] M. Müller, S. Bounouar, K. D. Jöns, M. Glässl, and P. Michler. On-demand generation of indistinguishable polarization-entangled photon pairs. *Nature Photonics*, 8(3):224–228, 2014. 240, 241

[134] P. Müller, T. Tentrup, M. Bienert, G. Morigi, and J. Eschner. Spectral properties of single photons from quantum emitters. *Physical Review A*, 96:023861, 2017. 238

[135] D. Najer, I. Söllner, P. Sekatski, V. Dolique, M. C. Löbl, D. Riedel, R. Schott, S. Starosielec, S. R. Valentin, A. D. Wieck, N. Sangouard, A. Ludwig, and R. J. Warburton. A gated quantum dot strongly coupled to an optical microcavity. *Nature*, 575(7784):622–627, 2019. 296

[136] C. Nawrath, R. Joos, S. Kolatschek, S. Bauer, P. Pruy, F. Hornung, J. Fischer, J. Huang, P. Vijayan, R. Sittig, M. Jetter, S. L. Portalupi, and P. Michler. High emission rate from a Purcell-enhanced, triggered source of pure single photons in the telecom C-band. Adv. Quantum Technol. 2300111, 6, 2023. 74, 140, 276, 295, 319, 326

[137] C. Nawrath, F. Olbrich, M. Paul, S. L. Portalupi, M. Jetter, and P. Michler. Coherence and indistinguishability of highly pure single photons from non-resonantly and resonantly excited telecom C-band quantum dots. *Applied Physics Letters*, 115(2):023103, 2019. 105

[138] W. Nie, N. L. Sharma, C. Weigelt, R. Keil, J. Yang, F. Ding, C. Hopfmann, and O. G. Schmidt. Experimental optimization of the fiber coupling efficiency of gaas quantum dot-based photon sources. *Applied Physics Letters*, 119(24):244003, 2021. 306, 323

[139] K. P. O'Donnell and X. Chen. Temperature dependence of semiconductor band gaps. *Applied Physics Letters*, 58(25):2924–2926, 1991. 207

[140] F. Olbrich, J. Kettler, M. Bayerbach, M. Paul, J. Höschele, S. L. Portalupi, M. Jetter, and P. Michler. Temperature-dependent properties of single long-wavelength InGaAs quantum dots embedded in a strain reducing layer. *Journal of Applied Physics*, 121(18):184302, 2017. 201, 202

[141] H. Ollivier, I. Maillette De Buy Wenniger, S. Thomas, S. C. Wein, A. Harouri, G. Coppola, P. Hilaire, C. Millet, A. Lemaître, I. Sagnes, O. Krebs, L. Lanco, J. C. Loredo, C. Antón, N. Somaschi, and P. Senellart. Reproducibility of high-performance quantum dot single-photon sources. *ACS Photonics*, 7(4):1050–1059, 2020. 235

[142] H. Ollivier, S. E. Thomas, S. C. Wein, I. Maillette de Buy Wenniger, N. Coste, J. C. Loredo, N. Somaschi, A. Harouri, A. Lemaitre, I. Sagnes, L. Lanco, C. Simon, C. Anton, O. Krebs, and P. Senellart. Hong-ou-mandel interference with imperfect single photon sources. *Physical Review Letters*, 126:063602, 2021. 101, 102

[143] G. Ortner, M. Schwab, M. Bayer, R. Pässler, S. Fafard, Z. Wasilewski, P. Hawrylak, and A. Forchel. Temperature dependence of the excitonic band gap in InxGa1-xAs GaAs self-assembled quantum dots. *Physical Review B*, 72:085328, 2005. 206

[144] R. Pässler. Dispersion-related description of temperature dependencies of band gaps in semiconductors. *Physical Review B*, 66:085201, 2002. 207

[145] R. B. Patel, A. J. Bennett, I. Farrer, C. A. Nicoll, D. A. Ritchie, and A. J. Shields. Two-photon interference of the emission from electrically tunable remote quantum dots. *Nature Photonics*, 4(9):632–635, 2010. 209, 210

[146] P. K. Pathak and S. Hughes. Coherent generation of time-bin entangled photon pairs using the biexciton cascade and cavity-assisted piecewise adiabatic passage. *Physical Review B*, 83:245301, 2011. 124

[147] M. Pelton, C. Santori, J. Vucković, B. Zhang, G. S. Solomon, J. Plant, and Y. Yamamoto. Efficient source of single photons: A single quantum dot in a micropost microcavity. *Physical Review Letters*, 89:233602, 2002. 77

[148] A. Peres. Separability criterion for density matrices. *Physical Review Letters*, 77:1413–1415, 1996. 114

[149] M. Petruzzella, T. Xia, F. Pagliano, S. Birindelli, L. Midolo, Z. Zobenica, L. H. Li, E. H. Linfield, and A. Fiore. Fully tuneable, Purcell-enhanced solid-state quantum emitters. *Applied Physics Letters*, 107(14):141109, 2015. 294

[150] S. L. Portalupi. *Light Confinement and Emission in Silicon Photonic Crystal Cavities*. PhD thesis, University of Pavia, 2011. ISBN 978-88-95767-51-2. 271, 273, 275

[151] S. L. Portalupi, M. Galli, M. Belotti, L. C. Andreani, T. F. Krauss, and L. O'Faolain. Deliberate versus intrinsic disorder in photonic crystal nanocavities investigated by resonant light scattering. *Physical Review B*, 84(4):045423, 2011. 294, 306

[152] S. L. Portalupi, M. Galli, C. Reardon, T. Krauss, L. O'Faolain, L. C. Andreani, and D. Gerace. Planar photonic crystal cavities with far-field optimization for high coupling efficiency and quality factor. *Optics Express*, 18(15):16064, 2010. 289, 301, 302, 303, 306

[153] S. L. Portalupi, G. Hornecker, V. Giesz, T. Grange, A. Lemaître, J. Demory, I. Sagnes, N. D. Lanzillotti-Kimura, L. Lanco, A. Auffèves, and P. Senellart. Bright phonon-tuned single-photon source. *Nano Letters*, 15(10):6290–6294, 2015. 314

[154] M. Prilmüller, T. Huber, M. Müller, P. Michler, G. Weihs, and A. Predojević. Hyperentanglement of photons emitted by a quantum Dot. *Physical Review Letters*, 121:110503, 2018. 164

[155] J. H. Quilter, A. J. Brash, F. Liu, M. Glässl, A. M. Barth, V. M. Axt, A. J. Ramsay, M. S. Skolnick, and A. M. Fox. Phonon-assisted population inversion of a Single InGaAs/GaAs quantum dot by pulsed laser excitation. *Physical Review Letters*, 114:137401, 2015. https://doi.org/10.1103/PhysRevLett.114.137401. 242

[156] A. J. Ramsay, T. M. Godden, S. J. Boyle, E. M. Gauger, A. Nazir, B. W. Lovett, A. M. Fox, and M. S. Skolnick. Phonon-induced Rabi-frequency renormalization of optically driven single InGaAs/GaAs quantum dots. *Physical Review Letters*, 105(17):177402, 2010. 188

[157] A. J. Ramsay, A. Venu Gopal, E. M. Gauger, A. Nazir, B. W. Lovett, A. M. Fox, and M. S. Skolnick. Damping of exciton rabi rotations by acoustic phonons in optically excited InGaAs/GaAs quantum dots. *Physical Review Letters*, 104:017402, 2010. 188

[158] A. Rastelli, S. Kiravittaya, M. Benyoucef, Y. Mei, and O. G. Schmidt. In situ tuning of optical modes in single semiconductor microcavities by laser heating. In *2007 9th International Conference on Transparent Optical Networks*, volume 3, pages 58–60. IEEE, 2007. 295

[159] M. E. Reimer, G. Bulgarini, N. Akopian, M. Hocevar, M. B. Bavinck, M. A. Verheijen, E. P. A. M. Bakkers, L. P. Kouwenhoven, and V. Zwiller. Bright single-photon sources in bottom-up tailored nanowires. *Nature Communications*, 3(1):737, 2012. 322

[160] U. Rengstl. *III-V Semiconductor Photonic Integrated Circuits with Quantum Dots as Single-Photon Emitters*. PhD thesis, Institut für Halbleiteroptik und Funktionelle Grenzflächen, Universität Stuttgart, 2017. Verlag Dr. Hut, München. 274, 331

[161] U. Rengstl, M. Schwartz, T. Herzog, F. Hargart, M. Paul, S. L. Portalupi, M. Jetter, and P. Michler. On-chip beamsplitter operation on single photons from quasi-resonantly excited quantum dots embedded in GaAs rib waveguides. *Applied Physics Letters*, 107(2):021101, 2015. 335

[162] K. Roy-Choudhury and S. Hughes. Quantum theory of the emission spectrum from quantum dots coupled to structured photonic reservoirs and acoustic phonons. *Physical Review B*, 92:205406, 2015. https://doi.org/10.1103/PhysRevB.92.205406. 198, 199

[163] B. E. A. Saleh and M. C. Teich. *Fundamentals of Photonics-Second edition*. John Wiley & Sons, Inc., 2007. 65

[164] C. Santori, D. Fattal, J. Vucković, G. S. Solomon, and Y. Yamamoto. Indistinguishable photons from a single-photon device. *Nature*, 419(6907):594–597, 2002. 100, 101

[165] C. Santori, D. Fattal, J. Vuckovic, G. S. Solomon, and Y. Yamamoto. Single-photon generation with InAs quantum dots. *New Journal of Physics*, 6(1):89, 2004. 260, 261

[166] C. Santori, M. Pelton, G. Solomon, Y. Dale, and Y. Yamamoto. Triggered single photons from a quantum dot. *Physical Review Letters*, 86(8):1502–1505, 2001. 193

[167] L. Sapienza, M. Davanço, A. Badolato, and K. Srinivasan. Nanoscale optical positioning of single quantum dots for bright and pure single-photon emission. *Nature Communications*, 6:7833, 2015. 279, 295

[168] M. Sartison. *Fabrication of efficient single-photon devices based on pre-selected quantum dots using deterministic optical lithography*. PhD thesis, Institut für Halbleiteroptik und Funktionelle Grenzflächen, Universität Stuttgart, 2019. Verlag Dr. Hut, München. 274, 275, 280, 281, 282, 283, 284, 285, 286

[169] M. Sartison, L. Engel, S. Kolatschek, F. Olbrich, C. Nawrath, S. Hepp, M. Jetter, P. Michler, and S. L. Portalupi. Deterministic integration and optical characterization of telecom O-band quantum dots embedded into wet-chemically etched Gaussian-shaped microlenses. *Applied Physics Letters*, 113(3):032103, 2018. 273, 277, 284, 290, 291, 323

[170] M. Sartison, S. L. Portalupi, T. Gissibl, M. Jetter, H. Giessen, and P. Michler. Combining in-situ lithography with 3D printed solid immersion lenses for single quantum dot spectroscopy. *Scientific Reports*, 7:39916, 2017. 279, 286, 290

[171] M. Sartison, S. Seyfferle, S. Kolatschek, S. Hepp, M. Jetter, P. Michler, and S. L. Portalupi. Single-photon light-emitting diodes based on preselected quantum dots using a deterministic lithography technique. *Applied Physics Letters*, 114(22):222101, 2019. 248, 284, 315

[172] M. Sartison, K. Weber, S. Thiele, L. Bremer, S. Fischbach, T. Herzog, S. Kolatschek, M. Jetter, S. Reitzenstein, A. Herkommer, P. Michler, S. L. Portalupi, and H. Giessen. 3D printed micro-optics for quantum technology: Optimised coupling of single quantum dot emission into a single-mode fibre. Light: Advanced Manufacturing, 2(LAM2020060010):103, 2021. https://doi.org/10.37188/lam.2021.006. 286, 290, 291

[173] C. Sauvan, P. Lalanne, and J. P. Hugonin. Slow-wave effect and mode-profile matching in photonic crystal microcavities. *Physical Review B*, 71:165118, 2005. 301

[174] C. Schimpf, M. Reindl, P. Klenovský, T. Fromherz, S. F. Covre Da Silva, J. Hofer, C. Schneider, S. Höfling, R. Trotta, and A. Rastelli. Resolving the temporal evolution of line broadening in single quantum emitters. *Optics Express*, 27(24):35290, 2019. 94, 95, 97, 190

[175] E. Schöll, L. Schweickert, L. Hanschke, K. D. Zeuner, F. Sbresny, T. Lettner, R. Trivedi, M. Reindl, S. F. Covre Da Silva, R. Trotta, J. J. Finley, J. Vučković, K. Müller, A. Rastelli, V. Zwiller, and K. D. Jöns. Crux of using the cascaded emission of a three-level quantum ladder system to generate indistinguishable photons. *Physical Review Letters*, 125:233605, 2020. https://doi.org/10.1103/PhysRevLett.125.233605. 165, 188

[176] I. Schwartz, D. Cogan, E. R. Schmidgall, Y. Don, L. Gantz, O. Kenneth, N. H. Lindner, and D. Gershoni. Deterministic generation of a cluster state of entangled photons. *Science*, 354(6311):434–437, 2016. 174

[177] M. Schwartz, U. Rengstl, T. Herzog, M. Paul, J. Kettler, S. L. Portalupi, M. Jetter, and P. Michler. Generation, guiding and splitting of triggered single photons from a resonantly excited quantum dot in a photonic circuit. *Optics Express*, 24(3):3089, 2016. 338

[178] M. Schwartz, E. Schmidt, U. Rengstl, F. Hornung, S. Hepp, S. L. Portalupi, K. Ilin, M. Jetter, M. Siegel, and P. Michler. Fully on-chip single-photon hanbury-brown and twiss experiment on a monolithic semiconductor-superconductor platform. *Nano Letters*, 18(11):6892–6897, 2018. 68, 337, 338

[179] L. Schweickert, K. D. Jöns, K. D. Zeuner, S. F. Covre Da Silva, H. Huang, T. Lettner, M. Reindl, J. Zichi, R. Trotta, A. Rastelli, and V. Zwiller. On-demand generation of background-free single photons from a solid-state source. *Applied Physics Letters*, 112(9):093106, 2018. 168

[180] K. Sebald, P. Michler, T. Passow, D. Hommel, G. Bacher, and A. Forchel. Single-photon emission of CdSe quantum dots at temperatures up to 200 K. *Applied Physics Letters*, 81(16):2920–2922, 2002. 184

[181] S. Seyfferle, F. Hargart, M. Jetter, E. Hu, and P. Michler. Signatures of single-photon interaction between two quantum dots located in different cavities of a weakly coupled double microdisk structure. *Physical Review B*, 97:035302, 2018. 327

[182] A. Shakoor, R. Lo Savio, P. Cardile, S. L. Portalupi, D. Gerace, K. Welna, S. Boninelli, G. Franzò, F. Priolo, T. F. Krauss, M. Galli, and L. O'Faolain. Room temperature all-silicon photonic crystal nanocavity light emitting diode at sub-bandgap wavelengths. *Laser & Photonics Reviews*, 7(1):114–121, 2013. 319

[183] G. Shooter, Z.-H. Xiang, J. R. A. Müller, J. Skiba-Szymanska, J. Huwer, J. Griffiths, T. Mitchell, M. Anderson, T. Müller, A. B. Krysa, R. M. Stevenson, J. Heffernan, D. A. Ritchie, and A. J. Shields. 1ghz clocked distribution of electrically generated entangled photon pairs. *Optics Express*, 28(24):36838–36848, 2020. 328

[184] C. Simon and J.-P. Poizat. Creating single time-bin-entangled photon pairs. *Physical Review Letters*, 94:030502, 2005. 124, 165

[185] H. Snijders, J. A. Frey, J. Norman, V. P. Post, A. C. Gossard, J. E. Bowers, M. P. van Exter, W. Löffler, and D. Bouwmeester. Fiber-coupled cavity-QED source of identical single photons. *Physical Review Applied*, 9(3):031002, 2018. https://doi.org/10.1103/PhysRevApplied.9.031002. 311, 312

[186] N. Somaschi, V. Giesz, L. De Santis, J. C. Loredo, M. P. Almeida, G. Hornecker, S. L. Portalupi, T. Grange, C. Antón, J. Demory, C. Gómez, I. Sagnes, N. D. Lanzillotti-Kimura, A. Lemaítre, A. Auffeves, A. G. White, L. Lanco, and P. Senellart. Near-optimal single-photon sources in the solid state. *Nature Photonics*, 10(5):340–345, 2016. 211, 258, 295, 312

[187] K. Srinivasan, P. E. Barclay, M. Borselli, and O. Painter. Optical-fiber-based measurement of an ultrasmall volume high-Q photonic crystal microcavity. *Physical Review B*, 70:081306, 2004. 304, 305

[188] P. Stepanov, A. Delga, N. Gregersen, E. Peinke, M. Munsch, J. Teissier, J. Mørk, M. Richard, J. Bleuse, J.-M. Gérard, and J. Claudon. Highly directive and Gaussian far-field emission from "giant" photonic trumpets. *Applied Physics Letters*, 107(14):141106, 2015. 290, 291, 323

[189] I. Stranski and L. Krastanow. Zur theorie der orientierten ausscheidung von ionenkristallen aufeinander. *Sitzungsber. Akad. Wiss Wien. Math.-Naturwiss.*, 146:797, 1938. 267

[190] A. Streltsov. *Quantum Correlations Beyond Entanglement*. SpringerBriefs in Physics, Springer International Publishing, Cham, 2015. 110

[191] T. Strobel, J. H. Weber, M. Schmidt, L. Wagner, L. Engel, M. Jetter, A. D. Wieck, S. L. Portalupi, A. Ludwig, and P. Michler. A unipolar quantum dot diode structure for advanced quantum light sources. *Nano Letters*, 23(14):6574–6580, 2023. 213, 266

[192] S. M. Sze. *Physics of Semiconductor Devices*. Wiley, 2007. 245

[193] H. Takesue and Y. Noguchi. Implementation of quantum state tomography for time-bin entangled photon pairs. *Optics Express*, 17(13):10976–10989, 2009. 117, 121, 122, 123

[194] S. Thomas and P. Senellart. The race for the ideal single-photon source is on. *Nature Nanotechnology*, 16(4):367–368, 2021. 72

[195] S. E. Thomas, M. Billard, N. Coste, S. C. Wein, Priya, H. Ollivier, O. Krebs, L. Tazaïrt, A. Harouri, A. Lemaitre, I. Sagnes, C. Anton, L. Lanco, N. Somaschi, J. C. Loredo, and P. Senellart. Bright polarized single-photon source based on a linear dipole. *Physical Review Letters*, 126:233601, 2021. 242, 243

[196] N. Tomm, A. Javadi, N. O. Antoniadis, D. Najer, M. C. Löbl, A. R. Korsch, R. Schott, S. R. Valentin, A. D. Wieck, A. Ludwig, and R. J. Warburton. A bright and fast source of coherent single photons. *Nature Nanotechnology*, 16(4):399–403, 2021. 197, 236, 258, 296, 321, 322

[197] N. V. Q. Tran, S. Combrié, and A. De Rossi. Directive emission from high-Q photonic crystal cavities through band folding. *Physical Review B*, 79:041101, 2009. https://doi.org/10.1103/PhysRevB.79.041101. 301, 302

[198] R. Trivedi, K. A. Fischer, J. Vučković, and K. Müller. Generation of non-classical light using semiconductor quantum dots. *Advanced Quantum Technologies*, 3(1):1900007, 2020. 174

[199] R. Trotta, J. Martín-Sánchez, J. S. Wildmann, G. Piredda, M. Reindl, C. Schimpf, E. Zallo, S. Stroj, J. Edlinger, and A. Rastelli. Wavelength-tunable sources of entangled photons interfaced with atomic vapours. *Nature Communications*, 7(1):10375, 2016. 222, 223, 224

[200] R. Trotta, E. Zallo, C. Ortix, P. Atkinson, J. D. Plumhof, J. Van Den Brink, A. Rastelli, and O. G. Schmidt. Universal recovery of the energy-level degeneracy of bright excitons in ingaas quantum dots without a structure symmetry. *Physical Review Letters*, 109:147401, 2012. https://doi.org/10.1103/PhysRevLett.109.147401. 220, 221

[201] W. Unrau, D. Quandt, J. H. Schulze, T. Heindel, T. D. Germann, O. Hitzemann, A. Strittmatter, S. Reitzenstein, U. W. Pohl, and D. Bimberg. Electrically driven single photon source based on a site-controlled quantum dot with self-aligned current injection. *Applied Physics Letters*, 101(21):211119, 2012. 248

[202] R. Uppu, H. T. Eriksen, H. Thyrrestrup, A. D. Uğurlu, Y. Wang, S. Scholz, A. D. Wieck, A. Ludwig, M. C. Löbl, R. J. Warburton, P. Lodahl, and L. Midolo. On-chip deterministic operation of quantum dots in dual-mode waveguides for a plug-and-play single-photon source. *Nature Communications*, 11(1):3782, 2020. https://doi.org/10.1038/s41467-020-17603-9. 321

[203] R. Ursin, F. Tiefenbacher, T. Schmitt-Manderbach, H. Weier, T. Scheidl, M. Lindenthal, B. Blauensteiner, T. Jennewein, J. Perdigues, P. Trojek, B. Ömer, M. Fürst, M. Meyenburg, J. Rarity, Z. Sodnik, C. Barbieri, H. Weinfurter, and A. Zeilinger. Entanglement-based quantum communication over 144 km. *Nature Physics*, 3(7):481–486, 2007. 118

[204] K. J. Vahala. Optical microcavities. *Nature*, 424(6950):839–846, 2003. 292, 293, 295

[205] Y. P. Varshni. Temperature dependence of the energy gap in semiconductors. *Physica*, 34(1):149–154, 1967. 207

[206] A. Vasanelli, R. Ferreira, and G. Bastard. Continuous absorption background and decoherence in quantum dots. *Physical Review Letters*, 89:216804, 2002. 177

[207] M. A. M. Versteegh, M. E. Reimer, K. D. Jöns, D. Dalacu, P. J. Poole, A. Gulinatti, A. Giudice, and V. Zwiller. Observation of strongly entangled photon pairs from a nanowire quantum dot. *Nature Communications*, 5(1):5298, 2014. 89, 329

[208] G. Vidal and R. F. Werner. Computable measure of entanglement. *Physical Review A*, 65:032314, 2002. 114

[209] S. Vignolini, F. Intonti, M. Zani, F. Riboli, D. S. Wiersma, L. H. Li, L. Balet, M. Francardi, A. Gerardino, A. Fiore, and M. Gurioli. Near-field imaging of coupled photonic-crystal microcavities. *Applied Physics Letters*, 94(15):151103, 2009. 306

[210] Y. A. Vlasov, X.-Z. Bo, J. C. Sturm, and D. J. Norris. On-chip natural assembly of silicon photonic bandgap crystals. *Nature*, 414(6861):289–293, 2001. 298

[211] M. Volmer and A. Weber. Keimbildung in übersättigten gebilden. *Zeitschrift für Physikalische Chemie*, 119U(1):277–301, 1926. 267

[212] H. Vural. *Optical and quantum optical properties of a quantum dot-atomic vapor interface*. PhD thesis, Institut für Halbleiteroptik und Funktionelle Grenzflächen, Universität Stuttgart, 2021. Verlag Dr. Hut, München. 90, 191, 252

[213] H. Vural, J. Maisch, I. Gerhardt, M. Jetter, S. L. Portalupi, and P. Michler. Characterization of spectral diffusion by slow-light photon-correlation spectroscopy. *Physical Review B*, 101(16):161401(R), 2020. 89, 90, 97, 190, 191, 252

[214] H. Vural, S. L. Portalupi, J. Maisch, S. Kern, J. H. Weber, M. Jetter, J. Wrachtrup, R. Löw, I. Gerhardt, and P. Michler. Two-photon interference in an atom-quantum dot hybrid system. *Optica*, 5(4):367, 2018. 191

[215] H. Vural, S. L. Portalupi, and P. Michler. Perspective of self-assembled InGaAs quantum-dots for multi-source quantum implementations. *Applied Physics Letters*, 117(3):030501, 2020. 89, 90, 91, 186, 187

[216] R. Wagner, L. Heerklotz, N. Kortenbruck, and F. Cichos. Back focal plane imaging spectroscopy of photonic crystals. *Applied Physics Letters*, 101(8):081904, 2012. 306

[217] H. Wang, Y.-M. He, T.-H. Chung, H. Hu, Y. Yu, S. Chen, X. Ding, M.-C. Chen, J. Qin, X. Yang, R.-Z. Liu, Z.-C. Duan, J.-P. Li, S. Gerhardt, K. Winkler, J. Jurkat, L.-J. Wang, N. Gregersen, Y.-H. Huo, Q. Dai, S. Yu, S. Höfling, C.-Y. Lu, and J.-W. Pan. Towards optimal single-photon sources from polarized microcavities. *Nature Photonics*, 13(11):770–775, 2019. 74, 235, 236, 314, 318

[218] H. Wang, H. Hu, T.-H. Chung, J. Qin, X. Yang, J.-P. Li, R.-Z. Liu, H.-S. Zhong, Y.-M. He, X. Ding, Y.-H. Deng, Q. Dai, Y.-H. Huo, S. Höfling, C.-Y. Lu, and J.-W. Pan. On-demand semiconductor source of entangled photons which simultaneously has high fidelity, efficiency, and indistinguishability. *Physical Review Letters*, 122(11):113602, 2019. 295, 326

[219] R. J. Warburton. Single spins in self-assembled quantum dots. *Nature Materials*, 12(6):483–493, 2013. 211, 212

[220] R. J. Warburton, C. Schaeferlein, D. Haft, F. Bickel, A. Lorke, K. Karrai, J. M. Garcia, W. Schoenfeld, and P. M. Petroff. Optical emission from a charge-tunable quantum ring. *Nature*, 405:926–929, 2000. 245

[221] J. H. Weber, B. Kambs, J. Kettler, S. Kern, J. Maisch, H. Vural, M. Jetter, S. L. Portalupi, C. Becher, and P. Michler. Two-photon interference in the telecom C-band after frequency conversion of photons from remote quantum emitters. *Nature Nanotechnology*, 14(1):23–26, 2019. 109

[222] J. H. Weber, J. Kettler, H. Vural, M. Müller, J. Maisch, M. Jetter, S. L. Portalupi, and P. Michler. Overcoming correlation fluctuations in two-photon interference experiments with differently bright and independently blinking remote quantum emitters. *Physical Review B*, 97(19):195414, 2018. 83, 106, 107, 108

[223] Y.-J. Wei, Y.-M. He, M.-C. Chen, Y.-N. Hu, Y. He, D. Wu, C. Schneider, M. Kamp, S. Höfling, C.-Y. Lu, and J.-W. Pan. Deterministic and robust generation of single photons from a single quantum dot with 99.5 % indistinguishability using adiabatic rapid passage. *Nano Letters*, 14(11):6515–6519, 2014. 236, 237

[224] K. Welna, S. L. Portalupi, M. Galli, L. O'Faolain, and T. F. Krauss. Novel dispersion-adapted photonic crystal cavity with improved disorder stability. *IEEE Journal of Quantum Electronics*, 48(9):1177–1183, 2012. 306

[225] A. G. White, D. F. V. James, W. J. Munro, and P. G. Kwiat. Exploring hilbert space: Accurate characterization of quantum information. *Physical Review A*, 65:012301, 2001. 114

[226] J. Wiersig, C. Gies, F. Jahnke, M. Aßmann, T. Berstermann, M. Bayer, C. Kistner, S. Reitzenstein, C. Schneider, S. Höfling, A. Forchel, C. Kruse, J. Kalden, and D. Hommel. Direct observation of correlations between individual photon emission events of a microcavity laser. *Nature*, 460(7252):245–249, 2009. 71

[227] M. Winger, T. Volz, G. Tarel, S. Portolan, A. Badolato, K. J. Hennessy, E. L. Hu, A. Beveratos, J. Finley, V. Savona, and A. Imamoğlu. Explanation of photon correlations in the far-off-resonance optical emission from a quantum-dot-cavity system. *Physical Review Letters*, 103(20):207403, 2009. https://doi.org/10.1103/PhysRevLett.103.207403. 198, 199, 200

[228] Z.-H. Xiang, J. Huwer, R. M. Stevenson, J. Skiba-Szymanska, M. B. Ward, I. Farrer, D. A. Ritchie, and A. J. Shields. Long-term transmission of entangled photons from a single quantum dot over deployed fiber. *Scientific Reports*, 9(1):4111, 2019. 119

[229] E. Yablonovitch, T. Gmitter, and K. Leung. Photonic band structure: The face-centered-cubic case employing nonspherical atoms. *Physical Review Letters*, 67(17):2295–2298, 1991. 298

[230] J. Yang, C. Nawrath, R. Keil, R. Joos, X. Zhang, B. Höfer, Y. Chen, M. Zopf, M. Jetter, S. L. Portalupi, F. Ding, P. Michler, and O. Schmidt. Quantum dot-based broadband optical antenna for efficient extraction of single photons in the telecom O-band. *Optics Express*, 28(13):19457–19468, 2020. 290

[231] J. Yin, Y. Cao, Y. H. Li, S. K. Liao, L. Zhang, J. G. Ren, W. Q. Cai, W. Y. Liu, B. Li, H. Dai, G. B. Li, Q. M. Lu, Y. H. Gong, Y. Xu, S. L. Li, F. Z. Li, Y. Y. Yin, Z. Q. Jiang, M. Li, J. J. Jia, G. Ren, D. He, Y. L. Zhou, X. X. Zhang, N. Wang, X. Chang, Z. C. Zhu, N. L. Liu, Y. A. Chen, C. Y. Lu, R. Shu, C. Z. Peng, J. Y. Wang, and J. W. Pan. Satellite-based entanglement distribution over 1200 kilometers. *Science*, 356(6343):1140–1144, 2017. 118

[232] Z. Yuan, B. E. Kardynal, R. M. Stevenson, A. J. Shields, C. J. Lobo, K. Cooper, N. S. Beattie, D. A. Ritchie, and M. Pepper. Electrically driven single-photon source. *Science*, 295:102–105, 2002. 246, 247

[233] T. Zander, A. Herklotz, S. Kiravittaya, M. Benyoucef, F. Ding, P. Atkinson, S. Kumar, J. D. Plumhof, K. Dörr, A. Rastelli, and O. G. Schmidt. Epitaxial quantum dots in stretchable optical microcavities. *Optics Express*, 17(25):22452, 2009. 295

[234] L. Zhai, M. C. Löbl, G. N. Nguyen, J. Ritzmann, A. Javadi, C. Spinnler, A. D. Wieck, A. Ludwig, and R. J. Warburton. Low-noise GaAs quantum dots for quantum photonics. *Nature Communications*, 11(1):4745, 2020. 196, 211

[235] L. Zhai, G. N. Nguyen, C. Spinnler, J. Ritzmann, M. C. Löbl, A. D. Wieck, A. Ludwig, A. Javadi, and R. J. Warburton. Quantum interference of identical photons from remote GaAs quantum dots. *Nature Nanotechnology*, 17(8):829–833, 2022. 105, 258

[236] A. Zrenner, E. Beham, S. Stufler, F. Findeis, M. Bichler, and G. Abstreiter. Coherent properties of a two-level system based on a quantum-dot photodiode. *Nature*, 418(6898):612–614, 2002. 210

Index

https://doi.org/10.1515/9783110703412-016

www.ingramcontent.com/pod-product-compliance
Lightning Source LLC
Chambersburg PA
CBHW080713220326
41598CB00033B/5405
9783110703405